AF369720

Entropy in Real-World Datasets and Its Impact on Machine Learning

Entropy in Real-World Datasets and Its Impact on Machine Learning

Editors

Jan Kozak
Przemysław Juszczuk

MDPI • Basel • Beijing • Wuhan • Barcelona • Belgrade • Manchester • Tokyo • Cluj • Tianjin

Editors

Jan Kozak
University of Economics
Katowice
Poland

Przemysław Juszczuk
University of Economics
Katowice
Poland

Editorial Office
MDPI
St. Alban-Anlage 66
4052 Basel, Switzerland

This is a reprint of articles from the Special Issue published online in the open access journal *Entropy* (ISSN 1099-4300) (available at: https://www.mdpi.com/journal/entropy/special_issues/Entropy_Dataset_Machine_Learning).

For citation purposes, cite each article independently as indicated on the article page online and as indicated below:

LastName, A.A.; LastName, B.B.; LastName, C.C. Article Title. *Journal Name* **Year**, *Volume Number*, Page Range.

ISBN 978-3-0365-7848-4 (Hbk)
ISBN 978-3-0365-7849-1 (PDF)

© 2023 by the authors. Articles in this book are Open Access and distributed under the Creative Commons Attribution (CC BY) license, which allows users to download, copy and build upon published articles, as long as the author and publisher are properly credited, which ensures maximum dissemination and a wider impact of our publications.

The book as a whole is distributed by MDPI under the terms and conditions of the Creative Commons license CC BY-NC-ND.

Contents

About the Editors

Jan Kozak

Jan Kozak earned his PhD in Computer Science in 2011 from the University of Silesia in Poland. Since 2013, he has been working in the Department of Machine Learning, University of Economics, in Katowice, Poland, Institute of Computer Science. From 2011 to 2016, Jan Kozak worked as an Assistant Professor at the University of Silesia, Institute of Computer Science. His research areas are related to data mining, decision trees, decision forests, and ant colony optimization. He proposed a new algorithm for constructing decision trees based on ant colony optimization. He also developed a new algorithm for the construction of a decision forest—ACDF. He has published a book and many articles concerning the subjects of ant colony optimization and data mining tasks. He has also been a speaker at many international conferences and has served as a member of several program committees of international conferences.

Przemysław Juszczuk

Przemysław Juszczuk earned his PhD in 2013 from the University of Silesia in Poland. Since 2020, he has been working in the Department of Decision Support in the Presence of Risk, Systems Research Institute, in Warsaw, Poland. His research interests include machine learning; multicriteria optimization; game theory; data analysis; financial markets; and decision support.

Preface to "Entropy in Real-World Datasets and Its Impact on Machine Learning"

Nowadays, machine learning is considered as a group of various methods used to solve the most complex real-world problems. Its usability is crucial in fields such as medicine, finance, text mining, image analysis, and more. Among the most prominent examples of machine-learning-related methods, we can find ensemble methods, multicriteria evolutionary algorithms, deep learning in neural networks, etc. Here, we are particularly interested in subjects connecting the entropy of datasets and the effectiveness of machine learning algorithms.

The main aspect of this book is devoted to entropy in the ever-growing amount of data available for users. Concepts such as big data and data streams are still increasingly gaining attention. The efficiency of classical methods seems to create debate amongst these types of data; thus, we believe that there is a necessity for continuous improvements in what is widely understood as machine learning. This book is dedicated to the analysis of real-world datasets, in particular, in terms of the entropy present in them and the impact on machine learning.

The topic of the book is very important nowadays, because ever-evolving machine learning techniques make it possible to obtain better real-world data. Therefore, this book contains information related to real data in fields such as automatic sign language translation, bike-sharing travel characteristics, stock index, sports data, fake news data, and more. However, it should be noted that the book also contains a lot of information on new developments in machine learning, new algorithms, algorithm modifications, and a new measure of classification quality assessment that also takes into account the preferences of the decision maker.

Jan Kozak and Przemysław Juszczuk
Editors

Article

Article

Learning to Classify DWDM Optical Channels from Tiny and Imbalanced Data

Paweł Cichosz [1], Stanisław Kozdrowski [1,*] and Sławomir Sujecki [2,3]

[1] Computer Science Institute, Warsaw University of Technology, Nowowiejska 15/19, 00-665 Warsaw, Poland; p.cichosz@elka.pw.edu.pl
[2] Faculty of Electronics, Military University of Technology, S. Kaliskiego 2, 00-908 Warsaw, Poland; slawomir.sujecki@wat.edu.pl
[3] Telecommunications and Teleinformatics Department, Wroclaw University of Science and Technology, Wyb. Wyspianskiego 27, 50-370 Wroclaw, Poland
* Correspondence: s.kozdrowski@elka.pw.edu.pl

Abstract: Applying machine learning algorithms for assessing the transmission quality in optical networks is associated with substantial challenges. Datasets that could provide training instances tend to be small and heavily imbalanced. This requires applying imbalanced compensation techniques when using binary classification algorithms, but it also makes one-class classification, learning only from instances of the majority class, a noteworthy alternative. This work examines the utility of both these approaches using a real dataset from a Dense Wavelength Division Multiplexing network operator, gathered through the network control plane. The dataset is indeed of a very small size and contains very few examples of 'bad" paths that do not deliver the required level of transmission quality. Two binary classification algorithms, random forest and extreme gradient boosting, are used in combination with two imbalance handling methods, instance weighting and synthetic minority class instance generation. Their predictive performance is compared with that of four one-class classification algorithms: One-class SVM, one-class naive Bayes classifier, isolation forest, and maximum entropy modeling. The one-class approach turns out to be clearly superior, particularly with respect to the level of classification precision, making it possible to obtain more practically useful models.

Keywords: machine learning; optical networks; imbalanced data; one-class classification

Citation: Cichosz, P.; Kozdrowski, S.; Sujecki, S. Learning to Classify DWDM Optical Channels from Tiny and Imbalanced Data. *Entropy* **2021**, *23*, 1504. https://doi.org/10.3390/e23111504

Academic Editors: Jan Kozak and Przemysław Juszczuk

Received: 8 October 2021
Accepted: 10 November 2021
Published: 13 November 2021

Publisher's Note: MDPI stays neutral with regard to jurisdictional claims in published maps and institutional affiliations.

Copyright: © 2021 by the authors. Licensee MDPI, Basel, Switzerland. This article is an open access article distributed under the terms and conditions of the Creative Commons Attribution (CC BY) license (https://creativecommons.org/licenses/by/4.0/).

1. Introduction

Constantly growing traffic in backbone networks makes dynamic and programmable optical networks increasingly important. This particularly applies to Dense Wavelength Division Multiplexing (DWDM) networks whereby efficient use of network resources is of paramount importance. Introducing automation, frequent network reconfiguration, re-optimization and network reliability monitoring allows DWDM network operators to minimize the capital expenditures (Capex) and operating expenditures (Opex) [1–6]. Currently, software-defined networking (SDN) is used to achieve all these objectives. SDN uses a logically centralized control plane in a DWDM network that is realized using purpose-built flexible hardware such as reconfigurable optical add/drop multiplexers (ROADMs), flexible line interfaces, etc. [7,8]. In modern DWDM optical networks, following the software defined network paradigm, DWDM network reconfiguration is becoming more frequent, making the evolving network more resilient and adapting faster to real changes in bandwidth demand so that network reconfigurations can closely match changes in bandwidth demand. However, bandwidth demand can change very quickly (fluctuations can occur within minutes), while network reconfigurations typically take much longer. This is mainly due to operational processes that are too slow to allow real-time network re-optimization. It is therefore important that DWDM network reconfigurations are automated and as fast as possible, without significantly increasing operational costs.

Frequent network reconfiguration and re-optimization necessary to make the best use of available resources has been facilitated by the introduction of software-defined networking (SDN) and knowledge-based networking (KDN) paradigms [7,9–11]. Central to SDN and KDN is automatic provisioning of optical channels (lightpaths), which is based on accurate quality estimation for them. Machine learning (ML) is a promising solution to this problem. Therefore, a number of algorithms have been proposed that first create a database using numerical modelling tools and then implement ML to estimate the quality of optical links ([6,7,12]). However, in the approach presented here, we apply ML to a database that has been extracted directly via the control plane from the DWDM network under analysis. This approach leads to an ML problem that is clearly different from the one addressed in [6,7,11,12], as there are significant challenges in using real optical network datasets, related to data representation, data size and class imbalance, which is intrinsic to data gathered via control plane from an operating DWDM network. The class imbalance follows from the fact that in an operating DWDM network there may be dozens or hundreds of operating connections but there is not much information available (if any) on connections that could not be realized due to excessive bit error rate. Therefore, a specially tailored ML approach is required to tackle this problem.

The main advantage of gathering data via the control plane is that it can be easily implemented by a DWDM network operator. As it will be further explained later, this approach imposes some constraints on the choice of appropriate ML methods due to above-mentioned class imbalance, which is an intrinsic feature of data collected via control plane. As already mentioned this makes the ML problem considered in this contribution clearly different from those considered so far in most of the available literature [6,7,12,13]. Expanding upon our previous work [14,15], we compare the predictive performance of the most successful binary classification algorithms combined with different techniques for class imbalance compensation and that of one-class classification algorithms that learn from majority class instances only.

1.1. Machine Learning Challenges

Successful applications of machine learning to support optical network design require real training data. While experiments on synthetic data may provide encouraging demonstrations, they are likely not to adequately represent the challenges that are associated with this application area and therefore provide overoptimistic predictive performance estimates or fail to identify potential obstacles and culprits. These challenges are mainly related to data size and quality.

DWDM network operators, particularly operating small or medium networks, may be unable to provide a dataset with more than several dozen or at best several hundred paths. More importantly, the vast majority if not all of those path configurations would usually correspond to correct, working channel designs. This is because unsuccessful path configurations are often discarded rather than archived, at least before the provider becomes aware of their utility as training data for machine learning. Before this awareness increases, available real datasets remain tiny and extremely imbalanced.

The data that has been made available for this study comes from a DWDM network operator providing services in Poland and is an excellent example of these issues. It contains just about a hundred of paths, including only three "bad" ones (i.e., such that could not be allocated due to a low quality of transmission). While it is still possible to use such data to train predictive models using classification learning algorithms, special care is needed to increase their sensitivity to the minority class and to reliably evaluate their quality. The extreme dominance of the "good" class makes it easy to come up with apparently accurate models with little or no actual predictive utility. To avoid this, we compensate the class imbalance using instance weighting and synthetic minority-class instance generation. However, the tiny size and extreme imbalance of the data may be still on the edge of the capabilities of standard binary classification, even with such compensation techniques. Therefore one-class classification, in which only "good" paths are used as training data, may be a viable and promising alternative. To compare the predictive performance of the binary and one-class classification approaches, their predictions are evaluated using ROC and precision-recall curves in combination with stratified cross-validation.

1.2. Related Work

To the authors' best knowledge this work is the first to apply one-class classification in the optical network domain and to compare its predictive performance to binary classification using different techniques of handling class imbalance. There is, however, some related prior work on applying other machine learning methods to optical networks as well as on using one-class classification as an alternative to binary classification for imbalanced data.

Considering work related to optical networks first, in [16] authors show that a routing and spectrum allocation (RSA) that monitors QoT in multiple slices significantly improves network performance. Rottondi et al. [12] extensively discuss and use ML techniques to optimise complex systems where analytical models fail. However, the network data in [12] was generated artificially, whereas in this contribution the data is collected by control plane from an operating network.

Similar problems related to lightpath QoT estimation are addressed by Mata et al. [17] but they mainly focus on the SVM classifier only. Barletta et al. [18] on the other hand, use mainly Random Forest algorithm that predicts whether the BER (Bit Error Rate) of unestablished lightpaths meets the required threshold based on traffic volume, desired route and modulation format. As in [12] the system is trained and tested on artificial data, which is different to the approach adopted in this contribution.

Japkowicz [19] compared different ways of handling class imbalance including one-class classification. Japkowicz [19] found binary classification with imbalanced compensation superior to one class classification but experiments performed in [19] used artificial data and neural network classifier (with one-class classification performed using an autoassociative network type).

Lee and Cho [20] advocated the use of one-class classification for imbalanced data and demonstrated that it can outperform binary classification if the imbalanced ratio is high. They experimented with the standard and one-class versions of the SVM algorithm.

Bellinger et al. [21] discuss the potential utility of one-class classification in binary classification tasks with extreme class imbalance, as in our case. Their results suggest that binary classification with class imbalance compensation methods may be more useful than one-class classification when dealing with data from complex multi-modal distributions. However their results are based on datasets where the number of minority class instances is bigger than in our case.

1.3. Article organization

The rest of the paper is organized as follows. In Section 2 the analyzed optical network data, the applied machine learning algorithms, and model evaluation methods are described. The results of the experimental study are presented in Section 3 and discussed in Section 4. Contributions of this work and future research directions are summarized in Section 5.

2. Materials and Methods

The data comes from a real DWDM optical network of a large telecom operator. The network uses 96 DWDM channels allocated in C-band and is physically located in Poland, with network nodes corresponding to Polish cities.

2.1. Data

The network is equipped exclusively with coherent transponders. This is a typical representative of a new network created by an operator. The coherent transponders belong to Ciena's 6500 family, with transmission rates of 100 G, 200 G and 400 G and four types of modulation: QPSK, 16QAM, 32QAM and 64QAM.

Data preparation process is depicted in Figure 1. In order to better understand the meaning of the various database attributes presented later in the subsection, in the context of DWDM technology, an example DWDM network topology is shown in Figure 2. Figure 3 illustrates the concepts of network node, hop, hop length, path, and transponder. The dataset contains 107 optical paths, 3 of which correspond to unsuccessful designs ("bad") and rest of them (104) are operational ("good").

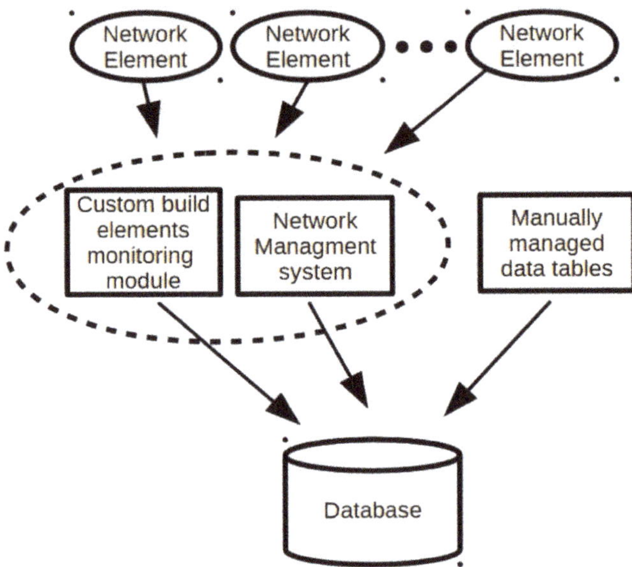

Figure 1. Data preparation process.

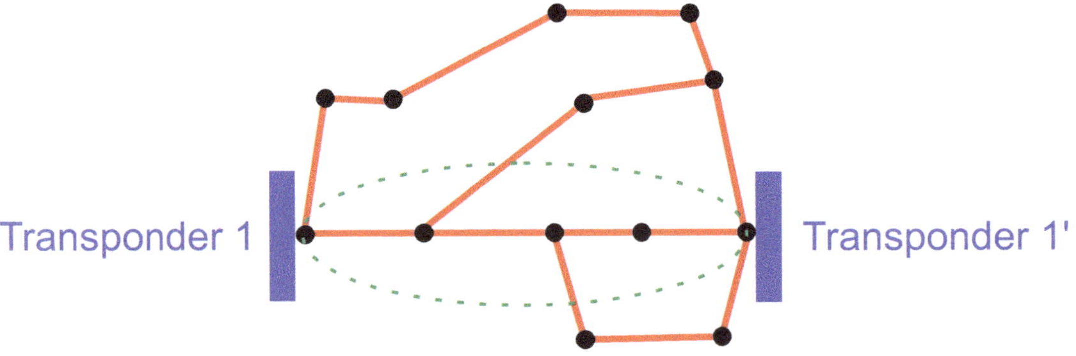

Figure 2. An example DWDM network topology.

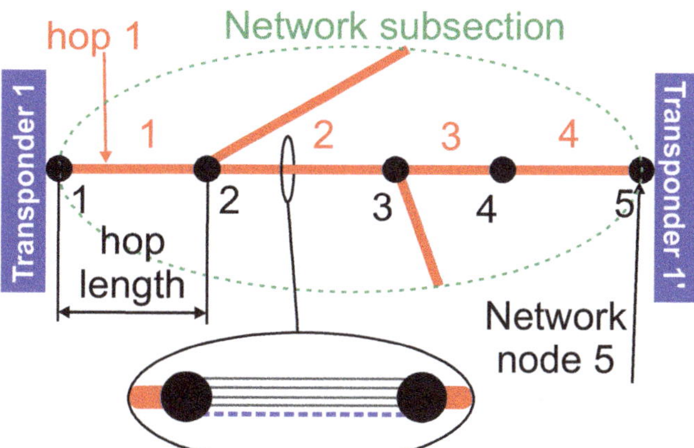

Figure 3. Network subsection illustrating the meaning of the specific channel attributes occurring in the studied database.

2.1.1. Path Description

Network paths are described by several properties that may be related to transmission quality and expected to be predictively useful. The `hop_lengths` property gives the length of each hop that forms a path from the initial transponder to the destination transponder. This property is important because the signal to noise ratio depends on the length of the fibre connecting both transponders. In each hop there are usually more wavelengths occupied. This is because these wavelengths are used by paths other than the one occupied by the considered path. All paths can interact through nonlinear phenomena like four wave mixing and thus affect the quality of transmission. Therefore, the `num_of_paths_in_hops` property, which gives the number of adjacent DWDM wavelengths in a given hop, is included. The `hop_losses` property gives the value of the optical loss for a given hop. Again, hop losses affect the signal to noise ratio and hence the corresponding property was included. Another property, `number_of_hops`, provides information on how many hops are present in a path from the initial to destination transponder. Since each hop corresponds to a signal passing through a DWDM node, the number of hops affects the signal to noise ratio due to optical regeneration taking place in a DWDM node. The last two properties are intrinsically related with a specific type of transponder used. The `transponder_modulation` property stores information on the transponder modulation format, e.g., QPSK or 16QAM. This property is important because modulation format is related to receiver sensitivity. Finally, the `transponder_bitrate` property is in essence self explanatory and gives the bit rate of a given transponder. Transponder bit rate also affects receiver sensitivity and hence it is included.

2.1.2. Vector Representation

Path descriptions were transformed to a vector representation, as expected by classification algorithms for tabular data, by a simple aggregation-based feature engineering technique. Each of the available edge properties (`hop_lengths`, `num_of_paths_in_hops`, `hop_losses`) was aggregated by calculating the mean and standard deviation over all edges in the path. This gives 6 attributes derived from edge properties (2 attributes for each of the 3 edge properties), in addition to the 3 path attributes unrelated to individual edges (`number_of_hops`, `transponder_modulation`, and `transponder_bitrate`).

Applying additional aggregation functions to edge properties, such as the minimum, the maximum, the median, the first quartile, the third quartile, or the linear correlation coefficient with the ordinal number of the edge in the path, as in our prior work [14], may create some additional predictively useful attributes. However, this would make the dimensionality of this representation relatively high in comparison to the size of the available dataset, considerably increasing the risk of overfitting.

2.2. Binary Classification

Any standard classification algorithm can be used to predict channel "good"/"bad" class labels or probabilities. In this work we limit our attention to the two algorithms that performed the best in our previous study [14]: Random forest and extreme gradient boosting. They belong to the most successful learning algorithms for tabular data and it is very unlikely that their performance could be beaten by other algorithms using the same vector path representation.

2.2.1. Random Forest

The random forest algorithm creates a model ensemble consisting of multiple decision trees [22]. They are grown on bootstrap samples from the training set by using a mostly standard decision tree growing algorithm [23,24]. However, since the expected improvement of the resulting model ensemble over a single model is contingent upon sufficient diversity of the individual models in the ensemble [25,26], the following modifications are applied to stimulate the diversity of decision trees that are supposed to constitute a random forest:

- large maximally fitted trees are grown (with splitting continued until reaching a uniform class, exhausting the set of instances, or exhausting the set of possible splits),
- whenever a split has to be selected for a tree node, a small subset of available attributes is selected randomly and only those attributes are considered for candidate splits.

To use a random forest model for prediction, simple unweighted voting of individual trees from the model is performed, and vote distribution is used to obtain class probability predictions. With dozens or (more typically) hundreds trees this voting mechanism usually makes random forests highly accurate and resistant to overfitting. An additional important advantage of the algorithm is its ease of use, resulting from limited sensitivity to parameter settings, which makes it possible to obtain high quality models without excessive tuning.

2.2.2. Extreme Gradient Boosting

Extreme gradient boosting or *xgboost* is is another highly successful ensemble modeling algorithm. As other boosting algorithms, it creates ensemble components sequentially in such a way that each subsequent model best combines with the previously created ones [27–30].

The *xgboost* algorithm internally uses regression trees for model representation and optimizes an ensemble quality measure that includes a loss term and a regularization term [31]. Each subsequent tree is grown to minimize the sum of loss and regularization terms of all trees so far. Split selection criteria, stop criteria, and leaf values are derived from this minimization by the Taylor expansion of the loss function, using its gradient and hessian decomposed to terms for particular training instances and then assigned to the corresponding nodes and leaves of the tree being grown.

Extreme gradient boosting applied to binary classification is typically used with logarithmic loss (the negated log-likelihood) and the summed up numeric predictions of individual regression trees are transformed by a logistic link function to obtain class probability predictions.

The extreme gradient boosting algorithm is capable of providing excellent prediction quality, sometimes outperforming random forest models. It can overfit, however, if the number of trees grown is too large.

2.2.3. Handling Class Imbalance

Techniques for compensating class imbalance can be divided in the following three main categories:

- internal compensation by the learning algorithm, controlled by its parameter settings,
- compensation by data resampling,
- compensation by synthetic minority class data generation.

Techniques of the first category are generally supposed to increase sensitivity to the minority class without modifying the training data. They are possible with many classification algorithms and often consist in specifying class weights or prior probabilities. The binary classification algorithms used by this work are both ensemble modeling algorithms, which tend to be quite robust with respect to class imbalance, but their model quality can be still improved by such compensation mechanisms.

In the case of the random forest algorithm there are actually two possible related techniques. One is drawing bootstrap samples in a stratified manner, with different selection probabilities for particular classes. In the extreme case, a bootstrap sample may contain all instances from the minority class and the sample of the same size from the dominating class. The other is to specify instance weights affecting split selection and stop criteria for tree growing. Since in our case classes are extremely imbalanced and there are very few instances of the minority class, the weighting technique is preferred to the stratified sampling technique, since the latter would have to severely undersample the dominating class, with a possibly negative effect on model performance. The same weighting technique is also used with the the *xgboost* algorithm. In this case instance weights are used when calculating the logarithmic loss, so that the contribution of minority class instances to the loss function minimized by the algorithm is increased.

Data resampling may be performed by minority class oversampling (replicating randomly selected minority class instances), majority class undersampling (selecting a sample of majority class instances), or a combination of both, so that the resampled training set has either fully balanced classes or at least considerably more balanced than originally. Unfortunately these techniques have very limited utility for datasets that are both small and extremely imbalanced, as in our case. Undersampling would remove most of the available training data, and oversampling would replicate the very few "bad" paths increasing the risk of overfitting to these specific instances. They can be therefore hardly expected to offer any advantages over internal imbalance compensation by weighting and are not used in this work.

Potentially more useful techniques of generating synthetic minority class instances can be considered more refined forms of oversampling in which minority class instances available in the training data are not directly replicated, but used to generate new synthetic instances. This is supposed to make the increased representation of the minority class in the modified training set more diverse and thus reduce the risk of overfitting. Two well known specific techniques based of this idea are SMOTE [32] and ROSE [33] and they are both used in our experimental study. SMOTE finds nearest neighbors of each minority class instance and then generates new synthetic instances by interpolating between the instance and its neighbors. ROSE adopts a smoothed bootstrap sampling approach, with new instances generated in the neighborhood of original instances by drawing from a conditional kernel density estimate of attribute values given the class. Both minority and majority class instances are generated, and the class distribution in the generated dataset can be controlled to achieve a desired level of balance.

2.3. One-Class Classification

One-class classification follows the following learning scenario [34,35]:

- the training contains only instances of a single class,
- the learned model is supposed to predict for any instance whether it belongs to the single class represented in the training set.

In our case the single class represented in the training set corresponds to "good" paths. When the obtained model is applied to prediction, it identifies paths which are likely to also be "good" (i.e., be of the same class as that represented in the training set) and those which are likely to be "bad" (i.e., not to be of the same class as that represented in the training set). It can be assumed that model predictions are provided in the form of decision function values (numeric scores) such that higher values are assigned to instances that are considered less likely to be if the same class as that represented in the training set, i.e., in our case, more likely to be "bad" paths.

One-class classification is most often applied to unsupervised or semi-supervised anomaly detection [36], where an unlabeled training set, assumed to contain only normal instances, sometimes with a small fraction of anomalous instances, is used to learn a model that can detect anomalous instances in new data. It can be also useful, however, for binary classification tasks with extreme class imbalance [21], particularly when the number of minority class instances is too small for standard binary classification algorithms, even combined with imbalance compensation techniques. This may be often the case with data for optical channel classification.

The best known and widely used one-class classification algorithm is one-class SVM. In our experimental study it is compared with three other algorithms: The one-class naive Bayes classifier, the isolation forest algorithm, and the maximum entropy modeling algorithm. The first of those is a straightforward modification of the standard naive Bayes classifier and probably the simplest potentially useful one-class learning algorithm. The second one, while designed specifically for anomaly detection applications, can also serve as a general-purpose one-class classification algorithm. The third one, while originally intended for creating models of species distribution in ecosystems, has been also found to be useful for one-class classification.

2.3.1. One-Class SVM

The one-class SVM algorithm uses a linear decision boundary, like standard SVM, but adopts a different principle to determine its parameters. Rather than maximizing the classification margin, which is not applicable to one-class classification, it maximizes the distance from the origin while separating the majority of training instances therefrom [37]. The side of the decision boundary opposite from the origin corresponds to the class represented in the training set. Only a small set of outlying training instances are permitted to be left behind, and the upper bound on the share of such outlying instances in the training set is specified via the ν parameter.

The principle of separating most of the training set from the origin of the space is typically combined with a translation-invariant kernel function (such as the radial kernel), sothat instances in the transformed representation lie on a sphere centered in the origin. The separating hyperplane then cuts off a segment of the sphere where most training instances are located.

One-class SVM predictions are signed distances from the decision boundary, positive on the "one-class" (normal) side and negative on the outlying side. The negated value of such signed distance can therefore serve as a numeric score for ranking new instances with respect to their likelihood of not belonging to the class represented in the training set.

2.3.2. One-Class Naive Bayes Classifier

The one-class modification of the naive Bayes classifier is particularly straightforward. Since only one class is represented in the training set, its prior probability is assumed to be 1, conditional attribute-value probabilities within this class are estimated on the full training set, and the probability of an instance belonging to this class is proportional to the product of such attribute-value probabilities [38]. For numeric attributes, Gaussian density function values, with the mean and standard deviation estimated on the training data, are used instead of discrete attribute-value probabilities.

Discrete class predictions can be made by comparing the product of attribute-value probabilities for a given instance to a threshold, set to or around the minimum value of this product over the training set. Numeric scores (decision function values) can be defined as the difference between such a threshold and the probability being compared.

2.3.3. Isolation Forest

The isolation forest algorithm was proposed as an anomaly detection method [39], but it can also serve as a one-class classification algorithm regardless of whether instances not belonging to the class represented in the training set are interpreted as anomalous. Its model representation consists of multiple isolation trees grown with random split selection. These are not standard decision or regression trees, since no labels or values are assigned to leaves, and they just partition the input space. Splitting is stopped whenever a single training instance is left or a specified maximum depth is reached.

In the prediction phase each isolation tree is used to determine the path length between the root node and the leaf at which the instance arrives after traversing down the tree along splits. Instances that do not belong to the class represented in the training set can be expected to be easier to isolate (have shorter paths) than those which do belong to the class. The average path length over all trees in the forest can then serve as a decision value function for determining whether an instance is likely to belong to this class or not. The original algorithm transforms this average path length into a standardized anomaly score for generating alerts in anomaly detection applications, using the expected depth of unsuccessful BST searches. This is not necessary for one-class classification, since the negated average path length is sufficient to rank new instances with respect to their likelihood of not belonging to the class represented by the training set.

An extended version of the isolation forest used for this work employs multivariate rather than univariate splits [40]. This eliminates a bias that resulted in the original algorithm from using axis-parallel hyperplanes for data splitting.

2.3.4. Maximum Entropy Modeling

The maximum entropy modeling or maxent algorithm was originally developed for ecological species geographical distribution prediction based on available presence data, i.e., locations where a given species has been found and their attributes, used to derive environmental features [41]. These features, besides raw continuous attribute values, include attributes derived by several transformations, as well as binary features obtained by comparing continuous attributes with threshold values and by one-hot encoding of discrete attributes, with an internal forward feature selection process employed based on nested model comparison [42].

The algorithm, following the maximum entropy principle [43], identifies a species occurrence probability distribution that has a maximum entropy (i.e., is most spread out) while preserving constraints on environmental features. These constraints require that the expected values of environmental features under the estimated species occurrence probability distribution should be close to their averages from the presence points. The obtained model can provide, for an arbitrary point, the prediction of the species occurrence probability.

Despite its original intended purpose, maxent has been found to be useful as a general-purpose one-class classification algorithm [44,45]. Training instances take the role of "presence points", and input attributes are used to derive "environmental features", whereas background points can be generated by uniformly sampling the attribute ranges. Model prediction for an arbitrary instance can be interpreted as the probability that it belongs to the class represented in the training set.

2.4. Model Evaluation

Both binary and one-class classification algorithms used by this work produce scoring predictive models – their predictions are numeric values ranking instances with respect to the likelihood of being a "bad" path (or not belonging to the class represented in the training data). When applying standard binary classification quality measures to evaluate these predictions using, we refer to to the "good" class (represented in the training data for one-class classification), as *negative*, and the "bad" class (not represented in the training data for one-class classification) as *positive*.

ROC and precision-recall (PR) curves are used to visualize the predictive performance of the obtained models. ROC curves make it possible to observe possible tradeoff points between the *true positive rate* (the share of positive instances correctly predicted to be positive) and the *false positive rate* (the share of negative instances incorrectly predicted to be positive) [46,47]. PR curves similarly present possible levels of tradeoff between the *precision* (the share of positive class predictions that are correct) and the *recall* (the same as the true positive rate). The overall predictive power is summarized using the area under the ROC curve (AUC) and the area under the precision-recall curves (PR AUC).

In our case the true positive rate and the recall is the share of "bad" paths that are correctly predicted to be "bad", the false positive rate is the share of "good" paths that are incorrectly predicted to be "bad", and the precision is the share of "bad" class predictions that are correct. The area under the ROC curve can be interpreted as the probability that a randomly selected "bad" path is scored higher by the model than a randomly selected "good" path and the area under the PR curve can be interpreted as the average precision across all recall values.

When using data with heavily imbalanced classes, where positive instances are extremely scarce, even numerous false positives do not substantially decrease the false positive rate, since the number of false positives may be still small relative to the dominating negative class count. This is not the case for the precision, though, which is much more sensitive to false positives. Therefore precision-recall curves may be expected to be more informative and better highlight differences in the predictive performance obtained using different algorithms. For a more complete picture, however, both ROC and PR curves are presented.

To make an effective use of the small available dataset for both model creation and evaluation as well as to keep the evaluation bias and variance at a minimum, the $n \times k$-fold cross-validation procedure ($n \times k$-CV) is applied [48]. The dataset is split into k equally sized subsets, each of which serves as a test set for evaluating the model created on the combined remaining subsets, and this process is repeated n times to further reduce the variance. To evaluated binary classification models, the random partitioning into k subsets is performed by stratified sampling, preserving roughly the same number of minority class instances in each subset. For the evaluation of one-class classification models a one-class version of the $n \times k$-CV procedure is used in which the few instances of the "bad" class are never used for training but always used for testing. The true class labels and predictions for all $n \times k$ iterations are then combined to determine ROC curves, PR curves, and the corresponding AUC values.

While it appears a common practice to use the leave-out-out procedure rather than k-fold cross-validation when working with small datasets, the only potential advantage of the former would be avoiding the pessimistic bias resulting from the fact that in each iteration of the latter $\frac{1}{k}$ of the data are not used for model creation. However, the leave-one-out procedure has high variance (that cannot be reduced by multiple repetitions since the procedure is fully deterministic) and excluding only a single instance from the training data may cause optimistic bias due to underrepresenting the differences between the training data and the test data. We find it therefore more justified to use the $n \times k$-fold cross-validation procedure where the variance is substantially reduced and accept the fact that it may be pessimistically biased. This means that our reported results may underestimate the actually possible predictive performance levels, which should be preferred to any risk of optimistic bias.

3. Results

In the experimental study presented in this section binary and one-class classification algorithms described in Sections 2.2 and 2.3 are applied to the small and imbalanced dataset described in Section 2.1. The objective of the study is to verify the level of optical channel classification quality that can be obtained using these two types of algorithms. For binary classification the effects of class imbalance compensation using instance weights and synthetic minority class instance generation are also examined.

3.1. Algorithm Implementations and Setup

The following algorithm implementations are used in the experiments:

- **random forest (RF):** the implementation provided by the ranger R package [49],
- **extreme gradient boosting (XGB):** The implementation provided by the xgboost R package [50],
- **SMOTE:** The implementation provided by the smotefamily R package [51],
- **ROSE:** The implementation provided by the ROSE R package [52],
- **one-class SVM (OCSVM):** The implementation provided by the e1071 R package [53],
- **one-class naive Bayes classifier (OCNB):** The implementation provided by the e1071 R package [53], with a custom prediction method to handle the one-class classification mode specifically implemented for this work,
- **isolation forest (IF):** The implementation provided by the isotree R package [54],
- **maximum entropy modeling (ME):** The implementation provided by the MIAmaxent R package [55], with background data generation by random sampling of attribute value ranges specifically implemented for this work.

Since the *xgboost* algorithm does not directly support discrete attributes and one attribute in the dataset is discrete, it was preprocessed by converting discrete values to binary indicator columns.

The tiny size of the dataset and, particularly, the number of "bad" path configurations makes it hardly possible to perform algorithm hyper-parameter tuning. While the performance evaluation obtained by $n \times k$-fold cross-validation could be used to adjust algorithm settings and improve the results, as demonstrated in our previous work [14], without the possibility to evaluate the expected predictive performance of the tuned configurations on new data it could lead to overoptimistic results. This is why the algorithms are used in the following mostly default configurations, with only a few parameters set manually where defaults are unavailable or clearly inadequate:

- **random forest:** A forest of 500 trees is grown, with the number of attributes drawn at random for split selection set to the square root of the number of all available attributes,
- **extreme gradient boosting:** 50 boosting iterations are performed, growing trees with a maximum depth of 6 and scaling the contribution of each tree by a learning rate factor of 0.3, and applying L_2 regularization on leaf values with a regularization coefficient of 1,
- **SMOTE:** The number of nearest neighbors of minority class instances is set to 1 (which is the only available choice given the fact there are just three minority class instances in the data two of which are available for model creation in each cross-validation fold),
- **ROSE:** The generated dataset size and the probability of the minority class are set so as to approximately preserve the number of majority class instances and increase the number of minority class instances,
- **one-class SVM:** The radial kernel function is used, with the γ kernel parameter set to the reciprocal of the input dimensionality, and the ν parameter specifying an upper bound on the share of training instances that may be considered outlying is equal 0.5,
- **isolation forest:** The extended version of the algorithm is used [40], with multivariate splits based on three attributes, a forest of 500 isolation trees is grown, and for each of them the data sample size is equal the training set size (which is a reasonable

setup for a small dataset), and the maximum three depth is the ceiling of the base-2 logarithm thereof,

- **maximum entropy modeling:** All available attribute transformations [42] are applied to derive environmental features (linear, monotone, deviation, forward hinge, reverse hinge, threshold, and binary one-hot encoding), a significance threshold used for internal feature selection is set to 0.001, and the generated background data size is 1000.

For imbalance compensation with instance weighting the majority class weight is fixed as 1 and the minority class weight is set to values from the following sequence: $1, 2, 5, 10, 20, 50, 100$ (where 1 corresponds to no weighting). When using synthetic instance generation, the number of generated minority class instances is set to $d - 1$ times the number of real minority class instances, where d is in the same sequence as above. This can be achieved exactly for SMOTE and only approximately for ROSE due to its probabilistic nature.

The $n \times k$-fold cross-validation procedure is used with $k = 3$ (since there are only 3 minority class instances) and $n = 50$ (to keep the evaluation variance at a minimum).

3.2. Classification Performance

For each of the binary and one-class classification algorithm configurations described above cross-validated ROC and PR curves, with the corresponding area under the curve values, are reported and briefly discussed below. A bootstrap test (with 2000 replicates drawn from the data) is used for verifying the statistical significance of the observed AUC differences.

3.2.1. Binary Classification

Figure 4 presents the ROC and PR curves obtained for binary classification with instance weighting. The numbers in the parentheses after algorithm acronyms in the plot legends specify the minority instance weight value. For readability, only the results without weighting and with the best weight value are included. All the observed differences are statistically significant according to the bootstrap test except for those between RF(1) and XGB(5), and between RF(20) and XGB(5). One can observe that:

- according to the ROC curves the prediction quality appears very good, with AUC values of 0.96–0.97,
- nearly perfect ROC operating points are possible, with the true positive rate of 1 and the false positive rate of 0.05 or less,
- the precision-recall curves reveal that the prediction quality is not actually perfect, with the average precision just above 0.3 at best,
- without instance weighting the random forest algorithm outperforms *xgboost*, but with instance weighting they both perform on roughly the same level,
- imbalance compensation with instance weighting improves the predictive performance of both the algorithms, with the effect more pronounced for extreme gradient boosting.

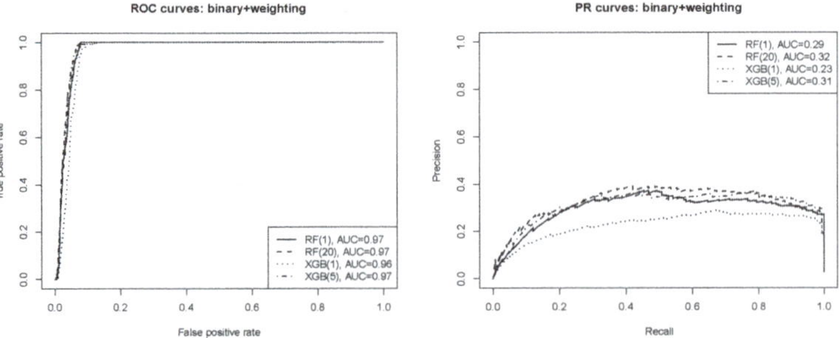

Figure 4. The ROC and PR curves for binary classification with instance weighting.

Figure 5 presents the ROC and PR curves obtained for binary classification with synthetic instance generation. The numbers in the parentheses after algorithm acronyms in the plot legends specify the minority class size multiplication coefficient. For readability, only the best results obtained when using SMOTE and ROSE are included and the results with no synthetic instance generation as a comparison baseline. All the observed differences are statistically significant according to the bootstrap test. One can observe that:

- synthetic instance generation reduces the prediction quality of the random forest algorithm, but provides an improvement for extreme gradient boosting,
- the effects of SMOTE and ROSE for *xgboost* are similar except for the fact that the latter works better with bigger minority class multiplication coefficients,
- the results for both SMOTE and ROSE are worse than those obtained with instance weighting.

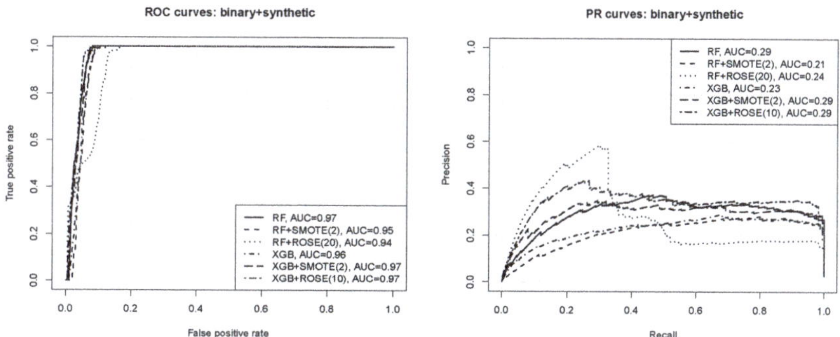

Figure 5. The ROC and PR curves for binary classification with synthetic instance generation.

3.2.2. One-Class Classification

The ROC and precision-recall curves for one-class classification are presented in Figure 6. All the observed differences are statistically significant according to the bootstrap test except for the one between IF and OCNB. One can observe that:

- all the algorithms produce models capable of successfully detecting out-of-class instances ("bad" paths), with AUC values between 0.96 and 0.98,
- the one-class naive Bayes and isolation forest algorithms achieve the maximum true positive rate value for a slightly less false positive rate value than the one-class SVM and maxent algorithms,
- the algorithms differ more substantially with respect to the average precision achieved, which is about 0.6 for one-class SVM and maxent, 0.66 for the one-class naive Bayes classifier, and 0.77 for the isolation forest algorithm,
- the isolation forest and one-class naive Bayes models maintain a high precision of 0.7 or above for a wide range of recall values (up to about 0.9), whereas the one-class SVM and maxent models can only maintain a precision level of 0.6 and 0.5, respectively, in the same range of recall values,
- all the one-class algorithms produce better models than those obtained by binary classification.

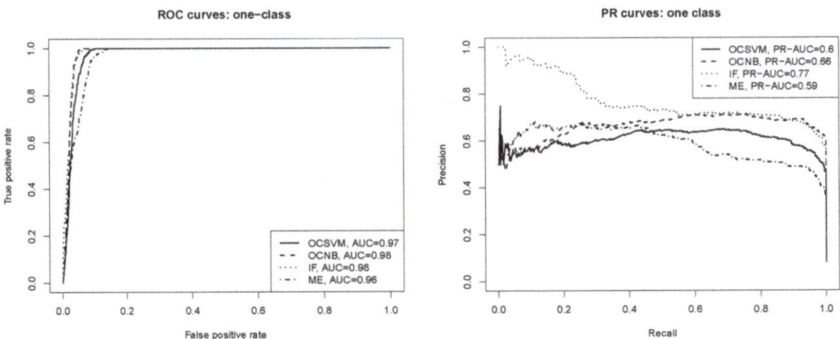

Figure 6. The ROC and PR curves for one-class classification.

4. Discussion

As discussed in Section 2.4, ROC curves do not provide a sufficient picture of model performance under severe class imbalance, because even with many false positives the false positive rate remains small due to the dominating overall negative class count. This is why they suggest that all the investigated algorithms achieve excellent prediction quality and their models exhibit only minor performance differences. Precision-recall curves indeed show a more useful view of the predictive quality of models obtained by particular algorithms and better highlight the differences between them.

For binary classification algorithms the simple instance weighting technique appears more useful than the more refined and computationally expensive synthetic instance generation techniques. This may be surprising at first, but actually neither SMOTE or ROSE are well suited to working with datasets not only heavily imbalanced but also very small. With just three minority class instances (two remaining for model creation within a single cross-validation fold) there is probably not enough real data to provide a reliable basis for synthetic data generation.

One-class classification algorithms, although using less input information (training data of the majority class only), all produce clearly better models than the best of those obtained using binary classification. The isolation forest algorithm turns out to deliver a superior overall predictive power and considerably more preferable operating points, with near-perfect detection of true positives ("bad" paths) without excessively many false positives. While all the algorithms deliver high quality models, the one-class naive Bayes and isolation forest algorithms clearly outperform the one-class SVM and maxent algorithms. It is particularly noteworthy that they can provide high precision in a wide range of recall values.

This study suggests that standard methods of handling class imbalance may be insufficient when the dataset is of a very small size. Indeed, it is not only the small share, but also the small absolute number of "bad" paths that prevents binary classification algorithms from creating more successful models. While the skewed class distribution can be compensated for by weighting, just a few training instances provide very poor basis for detecting generalizable patterns and for generating synthetic instances. Using only "good" paths for model creation leads to better results. The best obtained one-class models providing a precision level of about 0.7 are much more practically useful than the best binary classification models with precision just above 0.3.

5. Conclusions

The work has provided additional evidence that applying machine learning to optical channel classification is a promising work direction, but is associated with important challenges. To achieve models applicable in real-world conditions one has to use real-world datasets, but these suffer from severe imperfections, the most important of which are a small size and a heavy class imbalance. We have demonstrated that state-of-the-art binary classification algorithms may not achieve a very high level of prediction quality even when coupled with appropriate imbalance compensation techniques. The utility of the latter may be limited by the fact that it is not only the relative share of the minority class instances in the data that is small, but also their absolute count. The reported results confirm that one-class classification is a viable alternative, and models learned using majority class data only achieve better classification precision that those obtained using binary classification learning from all data.

Our findings provide an encouragement to continue this research direction by extending input representation with additional attributes, applying more one-class classification algorithms, and tuning their parameters to further improve the predictive performance. Gathering additional data not only would make the results of these enhanced future studies more reliable, but also make it possible to examine further ideas, such as model transfer between different networks or combining models trained on data from different networks. Expert knowledge on the physics of optical networks may permit defining alternative or additional path attributes, creating a more adequate input space representation for machine learning. Such knowledge could also be used to design a domain-specific data augmentation method that might be expected to perform better than general-purpose techniques of synthetic minority-class instance generation.

Author Contributions: Conceptualization, P.C., S.K. and S.S.; methodology, P.C., S.K. and S.S.; software, P.C.; validation, S.K., S.S. and P.C.; formal analysis, P.C., S.K. and S.S.; writing—original draft preparation, P.C., S.K. and S.S.; writing—review and editing, S.K., S.S. and P.C. All authors have read and agreed to the published version of the manuscript.

Funding: Not applicable.

Institutional Review Board Statement: Not applicable.

Informed Consent Statement: Not applicable.

Data Availability Statement: Not applicable.

Acknowledgments: The authors would like to thank the anonymous reviewers for their valuable comments.

Conflicts of Interest: The authors declare no conflict of interest.

References

1. Kozdrowski, S.; Żotkiewicz, M.; Sujecki, S. Ultra-Wideband WDM Optical Network Optimization. *Photonics* **2020**, *7*, 16. [CrossRef]
2. Klinkowski, M.; Żotkiewicz, M.; Walkowiak, K.; Pióro, M.; Ruiz, M.; Velasco, L. Solving large instances of the RSA problem in flexgrid elastic optical networks. *IEEE/OSA J. Opt. Commun. Netw.* **2016**, *8*, 320–330. [CrossRef]
3. Ruiz, M.; Pióro, M.; Żotkiewicz, M.; Klinkowski, M.; Velasco, L. Column generation algorithm for RSA problems in flexgrid optical networks. *Photonic Netw. Commun.* **2013**, *26*, 53–64. [CrossRef]
4. Dallaglio, M.; Giorgetti, A.; Sambo, N.; Velasco, L.; Castoldi, P. Routing, Spectrum, and Transponder Assignment in Elastic Optical Networks. *J. Lightw. Technol.* **2015**, *33*, 4648–4658. [CrossRef]
5. Kozdrowski, S.; Żotkiewicz, M.; Sujecki, S. Resource optimization in fully flexible optical node architectures. In Proceedings of the 20th International Conference on Transparent Optical Networks (ICTON), Bucharest, Romania, 1–5 July 2018.
6. Panayiotou, T.; Manousakis, K.; Chatzis, S.P.; Ellinas, G. A Data-Driven Bandwidth Allocation Framework With QoS Considerations for EONs. *J. Lightw. Technol.* **2019**, *37*, 1853–1864. [CrossRef]
7. Morais, R.M.; Pedro, J. Machine learning models for estimating quality of transmission in DWDM networks. *IEEE/OSA J. Opt. Commun. Netw.* **2018**, *10*, D84–D99. [CrossRef]
8. Kozdrowski, S.; Żotkiewicz, M.; Sujecki, S. Optimization of Optical Networks Based on CDC-ROADM Tech. *Appl. Sci.* **2019**, *9*, 399. [CrossRef]
9. Mestres, A.; Rodríguez-Natal, A.; Carner, J.; Barlet-Ros, P.; Alarcón, E.; Solé, M.; Muntés, V.; Meyer, D.; Barkai, S.; Hibbett, M.J.; et al. Knowledge-Defined Networking. *arXiv* **2016**, arXiv:1606.06222.
10. Musumeci, F.; Rottondi, C.; Nag, A.; Macaluso, I.; Zibar, D.; Ruffini, M.; Tornatore, M. An Overview on Application of Machine Learning Techniques in Optical Networks. *IEEE Commun. Surv. Tutor.* **2019**, *21*, 1383–1408. [CrossRef]

11. Żotkiewicz, M.; Szałyga, W.; Domaszewicz, J.; Bąk, A.; Kopertowski, Z.; Kozdrowski, S. Artificial Intelligence Control Logic in Next-Generation Programmable Networks. *Appl. Sci.* **2021**, *11*, 9163. [CrossRef]

12. Rottondi, C.; Barletta, L.; Giusti, A.; Tornatore, M. Machine-learning method for quality of transmission prediction of unestablished lightpaths. *IEEE/OSA J. Opt. Commun. Netw.* **2018**, *10*, A286–A297. [CrossRef]

13. Diaz-Montiel, A.A.; Aladin, S.; Tremblay, C.; Ruffini, M. Active Wavelength Load as a Feature for QoT Estimation Based on Support Vector Machine. In Proceedings of the ICC 2019—2019 IEEE International Conference on Communications (ICC), Shanghai, China, 20–24 May 2019; pp. 1–6.

14. Kozdrowski, S.; Cichosz, P.; Paziewski, P.; Sujecki, S. Machine Learning Algorithms for Prediction of the Quality of Transmission in Optical Networks. *Entropy* **2021**, *23*, 7. [CrossRef] [PubMed]

15. Cichosz, P.; Kozdrowski, S.; Sujecki, S. Application of ML Algorithms for Prediction of the QoT in Optical Networks with Imbalanced and Incomplete Data. In Proceedings of the 2021 International Conference on Software, Telecommunications and Computer Networks (SoftCOM), Split, Croatia, 23–25 September 2021; pp. 1–6. [CrossRef]

16. Panayiotou, T.; Savva, G.; Tomkos, I.; Ellinas, G. Decentralizing machine-learning-based QoT estimation for sliceable optical networks. *J. Opt. Commun. Netw.* **2020**, *12*, 146–162. [CrossRef]

17. Mata, J.; de Miguel, I.; Durán, R.J.; Aguado, J.C.; Merayo, N.; Ruiz, L.; Fernández, P.; Lorenzo, R.M.; Abril, E.J. A SVM approach for lightpath QoT estimation in optical transport networks. In Proceedings of the 2017 IEEE International Conference on Big Data (Big Data), Boston, MA, USA, 11–14 December 2017; pp. 4795–4797.

18. Barletta, L.; Giusti, A.; Rottondi, C.; Tornatore, M. QoT estimation for unestablished lighpaths using machine learning. In Proceedings of the 2017 Optical Fiber Communications Conference and Exhibition (OFC), Los Angeles, CA, USA, 19–23 March 2017; pp. 1–3.

19. Japkowicz, N. Learning from Imbalanced Data Sets: A Comparison of Various Strategies. In Proceedings of the AAAI Workshop on Learning from Imbalanced Data Sets, Austin, TX, USA, 31 July 2000; AAAI Press: Menlo Park, CA, USA, 2000.

20. Lee, H.; Cho, S. The Novelty Detection Approach for Different Degrees of Class Imbalance. In Proceedings of the Thirteenth International Conference on Neural Information Processing Systems, Hong Kong, China, 3–6 October 2006; Springer: Berlin, Germany, 2006.

21. Bellinger, C.; Sharma, S.; Zaïane, O.R.; Japkowicz, N. Sampling a Longer Life: Binary versus One-Class Classification Revisited. *Proc. Mach. Learn. Res.* **2017**, *74*, 64–78.

22. Breiman, L. Random Forests. *Mach. Learn.* **2001**, *45*, 5–32. [CrossRef]

23. Breiman, L.; Friedman, J.H.; Olshen, R.A.; Stone, C.J. *Classification and Regression Trees*; Chapman and Hall/CRC: Boca Raton, FL, USA, 1984.

24. Quinlan, J.R. Induction of Decision Trees. *Mach. Learn.* **1986**, *1*, 81–106. [CrossRef]

25. Breiman, L. Bagging Predictors. *Mach. Learn.* **1996**, *24*, 123–140. [CrossRef]

26. Dietterich, T.G. Ensemble Methods in Machine Learning. In Proceedings of the First International Workshop on Multiple Classifier Systems, Cagliari, Italy, 21–23 June 2000; Springer: Berlin, Germany, 2000.

27. Schapire, R.E. The Strength of Weak Learnability. *Mach. Learn.* **1990**, *5*, 197–227. [CrossRef]

28. Friedman, J.H. Greedy Function Approximation: A Gradient Boosting Machine. *Ann. Stat.* **2001**, *29*, 1189–1232. [CrossRef]

29. Friedman, J.H. Stochastic Gradient Boosting. *Comput. Stat. Data Anal.* **2002**, *38*, 367–378. [CrossRef]

30. Schapire, R.E.; Freund, Y. *Boosting: Foundations and Algorithms*; MIT Press: Cambridge, MA, USA, 2012.

31. Chen, T.; Guestrin, C. XGBoost: A Scalable Tree Boosting System. In Proceedings of the Twenty-Second ACM SIGKDD International Conference on Knowledge Discovery and Data Mining, San Francisco, CA, USA, 13–17 August 2016; ACM Press: New York, NY, USA, 2016.

32. Chawla, N.V.; Bowyer, K. W. Hall, L.O.; Kegelmeyer, W.P. SMOTE: Synthetic Minority Over-sampling Technique. *J. Artif. Intell. Res.* **2002**, *16*, 321–357. [CrossRef]

33. Menardi, G.; Torelli, N. Training and Assessing Classification Rules with Imbalanced Data. *Data Min. Knowl. Discov.* **2014**, *28*, 92–122. [CrossRef]

34. Moya, M.; Hush, D. Network Constraints and Multi-Objective Optimization for One-Class Classification. *Neural Netw.* **1996**, *9*, 463–474. [CrossRef]

35. Khan, S.S.; Madden, M.G. One-Class Classification: Taxonomy of Study and Review of Techniques. *Knowl. Eng. Rev.* **2014**, *29*, 345–374. [CrossRef]

36. Chandola, V.; Banerjee, A.; Kumar, V. Anomaly Detection: A Survey. *ACM Comput. Surv.* **2009**, *41*, 1–58. [CrossRef]

37. Schölkopf, B.; Platt, J.C.; Shawe-Taylor, J.C.; Smola, A.J.; Williamson, R.C. Estimating the Support of a High-Dimensional Distribution. *Neural Comput.* **2001**, *13*, 1443–1471. [CrossRef]

38. Datta, P. Characteristic Concept Representations. Ph.D. Thesis, University of California, Irvine, CA, USA, 1997.

39. Liu, F.T.; Ting, K.M.; Zhou, Z.H. Isolation-Based Anomaly Detection. *Acm Trans. Knowl. Discov. Data* **2012**, *6*, 3. [CrossRef]

40. Hariri, S.; Kind, M.C.; Brunner, R.J. Extended Isolation Forest. *arXiv* **2018**, arXiv:1811.02141.

41. Phillips, S.J.; Anderson, R.P.; Schapire, R.E. Maximum Entropy Modeling of Species Geographic Distributions. *Ecol. Nodelling* **2006**, *190*, 231–259. [CrossRef]

42. Halvorsen, R.; Mazzoni, S.; Bryn, A.; Bakkestuen, V. Opportunities for Improved Distribution Modelling Practice via a strict maximum likelihood interpretation of MaxEnt. *Ecography* **2015**, *38*, 172–183. [CrossRef]

43. Jaynes, E.T. Information Theory and Statistical Mechanics. *Phys. Rev.* **1957**, *106*, 620–630. [CrossRef]
44. Li, W.; Guo, O. A Maximum Entropy Approach to One-Class Classification of Remote Sensing Imagery. *Int. J. Remote Sens.* **2010**, *31*, 2227–2235. [CrossRef]
45. Liu, X.; Liu, H.; Gong, H.; Lin, Z.; Lv, S. Appling the One-Class Classification Method of Maxent to Detect an Invasive Plant Spartina alterniflora with Time-Series Analysis. *Remote Sens.* **2017**, *9*, 1120. [CrossRef]
46. Egan, J.P. *Signal Detection Theory and ROC Analysis*; Academic Press: Cambridge, MA, USA, 1975.
47. Fawcett, T. An Introduction to ROC Analysis. *Pattern Recognit. Lett.* **2006**, *27*, 861–874. [CrossRef]
48. Arlot, S.; Celisse, A. A Survey of Cross-Validation Procedures for Model Selection. *Stat. Surv.* **2010**, *4*, 40–79. [CrossRef]
49. Wright, M.N.; Ziegler, A. ranger: A Fast Implementation of Random Forests for High Dimensional Data in C++ and R. *J. Stat. Softw.* **2017**, *77*, 1–17. [CrossRef]
50. Chen, T.; He, T.; Benesty, M.; Khotilovich, V.; Tang, Y.; Cho, H.; Chen, K.; Mitchell, R.; Cano, I.; Zhou, T.; et al. *xgboost: Extreme Gradient Boosting*; R Package Version 1.1.1.1; 2020. Available online: https://CRAN.R-project.org/package=xgboost (accessed on 5 January 2021).
51. Siriseriwan, W. *smotefamily: A Collection of Oversampling Techniques for Class Imbalance Problem Based on SMOTE*; R Package Version 1.3.1; 2019. Available online: https://CRAN.R-project.org/package=smotefamily (accessed on 5 January 2021).
52. Lunardon, N.; Menardi, G.; Torelli, N. ROSE: A Package for Binary Imbalanced Learning. *R J.* **2014**, *6*, 82–92. [CrossRef]
53. Meyer, D.; Dimitriadou, E.; Hornik, K.; Weingessel, A.; Leisch, F. *e1071: Misc Functions of the Department of Statistics, Probability Theory Group (Formerly: E1071), TU Wien*; R Package Version 1.7-4; 2020. Available online: https://CRAN.R-project.org/package=e1071 (accessed on 5 January 2021).
54. Cortes, D. *isotree: Isolation-Based Outlier Detection*; R Package Version 0.1.20; 2020. Available online: https://CRAN.R-project.org/package=isotree (accessed on 5 January 2021).
55. Vollering, J.; Halvorsen, R.; Mazzoni, S. The MIAmaxent R package: Variable Transformation and Model Selection for Species Distribution Models. *Ecol. Evol.* **2019**, *9*, 12051–12068. [CrossRef]

Article

Real-World Data Difficulty Estimation with the Use of Entropy

Przemysław Juszczuk [1,*], Jan Kozak [2], Grzegorz Dziczkowski [2], Szymon Głowania [2], Tomasz Jach [2] and Barbara Probierz [2]

[1] Systems Research Institute, Polish Academy of Sciences, Newelska 6, 01-447 Warsaw, Poland
[2] Faculty of Informatics and Communication, Department of Machine Learning, University of Economics in Katowice, 1 Maja 50, 40-287 Katowice, Poland; jan.kozak@ue.katowice.pl (J.K.); grzegorz.dziczkowski@ue.katowice.pl (G.D.); szymon.glowania@ue.katowice.pl (S.G.); tomasz.jach@ue.katowice.pl (T.J.); barbara.probierz@ue.katowice.pl (B.P.)
* Correspondence: juszczuk@ibspan.waw.pl

Abstract: In the era of the Internet of Things and big data, we are faced with the management of a flood of information. The complexity and amount of data presented to the decision-maker are enormous, and existing methods often fail to derive nonredundant information quickly. Thus, the selection of the most satisfactory set of solutions is often a struggle. This article investigates the possibilities of using the entropy measure as an indicator of data difficulty. To do so, we focus on real-world data covering various fields related to markets (the real estate market and financial markets), sports data, fake news data, and more. The problem is twofold: First, since we deal with unprocessed, inconsistent data, it is necessary to perform additional preprocessing. Therefore, the second step of our research is using the entropy-based measure to capture the nonredundant, noncorrelated core information from the data. Research is conducted using well-known algorithms from the classification domain to investigate the quality of solutions derived based on initial preprocessing and the information indicated by the entropy measure. Eventually, the best 25% (in the sense of entropy measure) attributes are selected to perform the whole classification procedure once again, and the results are compared.

Keywords: entropy measure; real-world data; preprocessing; decision table; classification

Citation: Juszczuk, P.; Kozak, J.; Dziczkowski, G.; Głowania, S.; Jach, T.; Probierz, B. Real-World Data Difficulty Estimation with the Use of Entropy. *Entropy* **2021**, *23*, 1621. https://doi.org/10.3390/e23121621

Academic Editors: Geert Verdoolaege and Éloi Bossé

Received: 11 October 2021
Accepted: 26 November 2021
Published: 1 December 2021

Publisher's Note: MDPI stays neutral with regard to jurisdictional claims in published maps and institutional affiliations.

Copyright: © 2021 by the authors. Licensee MDPI, Basel, Switzerland. This article is an open access article distributed under the terms and conditions of the Creative Commons Attribution (CC BY) license (https://creativecommons.org/licenses/by/4.0/).

1. Introduction

In present times, we are facing the problem of a large amount of data flowing from different sources. In the era of the Internet of Things (IoT) and big data, the challenge is to effectively use and present the acquired data without generating redundant information. Due to the size of data available for decision-makers, it is nearly impossible to manually make any complex decisions. This difficulty is experienced even in machine learning algorithms, which must manage too many attributes, variables, and additional constraints, resulting in the whole process being lengthy and complicated [1]. As such, it is essential to simplify data in the cases where the decisions should be made very quickly, and a need exists to use a decision support system to maintain the decision-maker's sovereignty.

The main drawback of the existing datasets is their uniform structure. For the data related to a single domain, the distribution of attribute values, the size of data, or the overall difficulty of the given dataset classification is expected to be on a similar level. However, in the case of more general approaches, we often face inconsistency in data, including the need to use additional knowledge from the domain experts. In general, data available in repositories are mostly preprocessed and directed on a particular problem (like the classification or the regression). At the same time, the initially collected data may still be very complex.

The above problem had led to the construction of many complex algorithms and methods intending to decrease the complexity of the data used in the decision process.

Among these methods, we can emphasize approaches for reducing the number of variables included in the algorithm [2,3]. The idea of initially preprocessing the data related to the feature selection, removing the redundant data, or including more general attributes replacing the existing ones is not a new concept and it was deeply studied in many articles, where initial data limitation was needed. Examples of such feature selection methods can be found, for example, in extensions of the Principal Component Analysis method. One of the newest review articles in this subject can be found in [4]. A more general approach for future selection involving the swarm methods is presented in [5,6]. In comparison, one of the newest review articles related to the swarm methods is [7]. The second large set of algorithms used for the feature selection is related to the tree-based methods. In these methods, the attributes can be selected based on the importance of the attribute in the process of building the tree (classifier). An example of comparison for such algorithms can be found in [8].

For many cases, data dependencies are not linear. Thus, a complex method of variables elimination should be applied. For example, in the case of periodically important variables or in situations where the linear dependencies between elements are not obvious, different methods must be used to emphasize the crucial variables in the system. To avoid redundancy in the data, the selected variables should exhibit little or no mutual correlation. This requirement was described by [9], in which the phenomenon of the illusion of validity occurs: people have confidence in the results, which are based on redundant data. Thus, in decision support systems and during attribute selection, the role of decision-makers can also be marginalized.

A method that effectively identifies the crucial variables present in the complex data can be essential for the whole system's efficiency. However, in the case where the data structure and its complexity makes the data difficult or even impossible to process, the decision-maker faces a two-step problem: First, there is a need to adapt the data to fit the algorithm's input format. This can be achieved by some additional preprocessing methods, leading to a data format acceptable as the algorithm's input. However, the whole process may be lengthy and complex. It often covers concepts such as filling the missing data, discretization, and scalarization. Dealing with missing data cannot be solved with simple methods, and the literature covers various approaches to this problem [10–12].

Thus, today we observe many algorithms dedicated to a particular domain, which, opposite to the general approaches, can deal with the problems more efficiently. However, one should know that such available methods can still be beneficial, even as a starting point for emerging domains related to complex or big data. Our idea was to collect raw data from different fields and prepare it in a uniform, easy-to-analyze format based on decision tables. At the same time, we tried to use as general tools as possible, which unfortunately can lead to a decrease in classification quality. However, it maintains the generalized approach for all datasets.

Furthermore, we selected entropy as a concept, which allows us to describe the disorder of the data. By the disorder, we understand here the measure of complexity, where the more complex data (fewer dependencies between objects and attributes is visible) is defined by the higher entropy values. Therefore, we assumed that the increase in entropy could be equated with data difficulty. Furthermore, this assumption is verified by performing the actual classification on various datasets. Eventually, the results from the classification on the full set of attributes and subset generated on the basis of entropy can be compared. It is expected that high entropy should lead to less effective classification.

The entropy measure is considered from the point of view of all attributes. Thus, it is possible to identify the attributes with small disorder values (smaller entropy values). A subset of attributes with small entropy could be used to perform the classification while the data is limited.

In our data, a clear distinction exists between conditional attributes and decision class. Data from various fields cover several objects as well as different numbers of attributes. However, the common goal is to perform a classification task on the presented data. The second step of our research completely focused estimating the impact of the entropy-based measure on the classification task. First, we tried to determine if entropy can be effectively used to indicate data difficulty. Eventually, we investigated the results of the classification of the data. We expected that, initially, all conditional attributes analyzed in the dataset could be treated uniformly (i.e., have similar entropy values). Thus, the main questions were: is there a correlation between the entropy values and the quality of classification, and can the entropy-based measure be used to select the best-fitted attributes for the classification problem? To summarize, our research steps were as follows:

- initial preprocessing of real-world data;
- entropy calculation for different datasets;
- classification on all datasets;
- selection of the best-fitted 25% of attributes based on the lowest entropy measure;
- the comparison between the classification results for the full and limited set of attributes on different datasets.

To generalize our observations as much as possible, we tried to select data from various fields and describe the whole preprocessing framework with the use of domain knowledge presented by experts from different fields. Moreover, this preprocessing schema allowed us to use a general data format, which can be effectively used in entropy calculation and, finally, in classification problems.

The paper is organized as follows: In the next section, we present the related studies. In Section 3, we discuss the theoretical background related to the subject, including a description of entropy, decision tables, and efficiency measures used in classification tasks. Section 4 contains a description of the real-world data covering different domains. Section 5 presents the results of our experiments based on entropy calculation as well as the classification problem. Eventually, we conclude the study in Section 6.

2. Related Works

By classical entropy, we understand the measure of uncertainty related with some data. The idea was introduced by Shannon in 1948 [13] and further extended, for example, by Renyi and Tsallis [14,15], where Renyi entropy is the generalization of the Shannon entropy for specific parameters.

The classical entropy measure is used as a crucial element in many different algorithms and methods. Amongst the most prominent examples are the well-known classification algorithm C4.5 developed by Quinlan [16] as an extension of algorithm ID3 [17]. In both examples, entropy was used as a measure to generate a classifier (a decision tree). In C4.5, entropy was used for all algorithm steps to calculate the information gain based on the entropy for every attribute available in the dataset. A similar idea is used in greedy heuristic ID3, where, once again, the attribute used as a split criterion for the data is based on the highest information gain. Such an approach has been successfully used in machine learning [18] and signal processing [19].

Entropy is often used as an element of broader methods rather than a standalone measure. It has a role in novel metaheuristics such as an extension of classical particle swarm optimization [20]. In [21], it was used as an alternative approach to the concept of fuzzy sets to measure the uncertainty of the task in a task assignment problem. Entropy was used as an extension of the binary classification problem solved by particle swarm optimization [22]. In many articles, entropy has often been used as a replacement for classical measures such as variance [23].

Entropy mixed with the concept of fuzzy sets was included in an outlier detection approach [24]. In [25], entropy was included as a part of the feature selection mechanism based on fuzzy sets. Finally, a more complex approach, including the fuzzy multicriteria approach based on the TOPSIS method, was presented in [26].

Entropy was used in many different approaches to measure randomness in a clinical trial [27]. In [28], entropy was introduced to measure the uncertainty of ordered sets. In general, it can be used as an idea of measure for different fields such finance [29,30], chemistry [31], physics [32], and more. However, no works used entropy as a general measure for different domains simultaneously. A separate direction of research is devoted to various extensions of classical entropy. In [33], the idea of measuring an entropy on different scales (multiscale entropy) was presented. In the case of time-series data, the concept of approximate entropy is often used [34]. In [35], approximate entropy was extended, called sample entropy. This idea was further extended in [36]. Both methods were used in different applications to address various dynamic aspects of systems.

Another prevalent extension of the classical measure is permutation entropy, effectively used as a nonlinear measure in different fields such as cyber-security [37] and fault diagnosis in systems [38]. Some preliminary comparisons between the classical entropy measure and Pearson correlation were introduced [39]. In this example, the authors focused on the data derived from the system from the Internet of Things, focusing on spatio-temporal data.

The idea of using entropy as a complexity measure is well-known, and it has been recently studied by many researchers. Among interesting examples, we mention [40], where information entropy was used to measure the genetic diversity in colonies. Another example covers the general idea of measuring the complexity of time series [41].

Entropy as a measure of diversity was presented in [42], where the authors used Shannon entropy to measure the urban growth dynamics for a case study related to real-world data from the city of Sheffield in the U.K. More complex examples related to health and perception can be found in [43,44]. In the first case, the authors used entropy-based concepts for knowledge discovery in heart rate variability, whereas in the second example, approximate entropy was used for EEG data. Finally, among the newest works from the medical domain, Coates et al. [45] used entropy in the Parkinson's disease recognition process.

3. Methodology

For a set of objects X, every element can be described by a vector of n conditional attributes $\vec{x_{atr}} = \{x_{atr^1}, x_{atr^2}, \ldots, x_{atr^n}\}$ where n is a number of conditional attributes. A decision class is denoted as x_{class}. Thus, every object is described by a pair $(\vec{a_{atr}}, x_{class})$. For every conditional attribute, we have the attribute and value pair, and every attribute can have a numeric or symbolic value. In the case of attributes with continuous values, the discretization procedure, leading to limiting the number of values for a single attribute, is often performed.

In classification problems, the decision class x_{class}, including information about the decision class for a single object, has one of the values belonging to the decision class set of values.

In this article, we perform the preprocessing of real-world data, which allows transforming the initial raw data into a decision table defined as follows:

$$DS = (X, \vec{x_{atr}}, x_{class}). \tag{1}$$

All analyzed data differ in terms of the size of set X and the number of attributes in the vector of conditional attributes $\vec{x_{atr}}$. We did not assume simplifications related to the cardinality of the decision class. Thus, for some sets, this attribute is continuous, and an additional discretization procedure is needed. Eventually, for all datasets, the number of values in decision class x_{class} is discrete.

3.1. Entropy as a Measure of Classification Uncertainty

According to our aim, we wanted to explore the possibility of using entropy as an indicator of data difficulty. Therefore, we treated entropy as a measure of classification uncertainty. In addition, we explored how data can be simplified using only attributes selected in terms of entropy value. Therefore, we also examined the information attribute to assess the usefulness of entropy for data simplification.

Assuming that several different symbols describe information, entropy, in its basic form, can be calculated as follows:

$$E(DS) = -\sum_{i=1}^{|C|} p_i \cdot log(p_i),$$ (2)

where $|C|$ is the number of different decision classess, and p_i is the probability of occurence of the *i*-th decision class. With such a definition, entropy can be understood as a measure of data complexity. With an increasing number of decision classess available in the data, the overall complexity increases. In the most trivial case, for a single decision class, the p_i value is equal to 1, whereas $log(p_i)$ is zero (as well as the entropy). Thus, any increase in this value leads to higher entropy.

The value of the information attribute (Equation (3)) is determined for each conditional attribute to determine how it can change the entropy of the decision table DS. The resulting value determines the entropy that can be obtained by considering that attribute.

The information attribute is thus based on the calculation of entropy due to decision classes (Equation (2)), but this is performed due to the cases grouped by the values of the attribute being analyzed.

Formally, the information attribute is written as Equation (3), but note that these determinations are required for each attribute, where k is the number of attributes being analyzed, m is the number of possible values of the *k*-th attribute, and $|DS_i|$ is the number of instances having the *i*-th attribute value (analogously, DS_i is the subset of the decision table DS that has only the *i*-th attribute value on attribute k).

$$info\_att(k, DS) = \sum_{i=1}^{m} \frac{|DS_i|}{|DS|} \cdot E(DS_i)$$ (3)

In our considerations, $info\_att$ is crucial for simplifying the dataset. For each decision table DS with the number of conditional attributes n, values are determined based on Equation (4). This observation is used for further analysis.

$$all\_info\_att(DS) = \sum_{k=1}^{n} info\_att(k, DS)$$ (4)

3.2. Classification Measures

In our research, we wanted to examine the classification quality using state-of-the-art machine learning algorithms. We chose decision trees (CART algorithm) and ensemble methods: Random Forest, Bagging, and AdaBoost. To assess the quality of classification, in addition to the classical measures of classification quality (accuracy), we also used precision (called positive predictive value (PPV)) and recall (called true positive rate (TPR)). Notably, these are binary classification measures, i.e., for a dataset with only two decision classes. In real datasets, there are often more decision classes. Several methods can be used to generalize precision and recall. We wanted to provide as much information as possible in our solutions, so we computed precision and recall for each decision class.

Therefore, for PPV, the analyzed decision class is treated as positive and all others as negative, and analogously for TPR. So, in the definition of the measures of the quality of classification (accuracy in Equation (5), precision in Equation (6), and recall in Equation (7)), we denote:

TP: to identify all correctly classified cases of the analyzed class;
TN: to identify all cases outside the analyzed class that were not assigned to this class;
FP: to identify all cases outside the analyzed class that were assigned to this class;
FN: to identify all misclassified cases of the analyzed class.

$$accuracy = \frac{TP+TN}{TP+TN+FP+FN},$$ (5)

$$PPV = \frac{TP}{TP + FP}, \tag{6}$$

$$TPR = \frac{TP}{TP + FN}. \tag{7}$$

4. Data Preparation and Preprocessing

In this section, we provide details of the real-world data used in further experiments. The data were collected from external sources and cover various fields. We adapted the raw data into a decision table format, described in detail in the previous section, to perform the tests based on the classification problem. All necessary steps for data processing are described in this section.

However, despite the processing of all datasets, some general preprocessing steps were used. Below we indicate these steps in points with a short description.

- collect data in the raw format—the first step was to obtain the entire data. Please note that for some cases, these data were obtained from different sources; however, all information initially was presented as a table;
- join data tables from different sources—this step was used to merge all obtained data into a single table structure;
- eliminate all missing and incomplete data—no artificial methods allowing to repair missing data were included in this point;
- eliminate potential outliers in the data—by outlier, we mean observation outside the range $\langle Q1 - 3 \cdot IQ : Q3 + 3 \cdot IQ \rangle$ (where Q1 is the first quartile, Q3 is the third quartile, and IQ is the interquartile range);
- perform discretization for selected attributes (attributes pointed out by the domain expert having a relatively large number of values).

Please note that the last step was used for both conditional attributes as well as the decision attribute (if needed). Moreover, these were general steps adapted for all data. However, additional steps were explicitly performed for the selected data (for example, related to the natural language processing), described in detail in subsections related to different data.

4.1. Fake News Data

Universal access to the Internet created the possibility of the rapid creation and gaining of knowledge by users, which became a threat through the easy spread of false information in the form of fake news. Fake news aims to present users with a view that is not in line with reality or leads them to make wrong decisions or actions based on false information.

The problem of disinformation is best visible on social networking services and news sites, where fake news is spreading widely in the form of sharing, passing on to friends, or creating documents based on unreliable sources [46]. Therefore, it is essential to quickly classify the documents posted and adequately mark the articles as true or fake news. The subject matter of the documents from the fake news dataset is related to many different fields; in particular, it concerns political, media, and financial content, as well as current events [47,48].

Kannan et al. [49] claimed that preprocessing real text data for analysis using machine learning algorithms is always the longest stage and often amounts to around 80% of the total processing time. Therefore, to transform the fake news dataset into a decision table, we propose applying the statistical approach of natural language processing (NLP).

In the first step of NLP, the tokenization process is carried out, dividing a given text into the smallest unit (e.g., a sequence of words, bytes, syllables, or characters) called a token. The result is the creation of an n-gram model that is used to identify and analyze attributes used in natural language modeling and processing [50]. In our research, we used n-gram to define individual words from document titles, from which we additionally rejected words appearing on the stop word list. An example of a stop words list is presented in Figure 1.

```
a, an,
about,
are, be, is, was, will,
as, how,
by, for, of, from,
in, on,at,
or, and,
the, that, these, this,
too,
what, when, where, who,
```

Figure 1. A sample list of rejected words, the so-called Stop Words.

The next step in NLP is to perform the normalization process using two methods: stemming and lemmatization. The stemming method is used to extract the subject and the endings of the words. Eventually, similar words are replaced by the same base word [51]. The method of lemmatization consists of reducing the word to its basic form [52]. The purpose of the normalization process is to reduce the variability in the set of terms.

The final step in the NLP covered in this research is creating a word vector model as a document representation. Our vector model is presented as a matrix (Figure 2), where documents (dok_1–dok_n) are presented in the form of feature vectors representing particular attributes (at_1–at_n). In the model, we use a binary representation, where each value from the {0,1} set determines whether the word appears in a given document. In addition, the number of attributes is limited to the most common words in the title of the document. On this basis, the fake news dataset was transformed into a decision table consisting of the attributes of the most common words and a decision attribute (*decision*) containing two classes (true or fake).

	at_1	at_2	at_3	at_4	at_5	at_6	at_7	at_8	...	at_n	decision
dok_1	1	0	0	1	0	0	1	0	...	0	0
dok_2	1	0	1	0	0	0	0	0	...	0	0
dok_3	0	0	0	0	1	0	0	1	...	0	0
dok_4	0	0	0	1	0	0	0	0	...	1	1
dok_5	0	1	0	0	0	1	0	0	...	0	0
dok_6	0	0	1	0	1	0	0	0	...	0	0
dok_7	0	1	0	0	0	0	0	0	...	0	1
dok_8	0	0	0	0	0	1	0	0	...	0	0
dok_9	0	1	0	0	0	0	0	1	...	1	0
dok_10	0	0	0	1	0	0	0	0	...	0	0
dok_11	0	1	0	0	0	0	0	0	...	1	0
dok_12	0	0	0	0	0	0	0	0	...	0	0
dok_13	0	0	0	0	0	0	0	0	...	0	1
dok_14	0	0	0	0	0	0	0	0	...	1	0
dok_15	0	0	0	1	0	0	1	0	...	0	1
dok_16	0	0	0	1	0	0	1	0	...	0	0
dok_17	1	0	1	0	0	0	1	0	...	0	1
dok_18	0	0	0	1	0	0	0	0	...	1	0
dok_19	0	0	0	0	0	0	0	1	...	0	0
...	...	...	...	...	...	...	...	...	...	...	...
dok_n	0	0	0	0	0	0	0	0	...	0	0

Figure 2. The sample matrix of words occurrence (selected as conditional attributes) in documents.

The decision table structure consists of columns with conditional attributes and one decision, whereas rows include all documents from the set. Conditional attributes are words most often appearing in the text. The presence of specific words (in the decision table) is strictly dependent on the analyzed dataset. For this reason, the number of attributes is limited. Table 1 shows an example of the frequency of words (selected as conditional attributes) in the titles of true and fake news.

Table 1. The example frequency of words (selected as conditional attributes).

Attribute Name	True News	Fake News
word_1	608	463
word_2	592	715
word_3	1036	78
word_4	1151	655
word_5	840	585
word_6	631	692
word_7	2193	47
word_8	572	1441
word_9	2520	666
word_10	1227	654
word_11	859	975
word_12	371	970
word_13	471	1167
word_14	577	821
word_15	920	269
word_16	8843	5538
word_17	8369	40
word_18	592	703
word_19	1975	36
word_20	2874	815

Real text datasets are challenging to analyze due to the large number of attributes [53] that constitute single words for the fake news dataset. The distribution of attribute values due to decision classes (fake and true) is presented in Figure 3.

For each attribute, there is one histogram (Figure 3) consisting of two columns, which corresponds to the number of values for each attribute. The first column shows the number of objects (article content) in which the selected word does not appear (as an attribute value), while the second column shows the number of objects in which the selected word appears at least once. These numbers are shown in the chart. Additionally, each column shows the assignment of a word to the appropriate class: blue is the true class, and red is the fake class.

By such a distribution of attributes due to decision classes (fake and true), it can be seen that some words (such as *word_3*, *word_7*, *word_17*, *word_19*) do not appear at all in the fake class—the right column is entirely blue. However, in the case of the first column, the division into both classes is equal for almost all attributes.

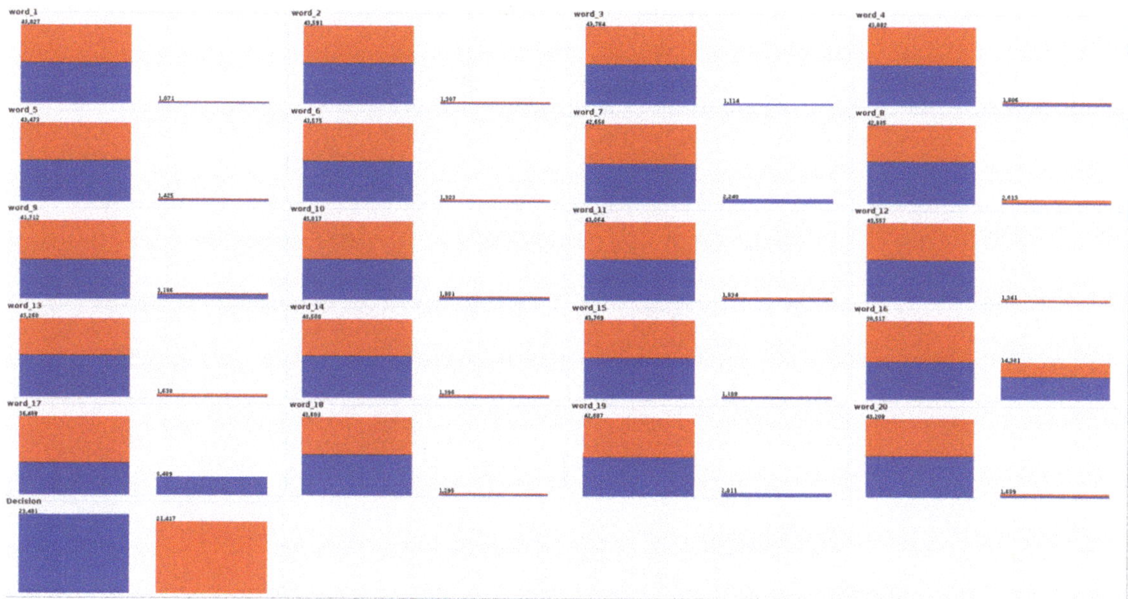

Figure 3. The distribution of attribute values due to decision classes for fake news data.

4.2. User Websites Navigation Data

Electronic commerce (e-commerce) has become popular as the Internet has grown, with many websites offering online sales, and e-commerce activity is undergoing a significant revolution. The major challenges in research are the collection, identification, and adoption of data supplied by Internet services to provide actionable marketing intelligence.

The main difficulty in web usage mining is the procurement of the desired database, as the only information we can collect from users visiting a website is through tracing the pages they have accessed.

Data collected from log files must be processed before data mining techniques (based on machine learning algorithms) can be used. Then, the personalization process is performed in the six main steps generally used in the field:

1. **Data collection**: Collecting the data from the server or the user side.
2. **Data filtering**: Removing or correcting undesirable data such as the log information obtained by crawlers.
3. **User identification**: Identification of user by IP address, cookies, and direct identification.
4. **Session identification**: Tracking the activity of the same user.
5. **Characteristics selection**: Selecting characteristics that can be useful for user behavior analysis.
6. **User behavior analysis**: Studying the behavior of different users for selecting dominant ones (i.e., the characteristics that change significantly from one behavior to another).

The main idea of analyzing the users' behavior during user navigation was to limit the users' sessions to 10 actions. Each action corresponds to a one-page view by the users. We chose the 10 actions limitation in the session because it was impossible to perform a pertinent clustering using less than 10 actions for the user session; the cluster was not significant enough, and differences between clusters were negligible.

Before the phase of navigation conditional attributes selection, the hierarchy of the website was derived. An example division of the site is as follows: First, we separated thematic websites to create universes. Websites from each universe were about the same topic. Then, we divided the entire site into seven different universes:

1. **Store**: the main universe (for example with products list),
2. **Quick order**: direct purchases by entering the catalog reference,
3. **Shopping cart (purchase)**,
4. **Sales**,
5. **Consulting**: customers questions and FAQs,
6. **Condition**: Terms of sale and shipping, and
7. **Various**: all others, such as home pages.

The universe store was divided into three levels of hierarchy: section, subsection, and subsubsection. Generally, the final product page corresponds to the subsubsection.

From this hierarchy, we selected conditional attributes that describe the user navigation of our commercial partner's website. The attributes are presented in Table 2.

Table 2. Session attributes.

User ID	Session ID
Day/Month/Year	Hour of begin
Hour of end	Purchase
Total amount	No. products bought
No. references bought	Discount code
New user	Source of navigation
Total time	Time universe (1–7)
No. total pages seen	No. pages universe (1–7) seen
No. universes changes	No. sections changes
No. subsect. changes	No. subsubsect. changes
No. of section seen	No. of subsection seen
No. product pages seen	No. of same product seen

The presented attributes are described as follows:

*   **User ID** describes the ID based on cookies, a unique ID for each user;
*   **Session ID** describes the session ID during one day each session is considered as closed after 30 min of inaction;
*   **Purchase** is a binary value that shows if the user made a purchase during their action;
*   **Discount code** is a boolean value describing the presence of a discount code during the purchase;
*   **New user** describes whether the user was recognized as a user who already made a purchase on the site;
*   **Source of navigation** describes whether the user is entered into our commercial partner's site voluntarily by using, for example, the search engine, or was pushed to visit the site by a mail company;
*   **Total time** describes the length of a session in seconds;
*   **Total universe (1–7)** represents the seven different attributes that describe the time that a visitor spends in each universe;
*   **Total no. of pages seen** describes the number of all the pages visited by the user during a session;
*   **No. pages universe (1–7) seen** represents the seven different attributes that describe the number of pages visited by a user in each universe;
*   **No. of universe, section, subsection, and subsubsection changes** are the four features that describe the number of changes the user makes during their navigation. If,

for example, the user switches universe and then returns to the previous one, the value of this attribute is equal to 2;

- **No. of sections or subsections seen** are the two attributes that describe the number of different sections or subsections seen during the user session;
- **No. of product pages seen** describes the number of product pages seen in total;
- **No. of same product seen** describes the sum of product pages that have been seen several times.

For the decision attribute, we chose the binary attribute *purchase*. Decisions classes were "yes" and "no". All the attributes were normalized.

The distribution of attribute values due to decision classes—*purchase* is presented in the Figure 4. Two colors correspond to the decision classes: blue indicates sessions not completed with the purchase, and red indicates the sections in which the purchase was made.

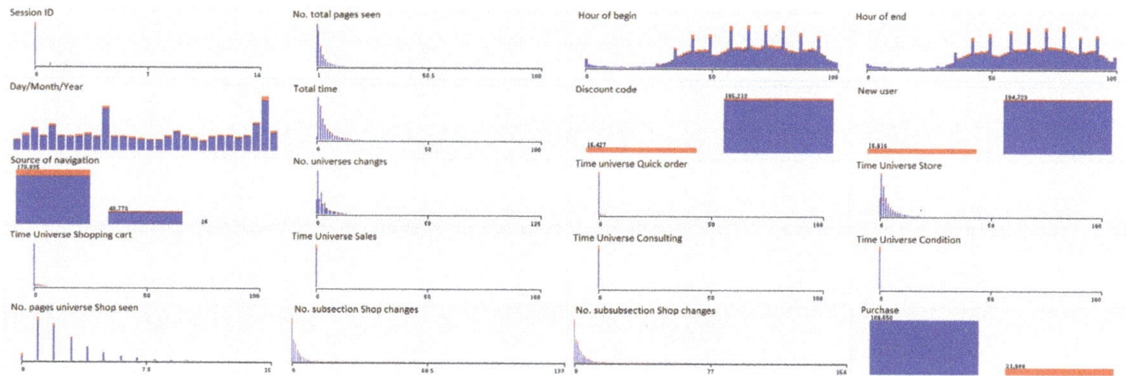

Figure 4. The distribution of attribute values due to decision classes for user websites navigation data.

As we can see in the Figure 4, some attributes do not discriminate the decision class. For example, the decision class distribution is identical for attributes such as *Day/Month/Year*, *Hour_of_end*, and *Source_of_navigation*. On the other hand, attributes such as *Discount_code*, *Total_time*, or *New_user*, clearly indicate the *purchase* class. According to the presented data distribution, we can determine that the user's session ending with purchase has the following attribute values: *avg._no_of_pages_viewed* and *average_amount_of_time_spent_on_navigation*, the customer is not the first time on the website, and he has a discount code, the customer does not spend a lot of time in the store section, but frequently changes subpages in this category.

4.3. Real Estate Market Data

The real estate market has grown rapidly during the recent years [54]. As such, both the volume of data and the number of processed details have increased. Investors are looking for attractive properties from which profit can easily be earned. As customer habits change, so do the features connected to a particular property that is essential for buyers.

The change in investor and end-consumer behavior has led to the inclusions of additional details in advertisements of properties. Each advertisement is currently filled with much additional information, some of it structured and some of it only provided in descriptive text. The real estate market data used in this paper originated from actual advertisements presented on multiple Polish market web pages. The details of the adverts are often hidden inside the text describing a particular property. However, many details are often presented in a structured form, allowing less sophisticated automatic scrapers to gather the data. For some of the conditional attributes, it is still necessary to perform more advanced processing. For instance, *the_floor_number* is usually provided as a number in the vast majority of cases. However, there are some occasions where it is stated verbally as "ground floor" or "higher than the 10th floor". Most of the advertising portals do not provide a good enough validation of this data, which is why, during the data acquisition, we had to construct more detailed methods to handle the special types of values and data. A similar process had to be performed for geo-encoding the spatial data. In almost every advertisement, the exact address of the property was not given; only the street name and the city were described. Sometimes the street names had spelling errors, were not correctly placed on a map, or used an old street name before the mandatory change of street names in Poland that recently occurred [55].

Notably, the process of acquiring data from web pages is complicated. The dataset used in the current study consists of the following conditional attributes:

- **Build date** is the year the property was built. This attribute needed extra preprocessing steps, as some of the records provided a textual representation such as "the late 1980s". Therefore, the dates were given as is without any numerical processing.
- **Total number of floors** in a whole building. As mentioned before, more advanced NLP methods were applied to clean up the data.
- **Building material**. As the materials change across the decades, a whole dictionary of construction materials was created using both automatic methods and expert knowledge. We also constructed a synonyms dictionary. The provided value was then compared to the dictionaries and cleaned up. This is a categorical attribute.
- **Floor number** on which the particular property is situated.
- Area] of the property. Here, the vast majority of data were provided in square meters, but some of the land properties provided this value in acres, which had to be converted to an SI-derived unit of measure.
- **Building type** is a categorical attribute denoting the building type (e.g., semi-detached building, loft, etc.). Here, we used similar preprocessing techniques to those used for building material.
- **Condition state** describes the overall condition of a property. As this is highly subjective, as there are virtually no norms that can standardize this attribute, we used a two-fold approach. As a starting point, the value presented in the advertisement was taken directly as-is. Next, this value was then compared to the dictionary of values and corrected for spelling errors and synonyms. In a second step, the description text was analyzed to find keywords that could decrease or increase the overall condition of a property. For instance, if the property was marked as "ready to move in", but the description mentioned that "painting needed" or "kitchen is not equipped with stove", the overall condition was decreased. Although this is considered a categorical attribute, current works involve introducing the order relation to items from the condition dictionary. Additionally, we are working on an image classifier that will automatically label the state of a property.
- **Windows** with which the property is equipped (wooden, PCV, etc.).
- **Private ad** is a dichotomous attribute discriminating if an advert was published by a professional dealer or a private party. As research has shown, these two types are constructed vastly differently. Most of the time, private advertisements have lower-quality photographs, but the description is more accurate and meaningful than in professional ads. The former often includes additional costs in the description (such as a mandatory extra-paid parking space).
- **Market type** has two values: primary and aftermarket.
- **Ownership type** describes the legal ownership type of a given property.

The last attribute, being the decision one, denotes the price per square meter. As this value can fluctuate widely, we transformed it using a simple discretization:

$$bucket = \lceil \tfrac{price\_per\_sq\_mt}{1000} \rceil.$$

(8)

Because of the nature of scrapped data and the frequent necessity for repairing or transforming the data (e.g., converting units of measurement between imperial and metrical), this data is rather difficult to analyze. Furthermore, many attributes, all interesting for the end-user, make this processing even more complicated.

The distribution of attribute values in accordance with decision classes was created, as shown in the Figure 5. Please note that due to many values in the decision class, there was no visible distinction related to color for each class.

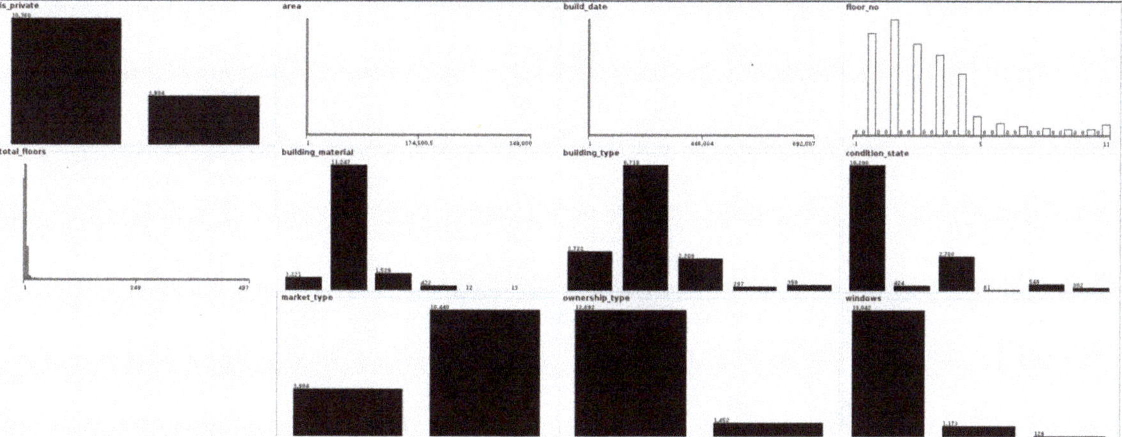

Figure 5. The distribution of attribute values due to decision classes for real estate market data.

Even though the data has been preprocessed extensively, some of the original values with mistakes were left intact. This is the case for *area* attribute, where one of the flat's areas is set to 349,000 square meters. This is clearly seen in the distribution plot, where the plot is heavily skewed. The same thing is happening with the *build_date* (a building has a date set to 892,007; there are also some spelling errors with a date like 19,000 or 20,014 where an individual probably inserted an additional 0). Because the number of records with such mistakes is relatively small (less than 0.02%), the authors included these outliers in the dataset to determine their influence on the overall entropy and classification results.

It is clearly seen that most of the properties are situated below fourth floor, which is expected, as it is far more easy to build such buildings in Poland compared to skyscrappers due to legal reasons. The owners tend to over-estimate the quality of interior, therefore the vast majority of apartments have the "ready to be moved" *condition_state*. Most of the analyzed apartments also have modern PVC windows.

4.4. Sport Data

Sport is a valuable part of many people's lives, understood both as physical activity and in terms following individual teams or athletes. Football is the most popular sport known, with the European leagues being some of the most famous in the world. Therefore, the top leagues from Germany, Italy, and Spain were selected for our analysis.

Numerous studies based on both expert analysis and machine learning techniques for predicting sports results can be found in the literature [56–59]. The most popular and accessible are predictions of match results in the form of win/loss/draw; however, both analyses and predictions may concern other elements such as the number of goals scored, the exact score, or the number of yellow cards [56,60].

The dataset was created from the tabular data available on a website [61]. For complete information, the data were extracted using the scraping method from two tables. The first one contains data about the league table. The second one consists of information about individual matches. The tables were then combined to obtain a full decision table that was divided into sets for each country. The conditional attributes included in the decision tables are presented below:

- **Season**: The season in which the games were played: a nominal variable using data for 10 consecutive seasons from 2011–2012 ("11/12") to 2020–2021 ("20/21").
- **Round**: The number of competition rounds. A quotient, integer variable ranging from six to 34 for Germany, and to 38 for Spain and Italy. Based on conducted experiments and the arguments indicated in the literature, the data for the first five rounds of each of the seasons were excluded from the analysis [62].
- **Team1**: The name of the first team. Categorical variable taking different values 28 for Germany 28, 34 for Italy 34, and 33 for Spain.
- **Position T1**: Position of Team1 in the competition table. A quotient, integer variable ranging from 1 to 18 for Germany, and 20 for Spain and Italy.
- **Match T1**: Match played by Team1 up to the current round. A quotient, integer variable ranging from six to 34 for Germany, and to 38 for Spain and Italy.
- **Winnings T1**: The number of matches won by Team1 up to the current round. A quotient, integer variable ranging from six to 34 for Germany, and to 38 for Spain and Italy.
- **Draws T1**: The number of draws for Team1 up to the current round. A quotient, integer variable ranging from six to 34 for Germany, and to 38 for Spain and Italy.
- **Losses T1**: The number of matches lost by Team1 up to the current round. A quotient, integer variable with values ranging from six to 34 for Germany, and to 38 for Spain and Italy.
- **Goals scored T1**: Goals scored by Team1 up to the current round. A quotient, integer variable.
- **Goals conceded T1**: Goals conceded by Team1 up to the current round. A quotient, integer variable.
- **Goal difference T1**: Difference between goals scored and lost by Team1. A quotient, integer variable.
- **Points T1**: The number of points gained by Team1 up to the current round.
- **Series T1**: Series of results match for Team1. A nominal variable, consisting of three symbols containing information about the results of the last three games played by the team. In the first position, there is the last played game, where W is team wins, R is a draw, P is team loss, and B indicates no data.

The same attributes are available for the second team as for the first team. The conditional attributes for the second team were marked by "T2". The last of the attributes is the decision class (*match_result*), which can have three values: 1 indicates a win for team 1, 2 indicates a win for team 2, and X is a draw. Team 1 is the team playing the game on its home field; team 2 is the team playing away.

The Figure 6 shows examples of distributions for the data of the German Bundesliga. A significant part of the data is characterized by right-hand asymmetry, which is naturally related to the domain specificity of the data. Representative examples of this fact are, among others, $Winnings\_T1$, $Draws\_T1$, $Goals\_scored\_T1$, $Goals\_conceded\_T1$, $Points\_T1$. A team starts with a value of 0 for the number of games won/lost, goals scored/conceded or the number of points. During the game, teams increase the values of these attributes, or they remain unchanged. This behavior contributes to the right asymmetry in the data. The distribution for $Goal\_difference\_T1$ is much closer to the normal distribution. In the decision class distribution, it can be seen that the most common values are related to the home team win (color = red), then the visiting team wins (color = cyan) and draw (color = blue). The last two classes have numbers much more similar to each other. The following rules are also observed for the "Team2" data and for other countries' leagues.

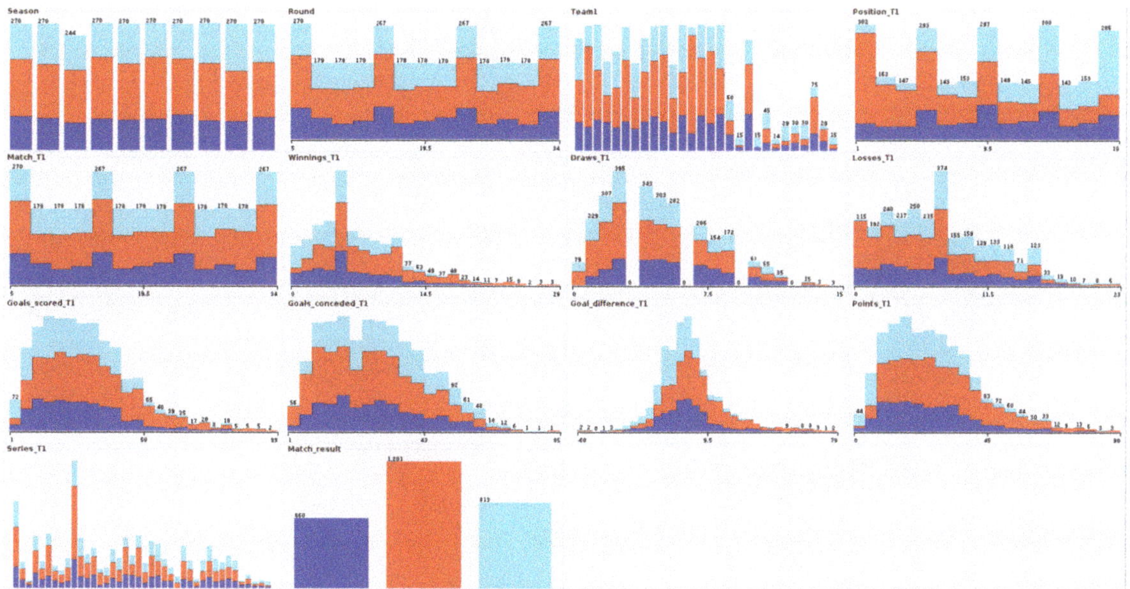

Figure 6. The distribution of attribute values due to decision classes for sport data.

4.5. Financial Data

From financial data, we can highlight two main groups of data. The first one is related to the well-known Markowitz model (and its extensions) and the portfolio selection problem, which is beyond the scope of this study. The second group is related to the price and indicator data from various markets. In this group, the most popular data are obtained from the financial markets (also known as forex market or foreign exchange market) and concerns the currency pairs.

A single market indicator (or group of indicators used jointly) is used in trading systems to generate buy signals. All indicator data were calculated according to market indicator formulas, which can be divided into two separate groups. The first covers trend-following indicators, which include the moving average (MA) market indicator. The MA for time t and s periods, denoted $MA_s(t)$, is calculated as:

$$MA_s(t) = \frac{\sum_{i=t-s}^{t-1} price_i}{s},\tag{9}$$

where $price_i$ is the value of the corresponding instrument at time i. In the above context, the period is the number of values considered when calculating the indicator. The second group of indicators covers the oscillators, whose primary purpose is to indicate rising or falling potential for the given currency pair. The indicator value is calculated using the currency value and can include the closing, opening, minimum, or maximum currency pair value from previous sessions (or any combination of the above). As an example, the oscillator Relative Strength Index (RSI) is calculated based on the last n periods in time t as follows:

$$RSI_s(t) = 100 - \frac{100}{1 - \frac{avg_{gain}}{avg_{loss}}},\tag{10}$$

where avg_{gain} is the sum of gains over the past s periods and avg_{loss} is the sum of losses over the past p periods.

All mentioned, indicators are calculated based on the currency pair value, which was included in the data. The decision (BUY or $SELL$) is based on the indicator value in time t

and its relation to the indicator value at time $t - 1$. Therefore, the general rule for opening the trade for indicators can be defined as follows:

$$cond_{Buy} = true \text{ if } (ind_s(t - 1) < c) \wedge (ind_s(t) > c), \tag{11}$$

where $ind_s(t)$ is the value of indicator ind in the present reading t considering the last s readings, $t - 1$ is the value in the previous reading, and c is the indicator level (different for each indicator), which should be crossed, to observe the signal.

As shown, the crucial aspect related to generating the signal by the indicator is the value difference between two successive readings. Thus, we decided to include this information in our data in some limited way (in the case of the MA indicator). For the remaining indicators, a discretization procedure was performed because, in the classification process performed in the experimental section, only a limited number of indicator values was accepted. The summary for each indicator is presented in Table 3.

Table 3. Discretization procedure for the market indicators. * in the rare cases, where indicator value exceeds the border value (cases with the word "above" or "below", the indicator value is set to the border value).

Indicator Name	Range *	Discretization Step
Bulls	$\langle 0 : 0.01 \rangle$	0.0005
Bulls	$\langle -0.01 : 0 \rangle$	0.0005
Bulls	Above 0.1	0.005
Bulls	Below -0.1	0.005
CCI	$\langle -200 : 200 \rangle$	20.0
DM	$\langle 0 : 1 \rangle$	0.1
OSMA	$\langle 0 : 0.01 \rangle$	0.0005
OSMA	$\langle -0.01 : 0 \rangle$	0.0005
OSMA	Above 0.1	0.005
OSMA	Below -0.1	0.005
RSI	$\langle 0 : 100 \rangle$	10.0
Stoch	$\langle 0 : 100 \rangle$	10.0

Each of our readings in data also included the decision taken as one of the following values: *STRONG BUY, BUY, WAIT, SELL*, or *STRONG SELL*. Each set's decision was based on calculating the difference between the present instrument value and the value observed after p readings. This schema is presented in Figure 7. In this study, we examined p equal to 5.

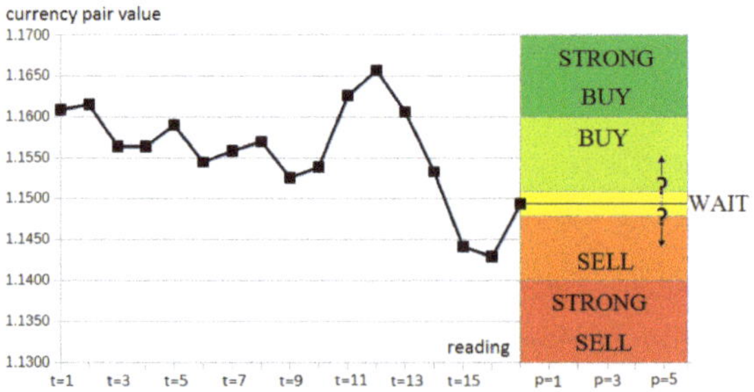

Figure 7. Decision calculation method for the financial data.

The distribution of attribute values in accordance to decision classes was created, as shown in Figure 8. We selected an example data for the AUDUSD instrument; however, a similar distribution of attribute values was noted for the remaining datasets. The blue color on the chart denotes the number of objects for which the STRONG BUY class was observed. Cyan color is related to the STRONG SELL class. Both classes cover the majority of all objects in the data. The red color shows the objects belonging to the SELL class. The two remaining classes are BUY and WAIT, respectively.

In general, we can divide the whole attribute set into three different categories. The first one is related to the instrument price (which is *Close* on the chart) and two indicators (*the_moving_average*) based on the price. For this category, we observe attributes, for which there are several values with a reasonably high number of objects assigned. The second category is related to the same indicators, where the difference between two successive readings was calculated. It gives us a distribution close to the normal distribution, where the minor differences (close to the 0) have a high number of objects assigned. Finally, the last category is related to the oscillator indicators like *Bulls* or *OSMA*, for which once again the approximation of the normal distribution is observed. Also, for these attributes, relative change between successive readings was included. The main problem in this data is that the slight differences (the middle part of attributes number 4 to 11) are frequently observed in the data. At the same time, most information comes from the relatively significant differences (tails of the distribution). Thus the most promising attribute values are the least observed in the data.

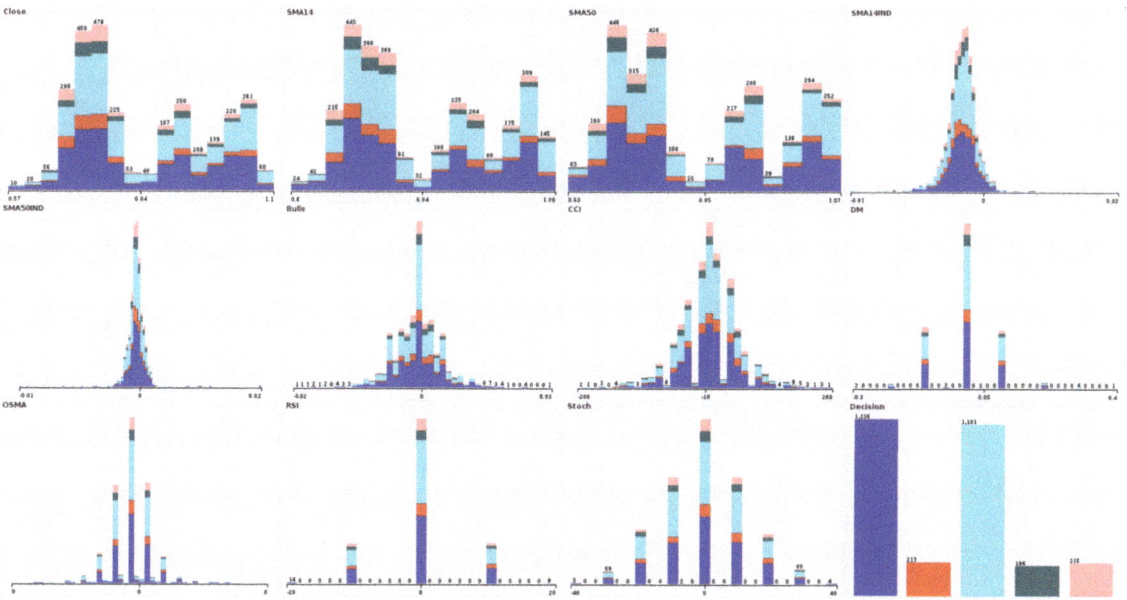

Figure 8. The distribution of attribute values due to decision classes for financial data.

5. Numerical Experiments

In this section, we describe the experiments we performed on different real-world datasets. For every set, the experiments consisted of four steps:

- calculation of the information for each conditional attribute (information attribute);
- classification of the obtained data;
- classification on the limited set of attributes (including the best 25% of the conditional attributes selected based on the information attributes) as well as the classification on the set of attributes selected by the correlation-based approach;

- sensitivity analysis on the parameter related to the percent of attributes included in the limited set of attributes.

We selected a group of well-known state-of-the-art algorithms for the classification: decision tree, Random Forest, Bagging, and AdaBoost. Two measures were used to estimate the quality of classification: the positive predictive value (PPV) and the true positive rate (TPR). Additionally, the accuracy of the classification was measured.

5.1. Fake News Data

The fake news detection research was conducted on the ISOT Fake News Dataset provided by the University of Victoria, Canada [63]. This collection includes 44,898 documents, of which 21,417 are real news cases and 23,481 are fake news. Each document in the set is described with the following attributes:

- *title,*
- *text,*
- *subject,*
- *date.*

Additionally, to determine the decision class, the main file was divided into two separate files:

- *FAKE*: documents that were detected and marked by Politifact.com as untrue sources;
- *TRUE*: real documents from Reuters.com, accessed on 31 August 2021.

In our fake news detection experiments, the dataset was limited to the *title,* and the *decision (true or fake news)* attributes only. This restriction allowed us to quickly mark the document based on the title without analyzing its content. In our previous research [64], we showed that the fake news detection model analyzing the titles produces accurate results and reduces the runtime of classification algorithms compared to the analysis of the entire content of the document.

In the first step of the experiments, we calculated the entropy of the decision class (see Equation (2)) and the information for each conditional attribute, which were the most common words in the documents. Notably, the values of the decision table (frequency of the occurrence of certain words) are strictly dependent on the documents that comprise the set on which the algorithm was trained. For this reason, the number of attributes was limited to 20. The results of this experiment are presented in Table 4. As can be seen, almost all information values for individual attributes are close to the maximum entropy value (1.0) and are in the range of 0.958–0.998. However, the last row in Table 4 shows the entropy value for the entire dataset.

In general, it is difficult to determine the set of attributes that most impact the classification results. Only attribute *word_17* has an advantage over other attributes because, for attribute *word_17*, the value of the information is visibly lower and amounts to 0.83. This means that after a single attribute—in this case, one word per document title—whether the document's full title is true or false cannot be determined. Moreover, the conditional attributes are different for a different set of documents, which entails the possibility of entirely different entropy values.

In the next step of the experiments, the values of the classification evaluation measures were calculated using selected machine learning algorithms, which were derived for each of two decision classes (true or fake news). Table 5 shows the results for the classification of fake news data by decision class for all twenty attributes.

In the case of decision class *FAKE*, PPV values were in the range of 91.38–98.88%, where the best result was obtained using a decision tree, where TPR values were in the range 46.05–58.67%, and the best result was obtained with Bagging. However, in the case of decision class *TRUE*, PPV values were in the range of 62.70–67.46% (Bagging was superior), and TPR values were in the range of 94.65–99.43% and the best results were obtained by the decision tree.

We also checked the influence of a limited number of attributes on the classification results. For this purpose, 25% of the attributes with the lowest value information attribute were selected (in this case, the top five attributes were selected). The obtained results are presented in Table 6, where the values are similar to those in Table 5. This proves that with a significantly limited number of attributes—in this case, up to 5 single words per document—the classification results for the algorithms used are the same as for a full set of conditional attributes.

The classification accuracy values for the entire set were calculated in terms of the number of attributes (five or 20 attributes), and the results are presented in Table 7. As can be seen, for the three algorithms, the accuracy was in the range of 74.17–75.49%, while for the decision tree, the accuracy was slightly lower at 71.51%.

When detecting fake news by title only, the classification accuracy measure determined how many documents were correctly classified. However, when using the PPV and TPR measures, it was possible to assess how many documents in a given class were correctly recalled and with what confidence (precision).

Table 4. Information attribute values for fake news data.

Attribute Name	Value Count	Information Attribute
word_1	2	0.998331
word_2	2	0.998048
word_3	2	0.983818
word_4	2	0.996864
word_5	2	0.998051
word_6	2	0.998287
word_7	2	0.957435
word_8	2	0.990542
word_9	2	0.981507
word_10	2	0.996318
word_11	2	0.998106
word_12	2	0.992925
word_13	2	0.992260
word_14	2	0.997342
word_15	2	0.993210
word_16	2	0.986877
word_17	2	0.803497
word_18	2	0.998101
word_19	2	0.961002
word_20	2	0.998470
Attribute Name	**Value Count**	**Entropy**
Decision	2	0.998473

Table 5. Classification results for fake news data by decision class for full set of attributes [in %] (all bold numbers correspond the best values obtained).

Decision Class	Decision Tree		Random Forest		Bagging		AdaBoost	
	PPV	**TPR**	**PPV**	**TPR**	**PPV**	**TPR**	**PPV**	**TPR**
FAKE	**98.88**	46.05	93.95	54.87	91.38	**58.67**	92.10	56.88
TRUE	62.70	**99.43**	66.02	96.12	**67.46**	93.93	**66.69**	94.65

Table 6. Classification results for fake news data by decision class for limited set of attributes (5 attributes selected) [in %] (all bold numbers correspond the best values obtained).

Decision Class	Decision Tree		Random Forest		Bagging		AdaBoost	
	PPV	**TPR**	**PPV**	**TPR**	**PPV**	**TPR**	**PPV**	**TPR**
FAKE	**98.88**	46.05	93.68	54.04	93.67	**54.28**	93.67	**54.28**
TRUE	62.70	**99.43**	65.58	96.00	**65.69**	95.97	**65.69**	95.97

Table 7. Accuracy results for the classification over fake news data [in %].

	Decision Tree	Random Forest	Bagging	AdaBoost
Accuracy (20 attributes)	71.51	74.55	75.49	74.89
Accuracy (5 attributes)	71.51	74.56	74.17	74.17

5.2. User Websites Navigation Data

The vital part of preprocessing the data is converting the raw data into a set of navigation attributes. During our research, we obtained the data of our commercial partner for one entire year. This data was more than 85 GB in size. For our learning base, we used a sample of data of one month. We chose the month of April due to avoid any marketing actions. The database for one month represents more than one million sessions with more than 10 actions performed. On account of the scale of the database, the treatment is time-consuming. After performing the limitation, we obtained 211,639 user sessions.

For entropy and classification analyses, we eliminated significantly correlated attributes such as *total_amount*. In the end, we obtained 31 attributes and one binary decision attribute, *purchase*.

The dataset for user behavior analysis consists of 211,639 unique rows. Each entry represents a unique user navigation session. First, the entropy value represents the entropy of a decision class of individual conditional attributes. Second, the results are shown in Table 8 along with the cardinality of the value set for each conditional attribute.

The entropy values for most attributes were near 0.5. For several attributes, the entropy value was lower than 0.5. For few attributes, the entropy was less than 0.2. An explanation may be the distribution of values for these attributes, which was strongly unbalanced. In most cases, the value of an attribute was equal to zero, only occasionally taking different values. Examples of these attributes are *discount_code* and *new_user*. When analyzing other attributes, the values of entropy were similar, indicating that most attributes carried an equivalent level of information. Intuitively, it seems that some attributes should be more discriminatory, but the analysis of the results did not confirm this. There were no highly biased attributes in the analyzed dataset.

Table 9 provides the classification results for the same dataset divided by each decision class value. The efficiency measures indicated relatively accurate results: PPV, TPR, and accuracy values were in the range of 0.89–1. However, both PPV and TPR were better for the decision class equal to "no". The results for the decision tree, Random Forest, and AdaBoost were similar. The results obtained using the Bagging algorithm were visibly worse than for the other algorithms. The PPV value for the class "yes" was around 0.5. Again, the reason seems to be the uneven distribution of the values of the target class.

Table 8. Information attribute values for user websites navigation data.

Attribute Name	Value Count	Information Attribute
No. of session	14	0.478533
No. total pages seen	65	0.465879
Hour of begin	101	0.478267
Hour of end	101	0.479259
Day/Month/Year	30	0.474476
Total time	77	0.402749
Discount code	2	0.173014
New customer	2	0.161386
Source of navigation	3	0.479204
No. universes changes	87	0.364767
Time Universe quick order	64	0.347537
Time Universe store	74	0.445645
Time Universe shopping cart	61	0.214535
Time Universe sales	27	0.481156
Time Universe consulting	36	0.481021
Time Universe condition	48	0.481082
Time Universe various	74	0.323663
No. pages universe quick order	38	0.339926
No. pages universe store	65	0.432955
No. pages universe shopping cart	60	0.217707
No. pages universe sales	20	0.481144
No. pages universe consulting	7	0.481129
No. pages universe condition	7	0.481159
No. pages universe various	64	0.349787
No. subsections seen	16	0.439947
No. of section changes	47	0.464543
No. product pages seen	64	0.478707
No. of same product seen	42	0.468668
No. of subsection seen	88	0.436594
No. subsection changes	78	0.471988
No. subsubsection changes	89	0.470698
Attribute Name	**Value Count**	**Entropy**
Purchase (Decision)	2	0.481233

Finally, we performed the limitation of the attributes used in classification. The limitation was based on the analysis of the value of entropy for each attribute. We selected the 25% most significant conditional attributes and performed the classification with a limited number of attributes. The classification results for user websites navigation data by decision class values for 25% of the attributes with the lowest information attribute are presented in Table 10.

The accuracy of the results for user websites navigation data is compared in Table 11. The number of all attributes participating in the classification process was 31. After limiting the set of attributes to seven, the results of the classifier efficiency increased, which may be counterintuitive. Depending on the classifier used, the improvement in efficiency ranges from 0% (DT) to 10% (Bagging). The presented analysis shows the importance of limiting the attributes at the data preprocessing stage and of classification parameterization.

Table 9. Classification results for user websites navigation data by decision class values for full set of attributes [in %].

	Decision Tree		Random Forest		Bagging		AdaBoost	
Decision Class	**PPV**	**TPR**	**PPV**	**TPR**	**PPV**	**TPR**	**PPV**	**TPR**
Purchase = yes	99.91	89.50	99.92	87.44	49.46	92.55	89.15	91.49
Purchase = no	98.80	99.99	98.56	99.99	99.03	89.03	99.01	98.70

Table 10. Classification results for user websites navigation data by decision class values for limited set of attributes (7 attributes selected) [in %].

	Decision Tree		Random Forest		Bagging		AdaBoost	
Decision Class	**PPV**	**TPR**	**PPV**	**TPR**	**PPV**	**TPR**	**PPV**	**TPR**
Purchase = yes	99.92	89.50	99.92	89.50	96.70	92.42	98.83	90.61
Purchase = no	98.80	99.99	98.80	99.99	99.13	99.63	98.92	99.88

Table 11. Accuracy results for user websites navigation data [in %].

	Decision Tree	Random Forest	Bagging	AdaBoost
Accuracy (31 attributes)	98.90	98.69	89.40	97.96
Accuracy (7 attributes)	98.90	98.90	98.89	98.91

5.3. Real Estate Market Data

The goal of the real estate market data experiment presented in this paper was to find which attributes are crucial and essential for AI model creation based on the presented decision table. To achieve this, the values of the information attributes were computed.

The dataset consisted of 14,344 unique rows. There were 13 conditional attributes (described earlier) and one decision (price bucket). In the first experiment, we computed the entropy of a decision class and the information of individual conditional attributes. The results are shown in Table 12, along with the cardinality of the value set for each attribute.

Because the data were obtained from actual advertisements, the cardinality of a decision class fell more or less in a normal distribution (Figure 9). The most frequent price fell into the PLN 6000–7000 per square meter bucket. The far-right side of the histogram plot shows the luxury properties that are part of the dataset. Remember that the property's region heavily influences the real estate market. A property located in the capital is far more expensive than the same property in a less rich part of the country. The overall decision entropy is relatively high, as the classification problem is rather difficult. Most of the attributes maintain a similar entropy value, with a single exception being the property area. Because of the cardinality of this attribute and the fact that the price of a property is usually heavily correlated with the location, this is to be expected. However, the surprising finding is that the value of entropy is also relatively high, which means that the price fluctuation between a property with a similar area is also significant. We found no noticeable changes in information for attributes such as *market_type* or *ownership_type*, indicating that such features have secondary importance for the selling price.

All attributes except the *area* obtained an information value close to the maximal entropy for the whole dataset. That means that no single conditional attribute was enough to predict the price bucket of a given property. Even the *area* conditional attribute, with a visibly lower information value equal to 2.07, was insufficient to correctly predict the price range. The price range agrees with intuition: a large but poorly located and unfurnished ruin might be cheaper than a downtown loft.

Table 13 provides the classification results for the same dataset divided by each decision class value. The Bagging algorithm produced the best results by far in nearly every decision class, both in terms of PPV and TPR. When using the limited set of attributes the following results were obtained (Table 14). Overall accuracy results were also superior using the Bagging algorithm (Table 15). Further research is required to determine whether a precise fine-tuning of hyper-parameters would increase the quality of results produced by the other algorithms.

Table 12. Information attribute values for real estate market data.

Attribute Name	Value Count	Information Attribute
Build date	166	3.294200
Total number of floors	35	3.418894
Building material	6	3.529588
Floor number	14	3.536040
Area	3,698	2.071138
Building type	5	3.455887
Condition state	6	3.485657
Windows	3	3.542212
Private ad	2	3.576844
Market type	2	3.552212
Ownership type	2	3.572712
Attribute Name	**Value Count**	**Entropy**
Price bucket (decision)	16	3.579787

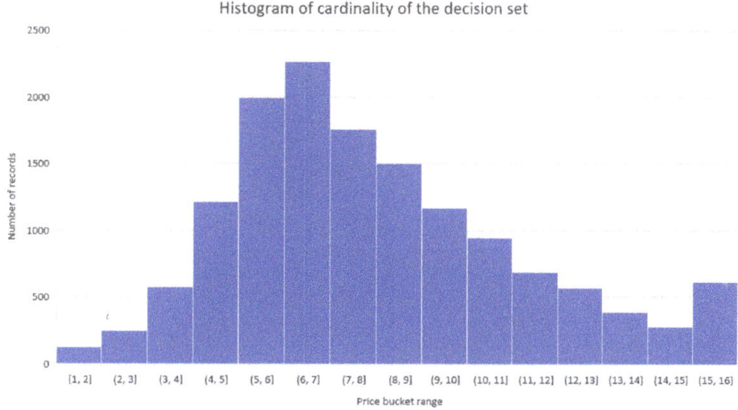

Figure 9. Histogram of cardinality of the decision set.

Table 13. Classification results for real estate market data by decision class for full set of attributes [in %] (all bold numbers correspond the best values obtained).

Decision Class	Decision Tree		Random Forest		Bagging		AdaBoost	
	PPV	TPR	PPV	TPR	PPV	TPR	PPV	TPR
1	**40.00**	9.52	0.00	0.00	25.00	**23.80**	0.16	**23.81**
2	0.00	0.00	**100.0**	0.00	99.47	**99.68**	24.61	10.02
3	0.00	0.00	0.00	0.00	**99.27**	**99.85**	0.00	0.00
4	0.00	0.00	0.00	0.00	**99.12**	**99.47**	0.00	0.00
5	0.00	0.00	0.00	0.00	**99.23**	**100.0**	0.00	0.00
6	0.00	0.00	0.00	0.00	**97.24**	**99.64**	0.00	0.00
7	0.00	0.00	49.08	21.62	**94.47**	**94.47**	27.44	87.80
8	0.00	0.00	0.00	0.00	**79.01**	**60.95**	0.67	9.52
9	0.00	0.00	0.00	0.00	**93.09**	**92.33**	0.00	0.00
10	99.57	**100.00**	100.00	7.10	99.57	99.74	99.57	79.98
11	0.00	0.00	0.00	0.00	**98.97**	**100.0**	0.00	0.00
12	21.55	**100.00**	31.25	21.79	**99.91**	**100.0**	67.78	10.03
13	**99.85**	**100.00**	32.13	91.97	**99.85**	99.90	0.00	0.00
14	**99.82**	0.99	73.49	**100.0**	**99.82**	99.65	38.48	**100.0**
15	**100.00**	**100.00**	87.70	**100.0**	**100.0**	**100.0**	39.41	9.97

Table 14. Classification results for real estate market data by decision class for a limited set of attributes (3 attributes selected) [in %]. (all bold numbers correspond the best values obtained).

Decision Class	Decision Tree		Random Forest		Bagging		AdaBoost	
	PPV	TPR	PPV	TPR	PPV	TPR	PPV	TPR
1	0.00	0.00	0.00	0.00	**27.78**	**23.81**	0.16	**23.81**
2	0.00	0.00	56.27	15.61	**99.58**	**99.79**	24.61	10.02
3	0.00	0.00	0.00	0.00	**99.14**	**100.00**	0.00	0.00
4	0.00	0.00	0.00	0.00	**98.79**	**99.82**	0.00	0.00
5	0.00	0.00	0.00	0.00	**99.24**	**100.00**	0.00	0.00
6	0.00	0.00	0.00	0.00	**96.91**	**99.64**	0.00	0.00
7	0.00	0.00	0.00	0.00	**89.01**	**90.89**	27.44	87.80
8	0.00	0.00	0.00	0.00	**60.38**	**30.48**	0.67	9.52
9	0.00	0.00	0.00	0.00	**89.37**	**91.53**	0.00	0.00
10	**99.57**	100.00	98.91	15.57	**99.57**	**99.66**	**99.57**	79.98
11	0.00	0.00	0.00	0.00	**98.79**	**99.82**	0.00	0.00
12	21.56	**100.00**	39.43	30.61	**99.92**	99.92	67.78	10.03
13	**99.85**	**100.00**	30.61	99.95	**99.85**	99.74	0.00	0.00
14	99.74	**100.00**	94.89	**100.00**	99.74	**99.82**	38.49	**100.00**
15	**100.00**	**100.00**	98.98	**100.00**	**100.00**	**100.00**	39.55	10.03

Table 15. Accuracy results for the classification over real estate data [in %].

	Decision Tree	Random Forest	Bagging	AdaBoost
Accuracy (15 attributes)	69.00	53.69	99.07	28.92
Accuracy (3 attributes)	69.00	56.71	98.71	28.92

5.4. Sport Data

Three datasets with 3362 unique rows for Spain, 2674 for Germany, and 3359 for Italy were analyzed. There was a total of 26 conditional attributes with *match_result* as a decision class. In the first stage, we calculated the entropy of a decision class and the information attribute values. The results are shown in Table 16, along with the cardinality of the value set for each attribute.

In all three analyzed datasets, the information attribute value was relatively small. It was the lowest for *Goal difference T1* and *Goal difference T2*, oscillating between 1.38 and 1.40. The highest information attribute value was recorded for *Season*. The next conditional attributes with high values were *Round* and *Matches T1 (T2)*. For the remaining measures, the values of attributes were similar. Table 16 presents the entropy of the datasets, all of which are similar (1.52–1.55).

Of the selected methods, random forest had the highest accuracy, followed by the AdaBoost algorithm. The decision tree performed the worst in the classification. None of the algorithms provided a significant advantage in terms of efficiency measures. A summary of the results is presented in Table 17.

Tests were also conducted using fewer attributes (from 24 to 6; 25% of the set based on the information attributes values). The results obtained are presented in Tables 18 and 19. As can be observed, similar results were obtained with a limited list of attributes. For some cases, the results obtained with a limited set of attributes were better. The best algorithms, in this case, were AdaBoost and random forest, whereas Bagging worked poorly.

Table 16. Information attribute values for sport data.

Attribute Name	Value Count	Information Attribute
	Germany	
Season	10	1.535055
Round	30	1.514509
Team1 (Team2)	28 (28)	1.471725 (1.464826)
Position T1 (T2)	18 (18)	1.417129 (1.438857)
Matches T1 (T2)	30 (30)	1.514509 (1.514509)
Winnings T1 (T2)	30 (30)	1.461058 (1.469589)
Draws T1 (T2)	16 (16)	1.506093 (1.502721)
Losers T1 (T2)	24 (25)	1.473808 (1.468121)
Goals scored T1 (T2)	92 (94)	1.457664 (1.454900)
Goals conceded T1 (T2)	74 (77)	1.476410 (1.470156)
Goal difference T1 (T2)	116 (117)	1.382856 (1.395159)
Points T1 (T2)	85 (86)	1.440930 (1.452137)
Series T1 (T2)	40 (40)	1.505347 (1.498025)
Match Result (Decision)	3	1.539089

Table 16. *Cont.*

Attribute Name	Value Count	Entropy
Match Result (Decision)	3	1.545029
	Italy	
Season	10	1.538402
Round	34	1.531829
*Team*1 *(Team*2*)*	34 (34)	1.458934 (1.460329)
Position T1 (T2)	20 (20)	1.415896 (1.424320)
Matches T1 (T2)	34 (34)	1.531829 (1.531829)
Winnings T1 (T2)	33 (32)	1.468639 (1.475596)
Draws T1 (T2)	19 (19)	1.514713 (1.512888)
Losers T1 (T2)	29 (30)	1.469099 (1.465326)
Goals scored T1 (T2)	92 (91)	1.481346 (1.475606)
Goals conceded T1 (T2)	86 (86)	1.484403 (1.488688)
Goal difference T1 (T2)	112 (110)	1.397699 (1.398332)
Points T1 (T2)	97 (97)	1.448064 (1.454139)
Series T1 (T2)	40 (40)	1.506473 (1.516376)
Attribute Name	**Value Count**	**Entropy**
Match Result (Decision)	3	1.545029
	Spain	
Season	10	1.519782
Round	34	1.514018
*Team*1 *(Team*2*)*	33 (33)	1.438616 (1.436341)
Position T1 (T2)	20 (20)	1.403615 (1.401159)
Matches T1 (T2)	34 (34)	1.514018 (1.514018)
Winnings T1 (T2)	32 (32)	1.451360 (1.455801)
Draws T1 (T2)	19 (19)	1.493165 (1.486429)
Losers T1 (T2)	27 (27)	1.462337 (1.453162)
Goals scored T1 (T2)	115 (111)	1.448292 (1.437169)
Goals conceded T1 (T2)	82 (81)	1.467792 (1.464831)
Goal difference T1 (T2)	130 (132)	1.376633 (1.375712)
Points T1 (T2)	95 (94)	1.438554 (1.435251)
Series T1 (T2)	40 (40)	1.493445 (1.488321)
Attribute Name	**Value Count**	**Entropy**
Match Result (Decision)	3	1.523545

Table 17. Classification results for sport data by decision class for full set of attributes [in %] (all bold numbers correspond the best values obtained).

Decision Class	Decision Tree		Random Forest		Bagging		AdaBoost	
	PPV	TPR	PPV	TPR	PPV	TPR	PPV	TPR
Germany								
1	53.66	85.51	53.96	**88.01**	56.94	70.36	**58.71**	68.44
2	55.74	48.34	**58.74**	**51.66**	51.92	49.94	55.05	50.92
X	**38.18**	3.18	0.00	0.00	33.09	20.45	**38.31**	**30.30**
Italy								
1	54.45	87.44	54.57	**89.79**	**60.39**	70.85	58.90	69.58
2	56.56	52.63	**59.12**	53.13	52.04	55.81	50.17	**57.10**
X	45.16	1.62	**100.00**	0.46	34.40	**21.21**	39.87	**20.97**
Spain								
1	55.78	88.11	55.57	**90.12**	**59.72**	69.23	**60.34**	69.79
2	52.00	**46.88**	**54.85**	44.87	49.55	**46.98**	50.73	40.53
X	0.00	0.00	**58.33**	0.85	30.29	22.83	31.73	**29.47**

Table 18. Accuracy results for the classification over sport data [in %].

	Decision Tree	Random Forest	Bagging	AdaBoost
Germany Accuracy 24 attributes	53.89	55.24	51.83	53.70
Germany Accuracy 6 attributes	53.31	54.84	49.01	55.78
Italy Accuracy 24 attributes	54.96	55.85	53.59	53.35
Italy Accuracy 6 attributes	54.47	54.85	50.09	53.54
Spain Accuracy 24 attributes	54.82	55.41	51.55	51.64
Spain Accuracy 6 attributes	55.01	55.49	51.74	55.31

The results (Table 17) show a problem with the prediction of class X (draw), which is best exemplified by the complete lack of prediction results by the Random Forest algorithm for data from Germany and Spain; for the remaining cases, this class had poor results. The unbalanced values in the decision class may be the reason for this finding. Note that a draw between teams seldom occurs.

The classification accuracy for the three sets and all selected algorithms oscillated between 51.55% and 55.85%, being higher than the random approach (for the three decision classes = 33.33%). The Random Forest algorithm achieved the highest classification accuracy on the Italy dataset and the lowest was achieved by Bagging on the Spain dataset. The exact results are presented in Table 18.

Table 19. Classification results for sport data by decision class values for for limited set of attributes (6 attributes selected) [in %] (all bold numbers correspond the best values obtained).

Decision Class	Decision Tree		Random Forest		Bagging		AdaBoost	
	PPV	TPR	PPV	TPR	PPV	TPR	PPV	TPR
Germany								
1	53.43	84.93	54.28	**86.59**	56.06	65.86	**57.08**	80.27
2	54.53	48.83	**56.27**	52.40	49.08	49.20	54.77	**60.76**
X	22.22	1.21	0.00	0.00	26.62	**18.06**	40.24	5.01
Italy								
1	54.47	86.83	53.73	**89.05**	**58.12**	67.61	55.92	80.31
2	55.76	54.32	**57.96**	51.34	49.85	50.94	52.44	**55.51**
X	2.00	0.12	0.00	0.00	27.26	**18.89**	28.39	5.10
Spain								
1	55.98	87.78	55.88	**89.48**	**60.12**	71.28	58.03	85.14
2	52.24	48.15	**54.28**	46.98	49.23	47.62	51.08	**50.26**
X	0.00	0.00	0.00	0.00	27.84	**18.96**	31.68	3.86

5.5. Financial Data Results

We used daily forex data in this study, which means that every new value was obtained at the beginning of the daily market session. We selected four different currency pairs as separate datasets: AUDUSD, EURUSD, GBPUSD, and NZDUSD, each containing 2865 readings. In addition, we used six different oscillator indicators: the Bulls indicator (*Bulls*), Commodity Channel Index (*CCI*), DeMarker indicator (*DM*), Oscillator of Moving Average (*OSMA*), Relative Strength Index (*RSI*), and the *stochastic_oscillator*. Additionally, the moving average (MA) indicator, calculated for 14 (*MA*14) and 50 (*MA*50) past readings, were included. For the results, we used the MA indicator and *MA* to denote the absolute difference between two successive readings for the indicator. It provided us with an overall number for 10 attributes.

In Table 20, we present the entropy of the decision class along with the information attributes values for the four different datasets. Firstly, there are no visible differences between the entropy values for the different datasets. However, a significant difference exists in the case of trend-following indicators (the first four attributes related to the MA indicator). This is obvious for *MA*14 and *MA*50. However, these attributes were not preprocessed and were used as was. Small entropy values suggest the strong predictive power of these indicators; however, their practical usability is lower due to a large number of different attribute values (in comparison to other oscillator indicators such as *RSI*).

In the case of oscillators, information attribute values were held on the same level instead, and it would not be easy to identify the best (in the sense of information) indicators. However, it is easy to find many examples of articles confirming that indicators' predictive capabilities are similar.

Table 21 presents the results of classification based on the PPV and TPR measures for the complete set of attributes available in the dataset. The decision class values were highly unbalanced, and for some cases, values such as BUY or SELL did not occur even once. For other cases (such as in the case of the GBPUSD dataset), the results were poor quality because we observed the STRONG BUY or STRONG SELL decision for most cases. However, in general, the AdaBoost algorithm for these rare cases with buying or selling values was slightly better than the Bagging algorithm. For the remaining cases, all four algorithms achieved similar results oscillating between 30% and 40%. Lower results for some cases (such as the STRONG BUY for the EURUSD dataset) could be related to the market situation and overall advantage of the bearish trend.

Table 20. Information attribute values for the financial data (all bold numbers correspond the best values obtained).

Attribute Name	AUDUSD		EURUSD		GBPUSD		NZDUSD	
	Value Count	Inf. Attribute	Value Count	Inf. Attribute	Value Count	Inf. Attribute	Value Count	Inf. Attribute
$SMA14$	2714	**0.077183**	2717	**0.078407**	2749	**0.050490**	2665	**0.098638**
$SMA50$	2686	**0.087042**	2673	**0.094808**	2734	**0.057822**	2653	**0.119658**
$SMA14'$	689	1.380616	767	1.258658	853	1.103368	623	1.417858
$SMA50'$	440	1.612055	497	1.467387	510	1.355281	389	1.652498
$Bulls$	341	1.747386	336	1.654923	296	1.584065	290	1.800077
CCI	78	1.936437	76	1.852890	80	1.752818	80	1.964629
DM	10	2.008175	11	1.928776	10	1.807938	9	2.035717
$OSMA$	154	1.899572	166	1.812773	214	1.653933	131	1.939371
RSI	5	2.014515	5	1.931348	6	1.812072	5	2.039597
$Stoch$	10	2.008801	11	1.928170	12	1.807163	11	2.034959
Attribute Name	Value Count	Entropy	Value Count	Entropy	Value Count	Entropy	Value Count	Entropy
$Decision$	5	2.321928	5	1.935982	5	1.815961	5	2.044563

Table 21. Classification results for the financial data by decision class for full set of attributes [in %].

Decision Class	Decision Tree		Random Forest		Bagging		AdaBoost	
	PPV	TPR	PPV	TPR	PPV	TPR	PPV	TPR
				AUDUSD				
BUY	4.05	1.53	-	-	1.52	2.55	-	-
$SELL$	-	-	-	-	1.99	3.23	6.25	0.46
$STRONG\ BUY$	35.55	34.77	33.22	49.91	31.03	30.81	33.15	37.32
$STRONG\ SELL$	35.08	53.50	30.57	32.15	26.75	21.89	35.76	49.59
$WAIT$	-	-	-	-	1.29	0.93	5.88	0.47
				EURUSD				
BUY	-	-	-	-	1.83	2.30	2.22	0.57
$SELL$	-	-	-	-	1.44	1.62	-	-
$STRONG\ BUY$	33.01	20.79	29.45	19.65	35.46	38.43	34.06	35.28
$STRONG\ SELL$	38.06	66.58	37.71	65.08	36.43	34.42	35.80	44.44
$WAIT$	-	-	-	-	1.49	0.61	4.76	0.61
				GBPUSD				
BUY	-	-	-	-	2.27	0.88	-	-
$SELL$	-	-	-	-	1.56	2.05	-	-
$STRONG\ BUY$	42.92	84.02	46.09	65.25	42.82	48.61	43.09	51.39
$STRONG\ SELL$	44.44	16.84	47.45	43.72	41.20	40.00	44.17	49.72
$WAIT$	-	-	-	-	2.22	0.67	-	-

Table 21. *Cont.*

Decision Class	Decision Tree		Random Forest		Bagging		AdaBoost	
	PPV	TPR	PPV	TPR	PPV	TPR	PPV	TPR
NZDUSD								
BUY	-	-	-	-	5.95	4.81	12.82	2.40
SELL	-	-	-	-	5.21	4.80	7.27	1.75
STRONG BUY	40.20	85.76	39.35	83.06	37.41	45.94	40.11	62.01
STRONG SELL	39.63	13.94	39.29	16.25	33.80	31.02	38.48	35.00
WAIT	-	-	-	-	5.81	2.50	-	-

Next, we performed the classification once again on the limited set of attributes. The results are presented in Table 22. For both measures (PPV and TPR), the quality of classification slightly worsened. However, the results improved for some rare cases (for example, EURUSD and GBPUSD and the TPR measure). This was achieved despite considerably reducing the number of conditional attributes included in the classification process.

Table 22. Classification results for the financial data by the decision class values for 25% of attributes with the lowest information attribute (in %).

Decision Class	Decision Tree		Random Forest		Bagging		AdaBoost	
	PPV	TPR	PPV	TPR	PPV	TPR	PPV	TPR
AUDUSD								
BUY	2.74	1.02	4.44	1.02	2.33	3.06	2.33	0.51
SELL	-	-	-	-	2.95	4.61	-	-
STRONG BUY	33.43	29.69	32.32	36.56	33.85	34.45	36.80	44.23
STRONG SELL	34.89	56.49	31.99	44.60	29.72	23.89	32.94	43.60
WAIT	-	-	-	-	4.85	5.12	-	-
EURUSD								
BUY	-	-	-	-	2.97	3.45	-	-
SELL	-	-	-	-	3.91	5.41	-	-
STRONG BUY	32.11	21.50	34.14	28.23	38.10	38.20	30.41	24.21
STRONG SELL	39.37	67.34	37.81	59.23	41.24	38.76	36.59	58.48
WAIT	-	-	-	-	3.73	3.05	-	-
GBPUSD								
BUY	-	-	-	-	2.70	2.65	16.67	0.88
SELL	-	-	-	-	2.11	3.42	-	-
STRONG BUY	41.28	37.05	44.40	54.92	42.92	42.95	42.87	48.77
STRONG SELL	42.52	60.73	43.32	47.53	42.21	41.05	44.39	52.55
WAIT	-	-	-	-	-	-	-	-

Table 22. *Cont.*

Decision Class	Decision Tree		Random Forest		Bagging		AdaBoost	
	PPV	**TPR**	**PPV**	**TPR**	**PPV**	**TPR**	**PPV**	**TPR**
			NZDUSD					
BUY	3.51	0.96	-	-	1.27	1.44	-	-
SELL	-	-	-	-	6.46	8.30	27.27	1.31
STRONG BUY	39.61	84.97	38.87	82.60	37.45	38.72	40.69	70.28
STRONG SELL	39.13	11.63	42.26	16.90	29.32	27.89	39.56	30.10
WAIT	-	-	-	-	2.50	1.50	5.13	1.00

Eventually, we analyzed the classical accuracy measure for two cases: with the full set of conditional attributes along with the limited set. These results are presented in Table 23. Surprisingly, the results do not indicate that the full set of attributes allows obtaining the highest accuracy values. These results are ambiguous; for some cases, (AUDUSD or EURUSD with the Bagging algorithm), accuracy was higher using the limited number of attributes.

These observations were also confirmed for the remaining sets. Thus, it can be assumed that some core sets of attributes can allow obtaining a relatively accurate classification. However, dependencies between these attributes are more sophisticated than simple linear correlations.

Table 23. Accuracy results for the classification over the financial data [in %].

	Decision Tree	**Random Forest**	**Bagging**	**AdaBoost**
AUDUSD	34.45	32.15	21.12	33.93
AUDUSD 2 atr.	33.55	31.70	23.78	34.32
EURUSD	36.13	35.04	30.02	32.74
EURUSD 2 atr.	36.73	36.03	32.19	34.11
GBPUSD	43.04	46.63	38.12	43.32
GBPUSD 2 atr.	41.97	43.89	36.28	43.47
NZDUSD	39.55	39.34	30.99	38.32
NZDUSD 2 atr.	38.41	39.39	26.89	39.63

5.6. Attributes Selection and the Sensitivity Analysis

To test and evaluate our results based on the attributes selection (based on the entropy values), we used the well-known correlation-based feature selection (CFS) method implemented in the WEKA system [65]. As a result, a subset of attributes, including the essential elements, were selected—comparison of a number of attributes obtained by our method and the WEKA system can be found in Table 24. As it can be noted, for most cases, the number of attributes in our approach is smaller than the number of attributes selected by the CFS method. For example, only the User Websites Navigation Data attribute selection is shown five instead of seven (out of 31 possible) attributes. In the case of the financial data, the number of attributes was the same for both methods. In contrast, for the remaining datasets, our proposed method allowed us to use a smaller number of attributes—extreme cases related to Real Estate Market Data indicated nine instead of three (out of 31) attributes.

Table 24. Number of attributes after selection.

	CFS	Proposed Approach	Original
Fake News Data	8	5	20
User Websites Navigation Data	5	7	31
Real-Estate Market Data	9	3	15
Sport Data (Germany)	8	6	24
Financial Data (GBPUSD)	2	2	11

A smaller number of attributes resulting from the use of our method does not affect the overall quality of classification. The results of classification after the selection are presented in Table 25 (names of datasets were written as an acronym). The table shows the difference in classification based on the attribute set calculated using the CFS method and our proposed approach. As can be observed, despite the smaller number of attributes indicated by the proposed method, the classification quality is similar—mostly does not exceed 0.3%. Only for the Random Forest method used for the Real Estate Market Data, an overall improvement close to 1% is observed—it is the case, where the number of attributes selected by the CFS method was equal to nine (instead of three in our proposed method). Similarly for the Sport Data, where there is improvement around 1%. While for the Financial Data, the highest differences (favoring our proposed method) were observed. In the case of the Random Forest and Bagging algorithms, the attributes selection worsens the results for over 2%. For the Financial Data for both cases, the classification was performed based on two attributes.

Table 25. Accuracy results for the classification over the data after selection [in %].

Data	Decision Tree	Random Forest	Bagging	AdaBoost
FN	71.51	74.24	74.25	74.27
Change:	—	−0.32	+0.08	+0.10
UWN	98.90	98.90	98.60	98.81
Change:	—	—	−0.29	−0.10
R-EM	69.00	57.67	98.93	28.92
Change:	—	+0.96	+0.22	—
SD	53.87	55.67	50.39	55.89
Change:	+0.56	+0.83	+1.38	+0.10
FD	42.44	41.55	33.52	43.51
Change:	+0.47	−2.34	−2.76	+0.04

In the case of the proposed method, we used the threshold of 25% of attributes included in the classification. It was shown to evaluate if the small subset of attributes allows maintaining the relatively high classification quality. Attributes were selected as the most important from the point of view of the entropy measure. This threshold was set experimentally, and it was based on several different indicators. Going below the 25% could limit the subset of attributes to two or even a single value in the case of analyzed data. At the same time, in the case of many attributes, it was possible to observe the visible decrease of classification quality. An example chart for the Sport Data (Germany) is presented in Figure 10, where the quality of classification (the Y-axis) is presented depending on the number of attributes (the X-axis). The vertical line points out the 25% of attributes used in the article.

Quality of classification

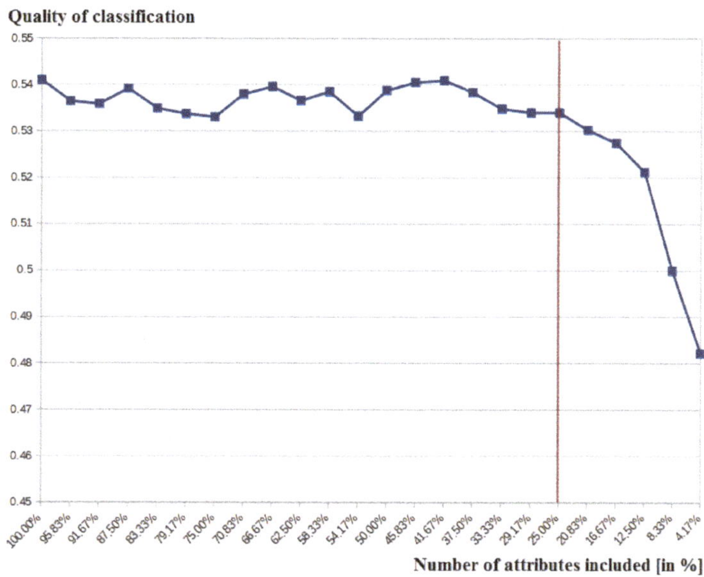

Figure 10. Classification accuracy depending on the number of attributes.

6. Conclusions and Future Works

In this study, we investigated the possibilities of using the entropy measure to select the best set of conditional attributes to be used in a classification problem. The general idea of the entropy, related works, and the problem background was introduced in the first part of the article. We also selected real-world data covering different fields. These data were retrieved and described with the use of domain knowledge experts. Finally, preprocessing was applied to all datasets, which were transformed into decision tables.

The datasets differed in their complexity, number of objects, number of conditional attributes, and the number of decision classes. Our goal was to calculate the entropy of decision classes and the information attribute values. Furthermore, we performed the classification with a set of well-known state-of-the-art algorithms. To estimate the quality of classification, we used the recall, precision, and accuracy measures. After the initial results, we selected the 25% best attributes (attributes with the best information attribute values) and performed the classification on the limited number of attributes.

For most of the cases, the algorithms obtained similar results. However, there were some examples, such as the real estate dataset, in which the Random Forest produced better results using only the limited attribute set. The Bagging algorithm showed slightly lower classification accuracy. The nature of the Random Forest algorithm, as the name implies, conducts each run providing similar but different results. The hyperparameters of Random Forest are the most prone to fine-tuning, but optimizing the parameter of each used algorithm for each used dataset was beyond the scope of this study. Notably, the value of real estate cannot be classified only using the significance of attributes but also must consider emotions and non-technical factors. For instance, we were unable to quantize the "cool" factor of a given property.

For the remaining datasets, the results were not uniform. It was difficult to identify the attributes with the best information attributes value. Differences in these values amongst the attributes in the single dataset were often negligible. However, eventually, we were able to select a subset of attributes with which the classification procedure was performed once again. Surprisingly, the limited set of attributes often allowed obtaining similar classification results. Unfortunately, it was impossible to capture the complex, nonlinear relations amongst the conditional attributes within the single dataset.

In the case of classification, we used the classical algorithms considered as a state-of-art approach. However, the multicriteria efficiency measure based on different entropy types could give much more useful information. This can be the case, especially for complex datasets without uniform structure (like Big Data). At the same time, we only investigated entropy in its basic form. An interesting approach could be related to introducing different entropy measures or even deriving estimates based on other entropy types.

In this article, we obtained some advantages over classical methods; however, the obtained results are not uniform. Therefore, our future goal could be related to extending the number of analyzed sets and emphasizing the quantitative results rather than focusing on the description of every single piece of data used in the experiments.

Author Contributions: Conceptualization, P.J. and J.K.; methodology, G.D., S.G., T.J., P.J., J.K. and B.P.; software, J.K.; validation, G.D., S.G., T.J. and B.P.; formal analysis, P.J. and J.K.; investigation, P.J.; resources, G.D., S.G., T.J., P.J., J.K. and B.P.; writing—original draft preparation, G.D., S.G., T.J., P.J., J.K. and B.P.; writing—review and editing, G.D., S.G., T.J., P.J., J.K. and B.P.; visualization, G.D., S.G., T.J., P.J., J.K. and B.P.; supervision, P.J. and J.K.; project administration, P.J. and J.K.; All authors have read and agreed to the published version of the manuscript.

Funding: This research received no external funding.

Institutional Review Board Statement: Not applicable.

Informed Consent Statement: Not applicable.

Conflicts of Interest: The authors declare no conflict of interest.

References

1. Zhang, J.Z.; Srivastava, P.R.; Sharma, D.; Eachempati, P. Big data analytics and machine learning: A retrospective overview and bibliometric analysis. *Expert Syst. Appl.* **2021**, *184*, 115561. [CrossRef]
2. Ayesha, S.; Hanif, M.K.; Talib, R. Overview and comparative study of dimensionality reduction techniques for high dimensional data. *Inf. Fusion* **2020**, *59*, 44–58. [CrossRef]
3. Yuan, Z.; Chen, H.; Xie, P.; Zhang, P.; Liu, J.; Li, T. Attribute reduction methods in fuzzy rough set theory: An overview, comparative experiments, and new directions. *Appl. Soft Comput.* **2021**, *107*, 107353. [CrossRef]
4. Jolliffe, I. A 50-year personal journey through time with principal component analysis. *J. Multivar. Anal.* **2021**, 104820. [CrossRef]
5. Wang, X.; Wang, Y.; Wong, K.C.; Li, X. A self-adaptive weighted differential evolution approach for large-scale feature selection. *Knowl.-Based Syst.* **2021**, *235*, 107633. [CrossRef]
6. Rostami, M.; Berahmand, K.; Nasiri, E.; Forouzandeh, S. Review of swarm intelligence-based feature selection methods. *Eng. Appl. Artif. Intell.* **2020**, *100*, 104210. [CrossRef]
7. Nguyen, B.H.; Xue, B.; Zhang, M. A survey on swarm intelligence approaches to feature selection in data mining. *Swarm Evol. Comput.* **2020**, *54*, 100663. [CrossRef]
8. Alsahaf, A.; Petkov, N.; Shenoy, V.; Azzopardi, G. A framework for feature selection through boosting. *Knowl.-Based Syst.* **2022**, *187*, 115895. [CrossRef]
9. Tversky, A.; Kahneman, D. Judgment under Uncertainty: Heuristics and Biases. *Science* **1974**, *184*, 1124–1131. [CrossRef]
10. Wang, S.; Celebi, M.E.; Zhang, Y.D.; Yu, X.; Lu, S.; Yao, X.; Zhou, Q.; Miguel, M.G.; Tian, Y.; Gorriz, J.M.; et al. Advances in Data Preprocessing for Biomedical Data Fusion: An Overview of the Methods, Challenges, and Prospects. *Inf. Fusion* **2021**, *76*, 376–421. [CrossRef]
11. Wang, M.C.; Tsai, C.F.; Lin, W.C. Towards missing electric power data imputation for energy management systems. *Expert Syst. Appl.* **2021**, *174*, 114743. [CrossRef]
12. Jia, X.; Dong, X.; Chen, M.; Yu, X. Missing data imputation for traffic congestion data based on joint matrix factorization. *Knowl.-Based Syst.* **2021**, *225*, 107114. [CrossRef]
13. Shannon, C. A mathematical theory of communications. *Bell Syst. Tech. J.* **1948**, *27*, 379–443. [CrossRef]
14. Rènyi, A. On measures of entropy and information. In Proceedings of the Fourth Berkeley Symposium on Mathematical Statistics and Probability, Berkeley, CA, USA, 20–30 June 1961; Volume 4.1, pp. 547–561.
15. Tsallis, C. Possible generalization of Boltzmann-Gibbs statistics. *J. Stat. Phys.* **1988**, *52*, 479–487. [CrossRef]
16. Quinlan, J.R. *C4.5: Programs for Machine Learning*; Morgan Kaufmann Publishers: San Mateo, CA, USA, 1993.
17. Quinlan, J.R. Induction of decision trees. *Mach. Learn.* **1986**, *1*, 81–106. [CrossRef]
18. Brown, G.; Pocock, A.; Zhao, M.J.; Luján, M. Conditional likelihood maximization: A unifying framework for information theoretic feature selection. *J. Mach. Learn.* **2012**, *13*, 27–66.

19. Chen, B.; Zhu, P.; Principe, J.C. Survival information potential: A new criterion for adaptive system training. *EEE Trans. Signal Process* **2012**, *60*, 1184–1194. [CrossRef]

20. Wan, M.; Gu, G.; Qian, W.; Ren, K.; Chen, Q.; Maldague, X. Particle swarm optimization-based local entropy weighted histogram equalization for infrared image enhancement. *Infrared Phys. Technol.* **2018**, *91*, 164–181. [CrossRef]

21. Lai, C.M.; Yeh, W.C.; Huang, Y.C. Entropic simplified swarm optimization for the task assignment problem. *Appl. Soft Comput.* **2017**, *58*, 115–127. [CrossRef]

22. Ganesh, M.R.; Krishna, R.; Manikantan, K.; Ramachandran, S. Entropy based Binary Particle Swarm Optimization and classification for ear detection. *Eng. Appl. Artif. Intell.* **2014**, *27*, 115–128. [CrossRef]

23. Principe, J.C. *Information Theoretic Learning: Rényi's Entropy and Kernel Perspectives*; Springer: New York, NY, USA, 2010.

24. Yuan, Z.; Chen, H.; Li, T.; Liu, J.; Wang, S. Fuzzy information entropy-based adaptive approach for hybrid feature outlier detection. *Fuzzy Sets Syst.* **2021**, *421*, 1–28. [CrossRef]

25. Li, Y.; Wang, S.; Yang, Y.; Deng, Z. Multiscale symbolic fuzzy entropy: An entropy denoising method for weak feature extraction of rotating machinery. *Mech. Syst. Signal Process.* **2022**, *162*, 108052. [CrossRef]

26. Kumar, R.; Gandotra, N.; Suman. A novel pythagorean fuzzy entropy measure using MCDM application in preference of the advertising company with TOPSIS approach. *Mater. Proc.* **2021**, in press. [CrossRef]

27. Hoberman, S.; Ivanova, A. The properties of entropy as a measure of randomness in a clinical trial. *J. Stat. Plan. Inference* **2022**, *216*, 182–193. [CrossRef]

28. Zhang, H.; Deng, Y. Entropy measure for orderable sets. *Inf. Sci.* **2021**, *561*, 141–151. [CrossRef]

29. Kuang, P.C. Measuring information flow among international stock markets: An approach of entropy-based networks on multi time-scales. *Phys. A Stat. Mech. Its Appl.* **2021**, *577*, 126068. [CrossRef]

30. Kozak, J.; Kania, K.; Juszczuk, P. Permutation entropy as a measure of information gain/loss in the different symbolic descriptions of financial data. *Entropy* **2020**, *22*, 330. [CrossRef] [PubMed]

31. Manzoor, S.; Siddiuqui, M.K.; Ahmad, S. On entropy measures of molecular graphs using topological indices. *Arab. J. Chem.* **2020**, *13*, 6285–6298. [CrossRef]

32. Kumar, K.; Prasad, V. Entropic measures of an atom confined in modified Hulthen potential. *Results Phys.* **2021**, *21*, 103796. [CrossRef]

33. Costa, M.; Peng, C.K.; Goldberger, A.L.; Hausdorff, J.M. Multiscale entropy analysis of human gait dynamics. *Phys. A Stat. Mech. Its Appl.* **2003**, *330*, 53–60. [CrossRef]

34. Pincus, S.M. Approximate entropy as a measure of system complexity. *Proc. Natl. Acad. Sci. USA* **1991**, *88*, 2297–2301. [CrossRef] [PubMed]

35. Richman, J.S.; Moorman, J.R. Physiological time-series analysis using approximate entropy and sample entropy. *Am. J. Physiol. Heart Circ. Physiol.* **2000**, *278*, H2039–H2049. [CrossRef] [PubMed]

36. Govindan, R.B.; Wilson, J.D.; Eswaran, H.; Lowery, C.L.; Preißl, H. Revisiting sample entropy analysis. *Phys. A Stat. Mech. Its Appl.* **2000**, *278*, H2039–H2049. [CrossRef]

37. Zhou, M.; Zhang, Z.; Xie, L. Permutation entropy based detection scheme of replay attacks in industrial cyber-physical systems. *J. Frankl. Inst.* **2021**, *358*, 4058–4076. [CrossRef]

38. Yan, R.; Liu, Y.; Gao, R.X. Permutation entropy: A nonlinear statistical measure for status characterization of rotary machines. *Mech. Syst. Signal Process.* **2012**, *29*, 474–484. [CrossRef]

39. Bermudez-Edo, M.; Barnaghi, P.; Moessner, K. Analysing real world data streams with spatio-temporal correlations: Entropy vs. Pearson correlation. *Autom. Constr.* **2018**, *88*, 87–100. [CrossRef]

40. Day, T. Information entropy as a measure of genetic diversity and evolvability in colonization. *Mol. Ecol.* **2015**, *24*, 2073–2083. [CrossRef]

41. Liu, X.; Jiang, A.; Xu, N.; Xue, J. Increment Entropy as a Measure of Complexity for Time Series. *Entropy* **2016**, *18*, 22. [CrossRef]

42. Zachary, D.; Dobson, S. Urban Development and Complexity: Shannon Entropy as a Measure of Diversity. *Plan. Pract. Res.* **2020**, *37*, 157–173. [CrossRef]

43. Mayer, C.; Bachler, M.; Hörtenhuber, M.; Stocker, C.; Holzinger, A.; Wassertheurer, S. Selection of entropy-measure parameters for knowledge discovery in heart rate variability data. *BMC Bioinform.* **2014**, *15*, S2. [CrossRef]

44. Chuckravanen, D. Approximate Entropy as a Measure of Cognitive Fatigue: An EEG Pilot Study. *Int. J. Emerg. Trends Sci. Technol.* **2020**, *20*, 1036–1042.

45. Coates, L.; Shi, J.; Rochester, L.; Del Din, S.; Pantall, A. Entropy of Real-World Gait in Parkinson's Disease Determined from Wearable Sensors as a Digital Marker of Altered Ambulatory Behavior. *Sensors* **2020**, *20*, 2631. [CrossRef]

46. Allcott, H.; Gentzkow, M. Social media and fake news in the 2016 election. *J. Econ. Perspect.* **2017**, *31*, 211–236. [CrossRef]

47. Guess, A.; Nagler, J.; Tucker, J. Less than you think: Prevalence and predictors of fake news dissemination on Facebook. *Sci. Adv.* **2019**, *5*, eaau4586. [CrossRef] [PubMed]

48. Lazer, D.M.; Baum, M.A.; Benkler, Y.; Berinsky, A.J.; Greenhill, K.M.; Menczer, F.; Metzger, M.J.; Nyhan, B.; Pennycook, G.; Rothschild, D.; et al. The science of fake news. *Science* **2018**, *359*, 1094–1096. [CrossRef]

49. Kannan, S.; Gurusamy, V.; Vijayarani, S.; Ilamathi, J.; Nithya, M. Preprocessing techniques for text mining. *Int. J. Comput. Sci. Commun. Netw.* **2014**, *5*, 7–16.

50. Wang, K.; Thrasher, C.; Viegas, E.; Li, X.; Hsu, B.J.P. An overview of Microsoft Web N-gram corpus and applications. In Proceedings of the NAACL HLT 2010 Demonstration Session, Los Angeles, CA, USA, 2–4 June 2010; pp. 45–48.
51. Amirhosseini, M.H.; Kazemian, H. Automating the process of identifying the preferred representational system in Neuro Linguistic Programming using Natural Language Processing. *Cogn. Process.* **2019**, *20*, 175–193. [CrossRef] [PubMed]
52. Straková, J.; Straka, M.; Hajic, J. Open-source tools for morphology, lemmatization, POS tagging and named entity recognition. In Proceedings of the 52nd Annual Meeting of the Association for Computational Linguistics: System Demonstrations, Baltimore, MD, USA, 23–24 June 2014; pp. 13–18.
53. Kalra, V.; Agrawal, R. Challenges of text analytics in opinion mining. In *Extracting Knowledge from Opinion Mining*; IGI Global: Hershey, PA, USA, 2019; pp. 268–282.
54. Koszel, M. The COVID-19 Pandemic and the Professional Situation on the Real Estate Market in Poland. *Hradec Econ. Days* **2021**, *11*, 412–425.
55. Wiktor, Z. Program, Strategy and Tactics of Communist Movement in Contemporary Epoche. *Real. Politics Estim.-Comments* **2020**, *11*, 83–95. [CrossRef]
56. Baboota, R.; Kaur, H. Predictive analysis and modelling football results using machine learning approach for English Premier League. *Int. J. Forecast.* **2019**, *35*, 741–755. [CrossRef]
57. Joseph, A.; Fenton, N.E.; Neil, M. Predicting football results using Bayesian nets and other machine learning techniques. *Knowl.-Based Syst.* **2006**, *19*, 544–553. [CrossRef]
58. Eryarsoy, E.; Delen, D. Predicting the Outcome of a Football Game: A Comparative Analysis of Single and Ensemble Analytics Methods. In Proceedings of the 52nd Hawaii International Conference on System Sciences, Maui, HI, USA, 8–11 January 2019; doi:10.24251/HICSS.2019.136. [CrossRef]
59. Schauberger, G.; Groll, A.; Tutz, G. Modeling football results in the German Bundesliga using match-specific covariates. *Engineering* **2016**. [CrossRef]
60. Schauberger, G.; Groll, A. Predicting matches in international football tournaments with random forests. *Stat. Model.* **2018**, *18*, 460–482. [CrossRef]
61. STS.PL. 2021. Available online: https://stats.sts.pl/pl (accessed on 31 August 2021).
62. Kozak, J.; Głowania, S. Heterogeneous ensembles of classifiers in predicting Bundesliga football results. *Procedia Comput. Sci.* **2021**, *192*, 1573–1582. [CrossRef]
63. Ahmed, H.; Traore, I.; Saad, S. Detecting opinion spams and fake news using text classification. *Secur. Priv.* **2018**, *1*, e9. [CrossRef]
64. Probierz, B.; Stefański, P.; Kozak, J. Rapid detection of fake news based on machine learning methods. *Procedia Comput. Sci.* **2021**, *192*, 2893–2902. [CrossRef]
65. Hall, M.A. Correlation-Based Feature Subset Selection for Machine Learning. Ph.D. Thesis, University of Waikato, Hamilton, New Zealand, 1998.

Article

Minimum Query Set for Decision Tree Construction

Wojciech Wieczorek [1], Jan Kozak [2,*], Łukasz Strąk [3] and Arkadiusz Nowakowski [3]

[1] Department of Computer Science and Automatics, University of Bielsko-Biala, Willowa 2,
43-309 Bielsko-Biała, Poland; wwieczorek@ath.bielsko.pl
[2] Department of Machine Learning, University of Economics in Katowice, 1 Maja 50, 40-287 Katowice, Poland
[3] Faculty of Science and Technology, University of Silesia in Katowice, Bankowa 14, 40-007 Katowice, Poland;
lukasz.strak@us.edu.pl (Ł.S.); arkadiusz.nowakowski@us.edu.pl (A.N.)
* Correspondence: jan.kozak@ue.katowice.pl

Abstract: A new two-stage method for the construction of a decision tree is developed. The first stage is based on the definition of a minimum query set, which is the smallest set of attribute-value pairs for which any two objects can be distinguished. To obtain this set, an appropriate linear programming model is proposed. The queries from this set are building blocks of the second stage in which we try to find an optimal decision tree using a genetic algorithm. In a series of experiments, we show that for some databases, our approach should be considered as an alternative method to classical ones (CART, C4.5) and other heuristic approaches in terms of classification quality.

Keywords: query set; decision tree; classification

Citation: Wieczorek, W.; Kozak, J.; Strąk, Ł.; Nowakowski, A. Minimum Query Set for Decision Tree Construction. *Entropy* **2021**, *23*, 1682. https://doi.org/10.3390/e23121682

Academic Editor: Geert Verdoolaege

Received: 14 November 2021
Accepted: 10 December 2021
Published: 14 December 2021

Publisher's Note: MDPI stays neutral with regard to jurisdictional claims in published maps and institutional affiliations.

Copyright: © 2021 by the authors. Licensee MDPI, Basel, Switzerland. This article is an open access article distributed under the terms and conditions of the Creative Commons Attribution (CC BY) license (https://creativecommons.org/licenses/by/4.0/).

1. Introduction

One of the main problems in machine learning is finding associations in empirical data in order to optimize certain quality measures. These associations may take different forms, such as Bayesian classifiers, artificial neural networks, rule sets, nearest-neighbor or decision tree classifiers [1]. Classical decision tree learning is performed using statistical methods. However, due to the large space of possible solutions and the graph representation of decision trees, stochastic methods can also be used.

Decision trees have been the subject of scientific research for many years [2]. The most recognized algorithms in that class are ID3 [3], C4.5 [4], and CART [5]. There are also works on the evolutionary approach to generating trees. The most popular ideas connected with this research direction are described in the article of Barros et al. [6]. Other approaches, for instance, the ant colony system, also have been studied [7]. To evaluate the performance of our approach, the following methods are selected for comparison: C4.5, CART (classification and regression trees), EVO-Tree (evolutionary algorithm for decision tree induction) [8], and ACDT (ant colony decision trees) [9]. We test the predictive performance of our method using publicly available UCI data sets.

The present proposal is about the building of decision trees which maximize the quality of classification measures, such as accuracy, precision, recall and F1-score, on a given data set. To this end, we introduce the notion of minimum query sets and provide a tree construction algorithm based on that concept. The purpose of the present proposal is fourfold:

1. Defining an integer linear programming model for the minimum query set problem. It entails preparing zero-one variables along with the set of linear inequalities and an objective function before starting the searching process.
2. Devising an algorithm for the construction of a decision tree with respect to the minimum query set. The second objective is also to implement this model through an available MIP (mixed integer programming) solver to get our approach working.
3. Performing experimental studies confirming the high classification quality of the proposed method. The third objective is also to investigate to what extent the power

of MIP solvers makes it possible to tackle the tree induction problem for large-size instances and to compare our approach with existing ones.

4. Sharing our program because of the possibility of future comparisons with other methods. The Crystal language implementation of our method is publicly available via GitHub. (https://github.com/w-wieczorek/mining, accessed on 8 December 2021).

This paper is organized into six sections. In Section 2, we present the necessary definitions and facts originated from the data structures and classification. Section 3 briefly introduces the related algorithms, while Section 4 describes our tree-construction algorithm based on solving an LP (linear programming) model and the genetic algorithm. Section 5 shows the experimental results of our approach with suitable statistical tests. Concluding comments and future plans are made in Section 6.

2. Preliminaries

In this section, we describe some definitions and facts about binary trees, decision trees, and the classification problem that are required for good understanding of our proposal. For further details about the topic, the reader is referred to the book by Japkowicz and Shah [10].

2.1. Observations and the Classification Problem

In supervised classification, we are given a training set called samples. This set consists of n *observations* (also called *objects*):

$$X = \{x_1, x_2, \ldots, x_n\}. \tag{1}$$

For each $1 \leq i \leq n$, an observation x_i is described by m *attributes* (also called *features*):

$$d(x_i) \in A_1 \times A_2 \times \cdots \times A_m, \tag{2}$$

where A_j $(1 \leq j \leq m)$ denotes the domain of the j-th attribute and $d \colon X \to A_1 \times \cdots \times A_m$ is a function. The values of the attributes can be quantitative (e.g., a salary) or categorical (e.g., sex—"female" or "male"). Furthermore, each observation belongs to one of $k \geq 2$ different *decision classes* defined by a function $c \colon X \to C$:

$$c(x_i) \in C = \{c_1, c_2, \ldots, c_k\}. \tag{3}$$

We assume that there are no two objects with the same description and different decision classes, that is, for any $1 \leq q, r \leq n, q \neq r,$

$$d(x_q) = d(x_r) \Rightarrow c(x_q) = c(x_r). \tag{4}$$

Based on the definitions given above, the *classification problem* can be defined as follows: assign an unseen object x to a class, knowing that there are k different decision classes $C = \{c_1, c_2, \ldots, c_k\}$, each object belongs to one of them, and that $d(x) = (a_1, a_2, \ldots, a_m)$. When $k = 2$, we are faced with the problem called *binary classification*. A learning algorithm $\mathcal{L}$ is first trained on a set of pre-classified samples S. In practice, a set S consists of independently obtained samples, according to a fixed—but unknown—probability distribution. The goal of an algorithm $\mathcal{L}$ is to produce a "classifier" which can be used to predict the value of the class variable for a new instance and to evaluate the classification performed on some test set V. Thus, we can say that in the learning process, a hypothesis h is proposed and its classification quality can be measured by means of accuracy, precision, recall, etc.

2.2. Decision Trees

We define a *binary tree* recursively as a tuple (S, L, R), where L and R are binary trees or the empty set, and S is a singleton set containing the value of the *root*. If L and R are empty sets, S is called a *leaf node* (or *leaf*); otherwise, S is called a *non-leaf node*. If (U, L_1, R_1)

is a binary tree and $L_1 = (V_L, L_2, R_2)$ or $R_1 = (V_R, L_2, R_2)$, then we say that there is an *edge* from U to V_L (or from U to V_R). Furthermore, V_L and V_R are called, respectively, left and right sons of U.

Let $Q = \{Q_1, Q_2, \ldots, Q_t\}$ be a collection of binary test (called *queries*) $Q_i: X \rightarrow \{0,1\}$, where X is a set of objects for which we define functions d and c as described in (2)–(4). A *decision tree*, T_X, is a binary tree in which each non-leaf node is labeled by a test from Q and has non-empty left and non-empty right subtrees; each leaf is labeled by a decision class; the edge from a non-leaf node to its left son is labeled 0 and the one to its right son is labeled 1. If $Q_{i_1}, O_{i_1}, Q_{i_2}, O_{i_2}, \ldots, Q_{i_h}, O_{i_h}$ is the sequence of node and edge labels on the path from the root to a leaf labeled by $c^* \in C$, then $c(x) = c^*$ for all objects $x \in X$ for which $Q_{i_j}(x) = O_{i_j}$ for all j $(1 \leq j \leq h)$. We also require that in this manner all leaves in a decision tree cover the whole set X, i.e., for all $x \in X$, there is at least one path from the root to a leaf corresponding to x.

The tree in Figure 1 is said to have a depth of 3. The *depth* (or *height*) of a tree is defined as the number of queries that have to be resolved down the longest path through the tree.

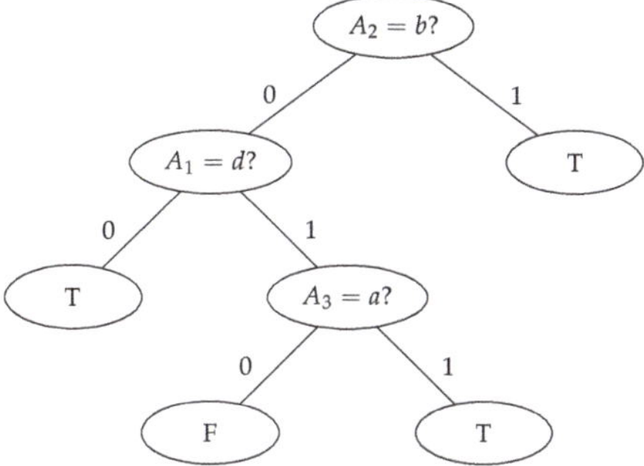

Figure 1. An exemplary decision tree.

Naturally, every decision tree T can play the role of a classifier as long as the queries can be resolved for other objects, i.e., those outside the training set. Having given a new object, let us say y, one may apply queries from the tree starting from the root and ending in a leaf ℓ that points out the predicted class p to which the object should belong. Every query in the tree directs us to a left or right son, toward a leaf ℓ. We denote such a prediction as $T(y) = p$.

2.3. Quality of Classification

To assess the quality of classification, we use the classical measures of classification quality: accuracy (5), precision (6), recall (7), and F1-score (8). Notably, these are binary classification measures, i.e., for a data set with only two decision classes. However, there are often more decision classes in data sets, so we use the so-called macro method to determine the values of these measures. Thus, in the definitions, we denote the following: TP_i to identify all correctly classified cases of the c_i class; TN_i to identify all cases outside the c_i class that are not assigned to this class; FP_i to identify all cases outside the c_i class that are assigned to this class; FN_i to identify all misclassified cases of the c_i class; and k as the number of decision classes.

$$acc = \frac{1}{k} \sum_{i=1}^{k} \frac{TP_i + TN_i}{TP_i + TN_i + FP_i + FN_i} \tag{5}$$

$$pre = \frac{1}{k} \sum_{i=1}^{k} \frac{TP_i}{TP_i + FP_i} \tag{6}$$

$$rec = \frac{1}{k} \sum_{i=1}^{k} \frac{TP_i}{TP_i + FN_i} \tag{7}$$

$$f1 = \frac{1}{k} \sum_{i=1}^{k} \frac{2 \cdot TP_i}{2 \cdot TP_i + FP_i + FN_i} \tag{8}$$

3. Related Works

This section describes the tree construction methods taken for our comparison. These are well-known, deterministic C4.5 and CART, and stochastic, population-based algorithms: EVO-Tree and ACDT.

3.1. C4.5

Developed initially by Ross Quinlan in 1993 [4], the C4.5 algorithm became one of the most popular decision tree-based algorithms [11] implemented as the standard in data mining tools, i.e., Weka (https://www.cs.waikato.ac.nz/~ml/weka/, accessed on 8 December 2021). Conceptually, the heuristic is a more advanced version of the ID3 algorithm proposed by the same author in 1986 [3]. The tree-building process recursively chooses the attribute with the highest information gain ratio. The higher the information gain the attribute has, the higher position in the tree from the root it has. Each selected feature splits a node's set of samples into subsets enriched in one class or the other [12]. To avoid over-fitting, the pruning technique is used to remove parts of the tree that minimally affect the estimated classification error. In contrast to ID3, some improvements can be made to handle missing values and continuous data [12].

3.2. CART

The classification and regression trees algorithm was co-authored by Leo Breiman, Jerome Friedman, Richard Olshen, and Charles Stone in 1984 [5], and is one of the most widely used decision tree making algorithms [11]. The CART is a binary (each node has two branches), recursive and non-parametric algorithm. It can be used for regression and classification problems. The decision tree making process uses the Gini impurity measure to determine attribute order in the tree [12]. The measure can be interpreted as the probability of incorrect classifying a randomly chosen observation from sample data if the attribute for the calculation is selected as the new decision tree node. The pruning mechanism is complex and produces a sequence of nested pruned trees, all candidate optimal trees. The best one is identified by evaluating the predictive performance of every tree in the pruning sequence by cross-validation.

3.3. EVO-Tree

The EVO-Tree algorithm [8] is an evolutionary algorithm that generates binary decision trees for classification. It uses the minimization of a multi-objective fitness function that utilizes the balance between the number of correctly classified instances and the size of the generated decision tree. The algorithm starts with the randomly initialized population of trees and uses two standard genetic operators: crossover and mutation. The crossover creates offspring by replacing a randomly selected sub-tree in the first parent with a sub-tree from the second parent. The parents' selection is made in a series of tournaments. In each tournament, a certain number of individuals from the population is randomly picked. Then, the best individual in terms of the fitness function value is chosen as a tournament winner to be put into the the pool of parents. The mutation randomly changes both attribute and split value of a decision tree. Finally, the algorithm stops if the maximum number of generations is reached or the fitness of the best individual does not improve after a fixed number of iterations.

3.4. ACDT

The ant colony decision tree (ACDT) algorithm [7] is an application of ant colony optimization algorithms [13] in the process of constructing decision trees. The good results typically achieved by the ant colony optimization algorithms when dealing with combinatorial optimization problems suggest the possibility of using that approach for the efficient construction of decision trees [14,15]. In the ACDT algorithm, each agent ant chooses an appropriate attribute for splitting the objects in each node of the constructed decision tree according to the heuristic function and pheromone values. The heuristic function is based on the twoing criterion (known from the CART algorithm) [5,16], which helps agent ants divide the objects into two groups. In this way, the attribute which best separates the objects is treated as the best condition for the analyzed node. Pheromone values represent the best way (connection) from the superior to the subordinate nodes—all possible combinations in the analyzed subtrees. For each node, the following values are calculated according to the objects classified, using the twoing criterion of the superior node.

4. Proposed Method

Our learning algorithm $\mathcal{L}$ receives as its input samples S, which are split into two subsets, the training set X and the test set Y (in experiments, we chose the proportions $4/7$ to X and $3/7$ to Y). Hypothesis space $H_{\mathcal{L}} = \{T_X^i\}_{i \in I}$ is searched in order to find a decision tree that approximates best the unknown true function. To this end, each tree is validated against Y: as a result, we output a tree T_X^* that minimizes $\mathrm{err} = |\{y \in Y : T_X^*(y) \neq c(y)\}|$. Unfortunately, in practice, we are not able to cover the whole hypothesis space. The selected hypothesis T_X^* can then be used to predict the class of unseen examples in the validation set, taken for the evaluation of $\mathcal{L}$. More exactly, $\mathcal{L}$ has two stages. In the first stage, by means of zero-one linear programming, a minimum query set Q is determined. In the second stage, by means of the genetic algorithm, the best ordering of Q—in the view of a decision tree construction—is settled. Let $x \in X$, $d(x) = (a_1, a_2, \ldots, a_m)$, and $v \in A_j$ ($1 \leq j \leq m$). In our approach, a *query* can be a function defined by $Q_i(x) = 1$ if $a_j = v$ and $Q_i(x) = 0$ if $a_j \neq v$. Thus, non-leaf nodes contain "questions" such as $A_j = v?$.

We require Q to be a minimum size query set satisfying the following condition: for each pair of distinct elements $u, w \in X$ with $c(u) \neq c(w)$, there is some query $q \in Q$ that $q(u) \neq q(w)$. We verified experimentally that this minimality is crucial in achieving good quality decision trees.

4.1. Linear Program for the Minimum Query Set Problem

Let us show how a collection of queries, Q, is determined via an integer program for the training set $X = \{x_1, x_2, \ldots, x_n\}$. The integer variables are $z_{jv} \in \{0, 1\}$, $1 \leq j \leq m$, $v \in A_j$, assuming that there are m attributes, $A_1, A_2, \ldots, A_m$. The value of z_{jv} is 1 if some query in Q is defined with A_j and $v \in A_j$; in other words, $A_j = v?$ is taken as a non-leaf node label representing the query and $z_{jv} = 0$ otherwise, i.e., there is no query based on A_j and v. Let us now see how to describe the constraints of the relationship between a set Q and a set X, with features and classes defined by functions d (as in (2)) and c (as in (3)), in terms of linear inequalities. For every pair of distinct elements $u, w \in X$ with $c(u) \neq c(w)$, we should have at least one query that distinguishes between the two. The following equation is the standard way of showing in a linear program that some elements (i.e., queries modeled as 0–1 variables) have to be included in the solution:

$$\sum_{\substack{1 \leq j \leq m \\ a_j \neq b_j}} z_{ja_j} + z_{jb_j} \geq 1, \tag{9}$$

where $(a_1, a_2, \ldots, a_m) = d(u)$ and $(b_1, b_2, \ldots, b_m) = d(w)$. Obviously, we are to find the minimum value of the linear expression

$$\sum_{\{(j,v):\, 1\leq j\leq m,\, v\in A_j\}} z_{jv}. \tag{10}$$

Please note that the above-mentioned problem is computationally complex (that is why we use an LP solver, specifically Gurobi optimizer) since Garey and Johnson's [17] NP-complete problem SP6 can be easily transformed to the decision version of the minimum query test problem.

4.2. The Construction of a Decision Tree with the Help of the Genetic Algorithm

After obtaining a minimum query set $Q = \{Q_1, Q_2, \ldots, Q_t\}$, we are ready to create a decision tree T_X by Algorithm 1.

Algorithm 1 A recursive algorithm for the construction of T_X.

---

   function BUILDTREE(X, Q) ▷ objects X as set, queries Q as array
      if all $x \in X$ have the same decision $c(x)$ **then**
         return $(\{c(x)\}, \varnothing, \varnothing)$ ▷ a leaf inside
      else
         find first i for which Q_i splits X into to non-empty sets
         $X_L = \{x \in X : Q_i(x) = 0\}$
         $X_R = \{x \in X : Q_i(x) = 1\}$
         return $(\{Q_i\}, \text{BUILDTREE}(X_L, Q), \text{BUILDTREE}(X_R, Q))$
      end if
   end function

Theorem 1. *Let X be a set of $n \geq 1$ observations and let $Q = \{Q_1, \ldots, Q_t\}$ be a set of such queries that for every pair of distinct elements $u, w \in X$ with $c(u) \neq c(w)$ there is some i $(1 \leq i \leq t)$ for which $Q_i(u) \neq Q_i(w)$. Then $\text{BUILDTREE}(X, Q)$ constructs a decision tree for X.*

Proof. Let T_X be a tree returned by BUILDTREE(X, Q). The conclusion of the theorem can be written as follows: $T_X(x) = c(x)$ for an arbitrary $x \in X$. We prove it by induction on n. Basis: We use $n = 1$ as the basis. The tree consisting of one leaf is returned, with the decision $c(x)$, so $T_X(x) = c(x)$, where x is the only element of X.
Induction: Suppose that the statement of the theorem holds for all $k < n$, where $k = |X|$. We want to show that for an arbitrary $x \in X$, where $|X| = n$, $T_X(x) = c(x)$ holds. Let us consider two cases: (i) all $x \in X$ have the same decision $c(x)$, and (ii) there is such $y \in X$ that $c(x) \neq c(y)$. In the former case, we can easily verify that $T_X(x) = c(x)$. In the latter case, there is some i $(1 \leq i \leq t)$ for which Q_i splits X into two non-empty sets, X_L and X_R. An element x is put into one of them. If it is X_L (i.e., $x \in X_L$), by the inductive hypothesis, we can claim that $T_{X_L}(x) = c(x)$, where T_{X_L} is the left subtree of a non-leaf node containing Q_i. Thus, $T_X(x) = c(x)$. For $x \in X_R$, we can repeat our reasoning.

Therefore, by strong induction, BUILDTREE(X, Q) constructs a decision tree for any set X of $n \geq 1$ observations. $\square$

Please notice that the shape of a tree T_X depends on the ordering of queries in an array Q. As a consequence, the order decides the quality of classification done by a tree returned by function BUILDTREE. That is why we apply the genetic algorithm (Algorithm 2) as a heuristic method to search such a large solution space [18]. Each individual is the permutation of the set $\{1, 2, \ldots, t\}$, which determines the order of $Q = \{Q_1, Q_2, \ldots, Q_t\}$.

Algorithm 2 The genetic algorithm for finding an optimal permutation.

function GENETICALGORITHM
    make an initial population P of POP_SIZE individuals
    iteration := 0
    while iteration < MAX_ITER and err(best_ind) > 0 **do**
        iteration := iteration + 1
        select T_SIZE elements from P
        recombine two best of them by means of PMX
        replace the worst selected element with the child
        mutate it with a probability PROB_MUTATION
    end while
    return best_ind
end function

The population size depends on the complexity of the problem, but usually contains several hundreds or thousands of possible solutions. We follow the advice of Chen et al. [19] and take POP_SIZE $= 2t \ln t$ (they suggested $|P| = O(\ln n)$, where n is the problem size, while our n is t!). The initial population is generated randomly, allowing the entire range of possible permutations.

During each successive iteration, a portion of the existing population (T_SIZE $= 3$ is chosen during preliminary experiments) is selected to breed a new individual. Solutions are selected through a fitness-based process, where fitter solutions (as measured by a fitness function) are chosen to be parents.

The fitness function is defined over the genetic representation and measures the quality of the represented solution. We use Algorithm 1 to decode a permutation. The number of misclassified objects for a test set Y is the fitness value.

For each new solution to be produced, a pair of "parent" solutions is selected for breeding from the pool selected previously. By producing a "child" solution using the crossover and mutation operations, a new solution is created which typically shares many of the characteristics of its "parents". We use partially mapped crossover (PMX for short) because it is the most recommended method for sequential ordering problems [18,20]. In the mutation operation, two randomly selected elements of a permutation are swapped with a probability PROB_MUTATION $= 0.01$. This process is repeated until one of the two termination condition is reached: (i) a solution is found that satisfies minimum criteria, or (ii) fixed number (MAX_ITER $= 500t$) of iterations reached. As a result, the best permutation encountered during all iterations is returned.

The final Algorithm 3 is depicted below. Note that heuristic search procedures that aspire to find globally optimal solutions to hard optimization problems usually require some diversification to overcome the local optimality. One way to achieve diversification is to restart the procedure many times [21]. We follow this advice and call the genetic algorithm 30 times, returning the best solution found over all starts.

Algorithm 3 The final algorithm.

Require: $S = X \cup Y$ the set of objects with functions d and c
Ensure: a decision tree T_X that tries to match a subset Y
    define a linear programming model according to (9) and (10)
    solve the model to obtain a minimum query set $Q = \{Q_1, \ldots, Q_t\}$
    multiple times run GENETICALGORITHM to obtain a permutation π
    return BUILDTREE(X, $[Q_{\pi(1)}, Q_{\pi(2)}, \ldots, Q_{\pi(t)}]$)

Because our algorithm relies on solving the minimum query set problem (finding the minimum set of attribute-value pairs that distinguishes every two objects) that is NP-hard, its overall complexity is exponential with respect to the size of input data. To tackle the problem, we use an integer linear programming solver. As modern ILP solvers are very ingenious, for practical data sets the computing time is not a big problem. Algorithms for solving ILP-problems and their NP-completeness were described in the book of [22].

5. Experiments

The section describes the comparison between selected referenced methods introduced in Section 3 and our proposed Algorithm 3 devised in the previous section.

5.1. Benchmark Data Sets

To verify our approach, we select 11 publicly available data sets with different numbers of objects, attributes, and decision classes. Used data sets are downloaded from the UCI data sets repository (https://archive.ics.uci.edu/, accessed on 8 December 2021) and are not subject to any modifications, except for possible ID removal. They are presented in Table 1, where the abbreviation used further in the paper is given in brackets, followed by the number of objects in the data set, the number of attributes, and the number of decision classes.

Table 1. Characteristics of data sets.

Data Set	Objects	Number of Attributes	Classes
balance-scale (bs)	625	4	3
breast-cancer-wisconsin (bcw)	699	9	2
car (car)	1728	6	4
dermatology (derm)	366	34	6
house-votes-84 (hv84)	435	16	2
lymphography (lymp)	148	18	4
monks-1 (monk1)	432	6	2
Somerville Happiness Survey 2015 (SHS)	143	6	2
soybean-large (soy-l)	307	35	19
tic-tac-toe (ttt)	958	9	2
zoo (zoo)	101	16	7

5.2. Performance Comparison

In this section, we describe some experiments comparing the performance of our approach implemented (https://github.com/w-wieczorek/mining, accessed on 8 December 2021) in Crystal language with ACDT implemented (https://github.com/jankozak/acdt_cpp, accessed on 8 December 2021) in C++, Weka's C4.5 implemented in Java, Scikit-learn's CART and EVO-Tree implemented (https://github.com/lazarow/dtree-experiments, accessed on 8 December 2021) in Python.

For the purpose of the experimental study, all data sets described in Section 5.1 are divided into three sets: training set (40%), test set (30%), and validation set (30%). For the classical algorithms (CART, C4.5) and EVO-Tree, the training and test sets are combined and used to learn the algorithm, while for the other algorithms, the training and test sets are used separately (according to the rule of the algorithm). In each case, the results are verified through the validation set. In this section, all given values are the results of classification performed on the validation set. So a train-and-test approach is used, but it is ensured that the data breakdowns are exactly the same in each case.

Additionally, for the algorithms that do not work deterministically (the proposed MQS and the compared EVO and ACDT) each experiment is repeated 30 times and the values presented in Tables 2 and 3 are the averages. The stability of the results obtained by these algorithms is also tested, which is presented in the form of box plots in Figures 2–4.

Table 2. The quality of classification depending on the approach (bold text is the best value).

Data Set	Measure	MQS	C4.5	CART	EVO	ACDT
bs	acc	0.7551	0.6809	**0.8085**	0.7730	0.7936
	pre	0.5559	0.4562	**0.5891**	0.5196	0.5482
	rec	0.5360	0.4843	**0.5739**	0.5505	0.5646
	f1	0.5436	0.4656	**0.5783**	0.5290	0.5538
bcw	acc	0.8817	**0.9333**	0.9190	0.9317	0.9192
	pre	0.8855	**0.9340**	0.9252	0.9270	0.9144
	rec	0.8808	0.9261	0.9059	**0.9313**	0.9173
	f1	0.8812	**0.9300**	0.9135	0.9290	0.9158
car	acc	0.9210	0.9056	**0.9730**	0.7069	0.9492
	pre	0.7946	0.7667	**0.9267**	0.3029	0.8511
	rec	0.8565	0.7600	**0.9329**	0.2609	0.9131
	f1	0.8205	0.7630	**0.9275**	0.2306	0.8714
derm	acc	0.8861	**0.9364**	0.9273	0.7879	0.9361
	pre	0.8605	**0.9334**	0.9152	0.7753	0.9276
	rec	0.8478	0.9244	0.9157	0.7225	**0.9248**
	f1	0.8488	**0.9278**	0.9142	0.7293	0.9253
hv84	acc	0.9078	0.9466	0.9313	**0.9603**	0.9450
	pre	0.8897	0.9300	0.9224	**0.9528**	0.9385
	rec	0.9096	0.9534	0.9326	**0.9641**	0.9442
	f1	0.8981	0.9436	0.9269	**0.9578**	0.9412
lymp	acc	**0.8222**	**0.8222**	**0.8222**	0.7896	0.8163
	pre	0.5411	**0.7613**	0.6677	0.6178	0.5764
	rec	0.6683	**0.9122**	0.6722	0.4980	0.5741
	f1	0.5837	**0.7912**	0.6679	0.5290	0.5718
monk1	acc	**1.0000**	0.8385	0.9538	0.7959	0.9331
	pre	**1.0000**	0.8807	0.9548	0.8469	0.9330
	rec	**1.0000**	0.8333	0.9548	0.7899	0.9330
	f1	**1.0000**	0.8323	0.9538	0.7857	0.9330
SHS	acc	**0.6125**	0.4419	0.4186	0.4682	0.4985
	pre	**0.6481**	0.5974	0.4378	0.5837	0.6118
	rec	**0.6500**	0.5428	0.4352	0.5532	0.5785
	f1	**0.6124**	0.4028	0.4173	0.4481	0.4844
soy-l	acc	0.5634	0.8478	**0.8495**	0.4706	0.7789
	pre	0.4974	**0.8565**	0.8560	0.4912	0.7173
	rec	0.6348	**0.8553**	0.8382	0.3224	0.6909
	f1	0.5294	0.8229	**0.8232**	0.3418	0.6367
ttt	acc	**0.9514**	0.8368	0.9132	0.7434	0.8927
	pre	**0.9626**	0.8092	0.8951	0.7387	0.8978
	rec	**0.9253**	0.8146	0.9066	0.6175	0.8485
	f1	**0.9412**	0.8118	0.9005	0.6217	0.8675
zoo	acc	0.8800	**0.9677**	**0.9677**	0.8720	0.9505
	pre	0.7381	**0.9524**	0.7857	0.7998	0.9080
	rec	0.8163	**0.9643**	0.8571	0.7539	0.8964
	f1	0.7636	**0.9510**	0.8095	0.7587	0.8857

Table 3. Decision tree characteristics depending on the approach.

Data Set	Parameter	MQS	C4.5	CART	EVO	ACDT
bs	time[s]	76.1	<0.1	<0.1	20.5	0.3
	size	257.1	31.0	241.0	15.1	79.4
	height	14.9	4.0	10.0	8.1	8.9
bcw	time[s]	11.7	<0.1	<0.1	12.5	0.2
	size	51.1	22.0	71.0	9.1	18.0
	height	8.0	3.0	12.0	5.4	5.7
car	time[s]	114.1	<0.1	<0.1	11.2	0.5
	size	318.9	134.0	163.0	1.7	109.4
	height	13.3	6.0	14.0	1.5	11.8
derm	time[s]	26.6	<0.1	<0.1	10.2	0.4
	size	64.7	25.0	27.0	10.8	16.6
	height	9.0	7.0	10.0	5.7	6.8
hv84	time[s]	2.6	<0.1	<0.1	4.3	0.1
	size	31.6	7.0	41.0	3.8	16.4
	height	5.9	3.0	6.0	2.5	4.2
lymp	time[s]	1.2	<0.1	<0.1	5.5	0.1
	size	34.0	20.0	49.0	11.0	20.0
	height	6.0	6.0	7.0	6.2	5.0
monk1	time[s]	0.1	<0.1	<0.1	3.9	0.1
	size	20.3	32.0	89.0	4.1	23.0
	height	5.0	5.0	10.0	2.6	6.1
SHS	time[s]	2.4	<0.1	<0.1	2.5	0.1
	size	64.3	9.0	87.0	8.0	15.6
	height	7.9	3.0	13.0	4.2	6.1
soy-l	time[s]	14.6	<0.1	<0.1	14.3	1.3
	size	151.8	67.0	75.0	14.0	45.8
	height	9.3	9.0	17.0	6.6	8.8
ttt	time[s]	24.9	<0.1	<0.1	17.5	0.4
	size	228.2	124.0	151.0	7.8	54.2
	height	9.0	7.0	11.0	4.1	8.0
zoo	time[s]	0.5	<0.1	<0.1	3.3	<0.1
	size	19.1	15.0	19.0	9.0	13.2
	height	4.9	6.0	7.0	5.1	4.9

5.3. Results of Experiments

The proposed algorithm is compared with two classical approaches and two heuristic algorithms (another genetic algorithm and the ant colony optimization algorithm). Our goal was to experimentally verify whether the MQS algorithm allows finding different (often better) solutions than the compared algorithms. The achieved results show that our assumption is confirmed.

The MQS algorithm, in terms of the analyzed metrics (see Section 2.3), allows for a significant improvement in the results for 3 out of 11 data sets. Thus, in the case of the monks-1 data set, the improvements in classification quality of almost 5% (with respect to CART), almost 7% (with respect to ACDT), about 16% (with respect to C4.5), and as much as about 20% with respect to another genetic algorithm (EVO-Tree) are obtained. There is an even greater improvement for the 2015 Somerville Happiness Survey data set and slightly less for tic-tac-toe.

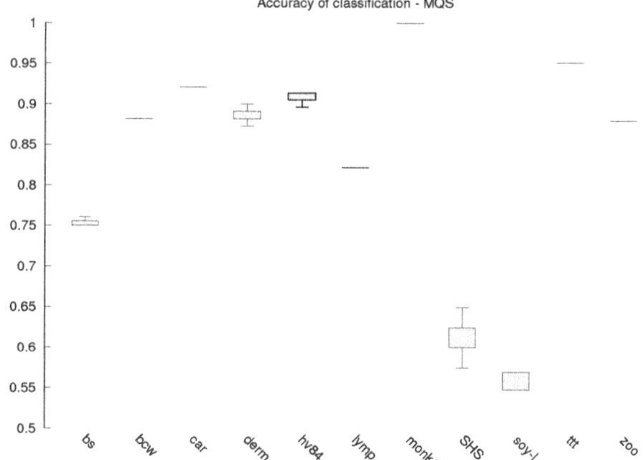

Figure 2. Box plot—accuracy of classification for the MQS algorithm.

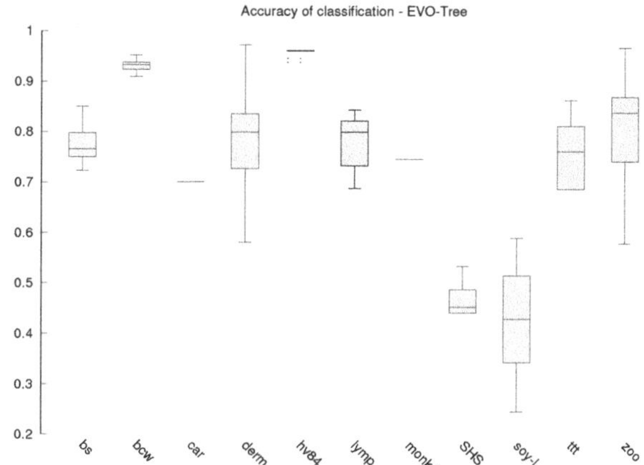

Figure 3. Box plot—accuracy of classification for the EVO-Tree algorithm.

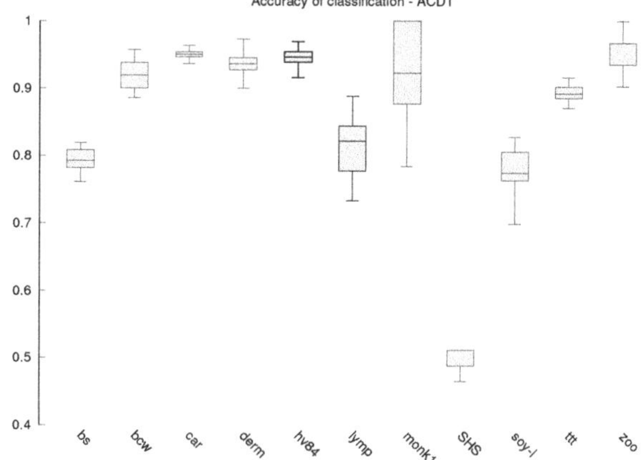

Figure 4. Box plot—accuracy of classification for the ACDT algorithm.

For the remaining data sets, the MQS algorithm obtains similar or slightly worse results, but only in one case the difference in classification quality is large—this is for the soybean-large data set. However, in two more cases, it is noticeable: dermatology and zoo. In each of these cases, the second GA algorithm has also poorer classification quality. As can be seen, the problem concerns sets with a large number of attributes (34 for dermatology, 16 for soybean-large, and 16 for zoo), so as the solution space increases (for classification, it depends on the number of attributes and the values of these attributes), the MQS algorithm has a harder time finding a suitable solution.

Our aim is to propose a new algorithm that will allow finding new optima in the solution space (in terms of classification quality). Thus, in some cases, it will allow to improve the quality of classification compared to other algorithms. Therefore, we do not try to improve either the size of the tree, the height of the tree, or the algorithm's running time, which is hard to compare between genetic and deterministic algorithms. However, we make a comparison of these decision tree-related parameters, and the results are shown in Table 3.

As can be seen, the MQS algorithm is similar in the decision tree learning time to another algorithm related to genetic algorithms (EVO-Tree). However, in terms of decision tree size and height, the proposed algorithm mostly constructs the largest trees. This is probably related to searching the solution space and covering the solution with the local optima. The size of the decision tree does not correlate with its classification quality (in relation to other algorithms) and so a significantly larger tree, e.g., in the case of the balance-scale data set, does not improve the results, while in the case of tic-tac-toe, the results are improved while increasing the decision tree.

The stability of the results obtained is also subject to our analysis, because the stability allows us to assume that the classifier will always be of similar quality. While in the case of classical algorithms, the results are deterministic, in the case of MQS, EVO-Tree and ACDT, a different classifier may emerge each time. Box plots are prepared with classification accuracy for each data set in case of MQS (Figure 2), EVO-Tree (Figure 3), and ACDT (Figure 4) algorithms. To prepare the graphs, the corresponding quantiles (minimum value is lowest on the OY axis, 1st quantile, 2nd quantile (median), 3rd quantile and maximum value that is highest on the OY axis) from all 30 repetitions of learning the decision tree are determined.

The MQS algorithm is the most stable; in Figure 2, we can see that only for the Somerville Happiness Survey 2015 and soybean-large data set, small (compared to the other algorithms) differences appear. For the other data sets, the results are very repeatable. For the other algorithms, the repeatability of the results is much lower, and so for EVO-Tree, we can see in Figure 3 that in seven cases, the differences are quite divergent; for the dermatology, soybean-large and tic-tac-toe databases, the classification accuracy in successive repetitions changes even by several dozen percentage points. In the case of the ACDT algorithm, the results are more reproducible (Figure 4)—significant differences appear in two to three cases, while for the monks-1 set, the difference can be as much as several dozen percentage points.

5.4. Statistical Analysis

The experimental results of the MQS approach are compared using a non-parametric statistical hypothesis test, i.e., the Friedman test [23,24] for $\alpha = 0.05$. Parameters of the Friedman test are shown in Table 4. The same table presents the average rank values for the compared algorithms for learning decision trees (in terms of classification quality). Results in terms of each of the classification quality measures analyzed are used for statistical testing.

The MQS algorithm obtains a rank of 3.1591, so it is significantly better than the EVO-Tree algorithm (the 5% critical difference is 0.6192); MQS is worse than the other algorithms, but this is by no means a critical difference. Therefore, we confirm that it is possible to use the MQS algorithm in the decision tree learning process, so it should always be considered and tested because it can output a significantly better classifier than the other algorithms. This is especially valid when we are given a data set with a small number of attributes. At the same time, we confirm that the proposed algorithm is significantly better than another genetic algorithm used for decision tree learning.

Table 4. The Friedman test results and mean ranks.

	Values
N	44
Chi-Square	24.0594
degrees of freedom	4
p value is less than	0.0001
5% critical difference	0.6192
Mean ranks	
MQS	3.1591
C4.5	2.6932
CART	2.5568
EVO	3.9545
ACDT	2.6364

As the EVO-Tree algorithm is found to be critically inferior to all other approaches analyzed, we perform a second round of statistical analysis. The results of the Friedman test and the mean ranks after rejecting the critically inferior method are recorded in Table 5. As can be seen, in this case, none of the methods is critically better or worse than all the others. The big difference remains only when contrasting MQS with CART.

Due to the lack of significant differences and the advantage of obtaining significantly higher results (when the MQS algorithm gets a rank of 1, it is better by several/dozen percentage points, where in other methods, the advantage is often negligible—see Table 2), the proposed method can be considered for use in selected classification problems.

Table 5. Friedman test results and mean ranks after rejection of the critically worse method.

	Values
N	44
Chi-Square	5.8
degrees of freedom	3
p value is less than	0.1218
5% critical difference	0.5305
Mean ranks	
MQS	2.8864
C4.5	2.4205
CART	2.2614
ACDT	2.4318

5.5. Discussion

To evaluate the proposed algorithm, we made comparisons with classical approaches and other non-deterministic algorithms. This is a new algorithm proposal, so we wanted to make a fair comparison. We used up to four different measures of classification quality. We also compared the size and height of the decision tree and the learning time of the classifier. Finally, we performed statistical tests.

As decision trees learned with non-deterministic methods often search a much larger solution space, this must affect their running time. It can also result in larger, more extensive decision trees. When proposing the MQS algorithm, we knew that the classifier learning time would require more time. Therefore, its application, like other stochastic methods, should be considered for classifiers that are built once in a while—not online classifiers. Our study confirmed that the MQS algorithm takes longer to learn than statistical methods. However, it is comparable to non-deterministic methods (especially another genetic algorithm).

In this case, the classification time is more important, and it depends primarily on the height of the decision tree. Our analyses indicated, for example, that the MQS algorithm is better than the CART algorithm in 10 out of 11 cases, remaining worse than the other algorithms in 7–9 cases. In terms of the size of the decision trees (this affects the memory occupation needed to store the finished classifier), the situation is similar. The MQS and CART algorithm learn larger decision trees than the others. However, it should be emphasized that no pruning of decision trees is performed for the proposed MQS algorithm. At this stage, we wanted to keep the complete decision trees.

However, our aim was to find new alternative classifiers with which a better classification could be achieved. Therefore, the most important analysis concerned the evaluation of classification quality. In this case, we were able to see that for some data sets, the MQS algorithm allows to build a classifier better than all other algorithms.

This is particularly important because often the differences (in classification quality assessment) between different algorithms are a few percentage points. However, for the monks-1, Somerville Happiness Survey 2015 and tic-tac-toe data sets, the MQS algorithm allows a very large improvement in each of the classification quality assessment measures.

We analyzed the exact structure of these data sets. Our observations show that the application of the proposed algorithm can be particularly beneficial for data sets with two decision classes and attributes with a small number of possible values (3–5 values of each attribute). However, the decision classes can be of different numbers. This does not mean, however, that the MQS algorithm obtains bad results with other sets—the suggestion described above indicates a situation where a classifier learned by MQL obtains results with much better classification quality.

Finally, we analyzed the stability of the results obtained. We did this to determine whether the classifiers learned by the MQS algorithm are always of similar quality. For this purpose, we performed 30 independent runs of the algorithm and obtained 30 independent classifiers. We performed the same tests with other stochastic algorithms (EVO-Tree and ACDT). The obtained results clearly indicate that the proposed algorithm is the most stable one, so it can be assumed that the classifier will always obtain similar results.

To confirm our observations, a statistical test was performed twice: the first time, for all approaches (and all classification quality values) and the second time, after rejecting the EVO-Tree algorithm (it obtained results with a critical difference with respect to other algorithms). This time, the critical difference of one algorithm against all others was not shown.

6. Conclusions

This paper deals with the construction of decision trees based on the finite set of observations (objects). In order to address the problem, we introduced the notion of minimum query set and made use of the genetic algorithm for suitable ordering of the found queries. As the result of the implemented algorithm, we achieved decision trees that perfectly match the training data set and have good classification quality on the test set. The conducted experiments and statistical inference showed that the new proposed, two-stage algorithm should be considered as an alternative method to classical ones (CART,

C4.5) and other heuristic approaches in terms of accuracy, precision, recall, and F1-score for all 11 UCI data sets.

Our method has also a few disadvantages. The most significant ones are that (i) the first stage of our approach relies on solving a computationally intractable problem, and (ii) for some cases, the obtained decision trees have too many nodes. In the near future, we are planning to adapt our approach to handle continuous attributes. In order to make it possible to reproduce our results or apply our method on new data, we share the source code of all algorithms via the Github platform.

Author Contributions: Conceptualization, W.W.; methodology, J.K., Ł.S. and A.N.; validation, J.K., Ł.S. and A.N.; formal analysis, W.W. and Ł.S.; investigation, J.K., Ł.S. and A.N.; resources, W.W., J.K., Ł.S. and A.N.; writing—original draft preparation, W.W., J.K., Ł.S. and A.N.; writing—review and editing, W.W., J.K., Ł.S. and A.N.; visualization, Ł.S.; supervision, W.W.; project administration, J.K. All authors have read and agreed to the published version of the manuscript.

Funding: This research received no external funding.

Institutional Review Board Statement: Not applicable.

Informed Consent Statement: Not applicable.

Data Availability Statement: Not applicable.

Conflicts of Interest: The authors declare no conflict of interest.

References

1. Kubat, M. *An Introduction to Machine Learning*; Springer: Berlin/Heidelberg, Germany, 2017.
2. Kotsiantis, S.B. Decision trees: A recent overview. *Artif. Intell. Rev.* **2013**, *39*, 261–283. [CrossRef]
3. Quinlan, J.R. Induction of decision trees. *Mach. Learn.* **1986**, *1*, 81–106. [CrossRef]
4. Quinlan, J.R. *C4.5: Programs for Machine Learning*; Elsevier: Amsterdam, The Netherlands, 2014.
5. Breiman, L.; Friedman, J.H.; Olshen, R.A.; Stone, C.J. *Classification and Regression Trees*; Elsevier: Oxford, UK, 1984; p. 358.
6. Barros, R.C.; Basgalupp, M.P.; de Carvalho, A.C.P.L.F.; Freitas, A.A. A Survey of Evolutionary Algorithms for Decision-Tree Induction. *IEEE Trans. Syst. Man Cybern. Part C* **2012**, *42*, 291–312. [CrossRef]
7. Kozak, J. Ant Colony Decision Forest Approach. In *Decision Tree and Ensemble Learning Based on Ant Colony Optimization*; Springer: Berlin/Heidelberg, Germany, 2019; pp. 119–134.
8. Jankowski, D.; Jackowski, K. Evolutionary Algorithm for Decision Tree Induction. In *Computer Information Systems and Industrial Management*; Saeed, K., Snášel, V., Eds.; Springer: Berlin/Heidelberg, Germany, 2014; pp. 23–32.
9. Kozak, J.; Boryczka, U. Collective data mining in the ant colony decision tree approach. *Inf. Sci.* **2016**, *372*, 126–147. [CrossRef]
10. Japkowicz, N.; Shah, M. *Evaluating Learning Algorithms: A Classification Perspective*; Cambridge University Press: Cambridge, UK, 2011.
11. Wu, X.; Kumar, V.; Quinlan, J.R.; Ghosh, J.; Yang, Q.; Motoda, H.; McLachlan, G.J.; Ng, A.; Liu, B.; Philip, S.Y.; et al. Top 10 algorithms in data mining. *Knowl. Inf. Syst.* **2008**, *14*, 1–37. [CrossRef]
12. Hssina, B.; Merbouha, A.; Ezzikouri, H.; Erritali, M. A comparative study of decision tree ID3 and C4.5. *Int. J. Adv. Comput. Sci. Appl.* **2014**, *4*, 13–19.
13. Dorigo, M.; Stützle, T. *Ant Colony Optimization*; MIT Press: Cambridge, UK, 2004.
14. Jiang, W.; Xu, Y.; Xu, Y. A Novel Data Mining Method Based on Ant Colony Algorithm. In *Advanced Data Mining and Applications*; Li, X., Wang, S., Dong, Z.Y., Eds.; Springer: Berlin/Heidelberg, Germany, 2005; Volume 3584, pp. 284–291.
15. Dorigo, M.; Birattari, M.; Stützle, T. Ant Colony Optimization—Artificial Ants as a Computational Intelligence Technique. *IEEE Comput. Intell. Mag.* **2006**, *1*, 28–39. [CrossRef]
16. Timofeev, R. Classification and Regression Trees (CART) Theory and Applications. Master's Thesis, Humboldt University, Berlin, Germany, 2004.
17. Garey, M.R.; Johnson, D.S. *Computers and Intractability: A Guide to the Theory of NP-Completeness*; W. H. Freeman: New York, NY, USA, 1979.
18. Salhi, S. *Heuristic Search: The Emerging Science of Problem Solving*; Springer International Publishing: Berlin/Heidelberg, Germany, 2017.
19. Chen, T.; Tang, K.; Chen, G.; Yao, X. A large population size can be unhelpful in evolutionary algorithms. *Theor. Comput. Sci.* **2012**, *436*, 54–70. [CrossRef]
20. Michalewicz, Z. *Genetic Algorithms + Data Structures = Evolution Programs*; Springer: New York, NY, USA, 1996.
21. Martí, R.; Lozano, J.A.; Mendiburu, A.; Hernando, L. *Handbook of Heuristics*; Chapter Multi-Start Methods; Springer International Publishing: Cham, Switzerland, 2018; pp. 155–175.
22. Schrijver, A. *Theory of Linear and Integer Programming*; John Wiley & Sons: Hoboken, NJ, USA, 1998.

23. Friedman, M. The use of ranks to avoid the assumption of normality implicit in the analysis of variance. *J. Am. Stat. Assoc.* **1937**, *32*, 675–701. [CrossRef]
24. Kanji, G.K. *100 Statistical Tests*; Sage: Thousand Oaks, CA, USA, 2006.

Article

Breaking Data Encryption Standard with a Reduced Number of Rounds Using Metaheuristics Differential Cryptanalysis

Kamil Dworak * and Urszula Boryczka

Faculty of Science and Technology, University of Silesia in Katowice, Będzińska 39, 41-200 Sosnowiec, Poland; urszula.boryczka@us.edu.pl
* Correspondence: kamil.dworak@us.edu.pl

Abstract: This article presents the author's own metaheuristic cryptanalytic attack based on the use of differential cryptanalysis (DC) methods and memetic algorithms (MA) that improve the local search process through simulated annealing (SA). The suggested attack will be verified on a set of ciphertexts generated with the well-known DES (data encryption standard) reduced to six rounds. The aim of the attack is to guess the last encryption subkey, for each of the two characteristics Ω. Knowing the last subkey, it is possible to recreate the complete encryption key and thus decrypt the cryptogram. The suggested approach makes it possible to automatically reject solutions (keys) that represent the worst fitness function, owing to which we are able to significantly reduce the attack search space. The memetic algorithm (MASA) created in such a way will be compared with other metaheuristic techniques suggested in literature, in particular, with the genetic algorithm (NGA) and the classical differential cryptanalysis attack, in terms of consumption of memory and time needed to guess the key. The article also investigated the entropy of MASA and NGA attacks.

Keywords: differential cryptanalysis; metaheuristics; symmetric block ciphers; memetic algorithms; DES; simulated annealing

Citation: Dworak, K.; Boryczka, U. Breaking Data Encryption Standard with a Reduced Number of Rounds Using Metaheuristics Differential Cryptanalysis. *Entropy* **2021**, *23*, 1697. https://doi.org/10.3390/e23121697

Academic Editor: Masahito Hayashi

Received: 15 November 2021
Accepted: 15 December 2021
Published: 18 December 2021

Publisher's Note: MDPI stays neutral with regard to jurisdictional claims in published maps and institutional affiliations.

Copyright: © 2021 by the authors. Licensee MDPI, Basel, Switzerland. This article is an open access article distributed under the terms and conditions of the Creative Commons Attribution (CC BY) license (https://creativecommons.org/licenses/by/4.0/).

1. Introduction

The growing popularity of computerisation, and at the same time the Internet itself, results in a growing demand for more and more advanced security methods. Restrictions such as individual user access control or basic authentication have become insufficient today. For several decades, engineers concentrating on the topic of information security have designed special cryptographic algorithms that meet the most important security aspects.

The main assumption of cryptography is not to hide the fact of the existence of information, but to keep its real image secret. The message is transformed in such a way that it is readable only to its author and the recipient it is dedicated to [1,2].

Contemporary symmetric block ciphers implement the process of transformation of the plain text using the Feistel cipher and the generalized substitution-permutation network [2]. In 1990, a completely new cryptanalytical method was made public, namely differential cryptanalysis [3]. In the case of the most modern and advanced encryption algorithms, the differential cryptanalysis itself turns out to be ineffective. In order to improve the attack performance, it was proposed to combine metaheuristic algorithms with the differential cryptanalysis algorithm.

In general, metaheuristic algorithms are used to obtain approximate solutions. In the case of cryptanalysis, it is necessary to guess the ideal decryption key—an approximate solution is unacceptable. Due to the avalanche effect present in every encryption algorithm today, changing any bit at the input causes a complete mixing of all bits at the output, which in fact results in the generation of a completely new ciphertext [1]. The developed algorithm enables automatic sifting of the keys with the worst value of the fitness function, owing to which the set of potential solutions will be significantly reduced.

Additional analytical properties of memetic algorithms improve the local search process in such a way as to achieve the best solution in the shortest possible time.

Metaheuristic algorithms are more and more often used in computer science, and thus in the domain of computer security. In the literature, we can find publications describing all kinds of metaheuristic attacks targeting both classical ciphers, contemporary symmetric block ciphers and stream ciphers. A literature review of publications is presented in Table 1.

Table 1. Literature review of researches on metaheuristics cryptanalysis.

Year	Authors	Algorithm	Cipher
2007	Song et al. [4]	GA	Four-Round DES
2007	Tadros et al. [5]	GA	Four-Rounded DES
2009	Garg [6]	GA and MA	Simplified Data Encryption Standard (SDES)
2010	Hu [7]	GA	Tiny Encryption Algorithm (TEA)
2011	Abd-Elmonim [8]	PSO	DES
2011	Vimalathithan and Valarmathi [9]	GA, PSO and Genetic Swarm Optimization (GSO)	Simplified Data Encryption Standard (SDES)
2012	Jadon et al. [10]	Binary PSO	DES
2012	Pandey and Mishra [11]	PSO	DES
2013	Ali [12]	Bees algorithm	Substitution Ciphers
2014	Boryczka and Dworak [13]	EA	Transposition Cipher
2014	Mekhaznia and Menai [14]	ACO and PSO	Feistel, Vigenere, and substitution ciphers
2015	Bhateja et al. [15]	Cuckoo Search	Vigenere cipher
2015	Jain et al. [16]	Cuckoo Search	Substitution Ciphers
2016	Amic et al. [17]	Binary Firefly Algorithm	DES
2016	Dworak et al. [18]	GA and MA	Simplified Data Encryption Standard (SDES)
2016	Dworak and Boryczka [19]	EA	Four-Rounded Fast Data Encipherment Algorithm (FEAL)
2017	Amic et al. [20]	Binary Cat Swarm Optimization (BCSO)	DES
2017	Jain et al. [21]	Cuckoo Search	Knapsack Cryptosystem
2017	Dworak and Boryczka [22]	GA	Six-Rounded DES
2018	Polak and Boryczka [23]	Tabu Search	RC4 and VMPC
2019	Amic et al. [20]	Dolphin Swarm Algorithm (DSA)	DES
2019	Kamal et al. [24]	Binary Cuckoo Search	Simplified Data Encryption Standard (SDES)
2019	Polak and Boryczka [25]	Tabu Search	RC4+
2020	Sabonchi et al. [26]	DE, GA and PSO	Vigenere cipher
2021	Grari et al. [27]	ACO	Merkle-Hellman cipher

In [4], the authors focused on evolutionary cryptanalysis using GA on DES4 ciphers by comparing the same bits between original and encrypted ciphertexts. Tadros in [5] presented another GA used to break FEAL8 and DES4 ciphers. Garg in [6] included a comparison between MA and GA during cryptanalysis of SDES encryption algorithm relying on n-gram statistics and frequency analysis method. Another approach was present by Hu in [7], quantum-inspired GA has been applied to break TEA. Abd-Elmonim described another attack, based on the PSO algorithm, responsible to break the full 16-rounded DES cipher in [8]. Vimalathithan and Valarmathi presented their researches about combining the effectiveness of GA and PSO as a new Generic Swarm Optimization algorithm to attack SDES cipher. In 2012, Jadon [10] and Pandey, with Mishra published interesting approaches related to Binary PSO and original PSO algorithms used in cryptanalysis attacks dedicated to DES cipher.

In the following years, Ali [12], Mekhaznia and Menai [14], Bhateja [15], Jain [16,21], and Sabonchi [26] focused on cryptanalysis of classical ciphers such as substitution, transposition, and Vigenere ciphers using many popular metaheuristics like Bees, EA, ACO, PSO and Cuckoo Search algorithms.

Amic in [17,20,28] presented Binary Firefly, Binary Cat Swarm Optimisation (BCSO), and Dolphin Swarm (DSA) algorithm—all directed against DES cipher. In [24] Kamal described the Binary Cuckoo Search algorithm used on ciphertext generated by SDES cipher.

Polak and Boryczka presented new cryptanalysis attacks dedicated to another subset of encryption algorithms—stream ciphers (RC4, VMPC, and RC4+), using Tabu Search in [23,25]. In 2021, Grari [27] published ACO algorithm dedicated Markle-Hellman cipher.

The next chapter is dedicated to a brief introduction to symmetric block ciphers and the DES cipher. The third chapter presents the basic assumptions of differential cryptanalysis, which were used and which constituted a basis for the design work on the MASA algorithm. Chapter four contains a detailed description of the developed metaheuristic attack carried out with the use of MA. The next chapter focuses on describing the runtime environment, including presenting all the parameters selected for each attack. This chapter also presents the results of the experiments, including the entropy studies for the MASA and NGA algorithms. The second to last chapter presents a detailed analysis of the effectiveness of the attacks presented, both in terms of the number of proven solutions and the time of decryption of the cryptogram. The article is concluded with a brief summary of the various stages of the research. This chapter also suggests further research directions. Appendix A is attached to this article, detailing the results for the Ω_2 characteristic.

2. Symmetric Block Ciphers

Symmetric ciphers are still one of the most popular encryption algorithms. In this type of ciphers, only one main key is used, which simultaneously takes on a role of an encryption and decryption key, which can be written as $K_E = K_D$. In the case of block ciphers, each message is divided into a finite number of blocks of the same length—for example, 64-bit blocks. Then they are transferred to the appropriate encryption function. Exactly one block of the ciphertext is generated from one block of plain text. If the message cannot be divided into even blocks, an additional block is created to store the last, incomplete, fragment of data. Then, for consistency, it is supplemented with default values or zeros.

These algorithms are perfect for encrypting larger volumes of data stored, that is, in all kinds of warehouses, wholesalers or databases. The most popular block cipher schemes include ciphers such as: DES and AES.

Data Encryption Standard

The DES cipher has been designed in such a way that the avalanche effect occurs from the very beginning of the algorithm [1]. Changing any input bit forces us to change at least half, and sometimes even all, of the output bits. The state of each bit at the output depends on each bit specified at the input [29].

The basic version of the cipher converts 64-bit plain text blocks into 64-bit ciphertext blocks, using a 64-bit encryption key K [2,30]. After running the algorithm, the primary key is reduced to 56 bits by removing every eighth parity bit. K is then subjected to breaking into six 48-bit subkeys, used in each of the cipher rounds, $K_1, ..., K_6$—A description of the primary key distribution process is presented in detail in [1,2,29–32]. Figure 1 shows a 6-round DES algorithm.

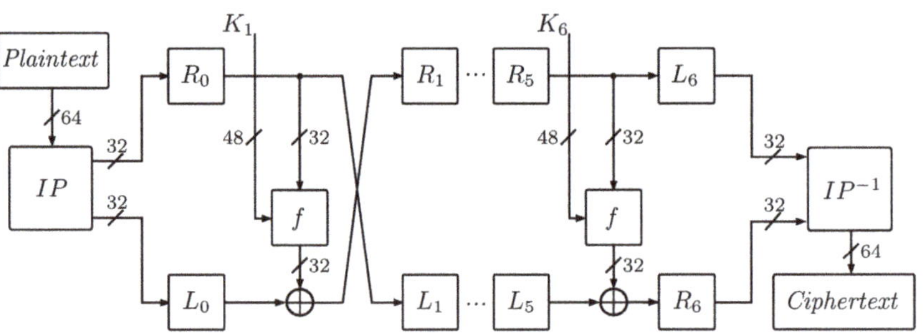

Figure 1. Simplified diagram of the six-rounded Data Encryption Standard DES algorithm.

The plain text block is passed to the initial IP permutation. Then, the generated block is divided into two regular 32-bit parts, R and L. In the next steps, six identical encryption cycles will be run, in which the right part of the R_i is passed to the f-round function along with the corresponding subkey K_i. Then, the generated data block is subjected to the exclusive disjunction operation with the left part of the L_i, resulting in a new right part of the R_{i+1}. The new left part of the L_{i+1} is copied from the right part of the previous R_i cycle.

After all the cipher rounds have been completed, parts of the L_6 and R_6 are combined into a 64-bit block, which will undergo the last transformation by the IP^{-1} inverse permutation function. The result of transposition of individual bits will be a 64-bit cryptogram block.

The f round function has been visualized in Figure 2. As an input parameter, a 32-bit data block is given, which at the very beginning will be extended via permutation E. The aim of this transformation is to align the length of the transferred block with the size of the subkey by duplicating the selected bits. By allowing one bit to influence two substitutions, the avalanche effect is increased [1]. The generated sequence is modulo two sum with subkey bits and then divided into eight 6-bit B_1–B_8 blocks.

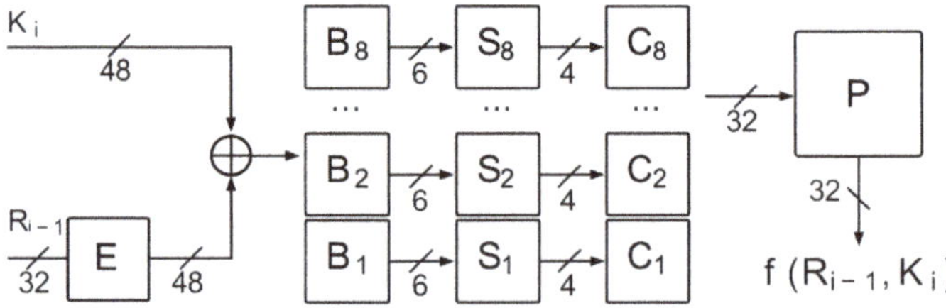

Figure 2. The f round function of the Data Encryption Standard DES algorithm. The only one, nonlinear, element of the DES cipher.

Each of the B_j blocks will be transferred to the so-called substitution matrix called S-blocks S_j. The main aim of this transform is to compress the input data. 6-bit data blocks will be converted into 4-bit blocks. S_j consist of integers between 0 and 15, stored in matrices of sixteen columns and four rows. The first and last bits of a 6-bit sequence B_j determine the line number. The remaining four bits represent the number of the column from which the return value will be selected [1,2,30].

S_j are the only nonlinear element of the DES standard. Changing one bit in an input sequence can lead to a complete mixing of all generated bits at the output. Modifications carried out in them have a significant impact on the level of complexity of cryptanalysis of the entire cipher. At the end of the f function, the generated sequences are combined into one 32-bit block, which will be passed to the permutation P—aimed at mapping each of the input bits to exactly one output bit without duplicating or omitting any of them [1].

3. Differential Cryptanalysis

The suggested algorithm is based on an attack with selected plain text. At the beginning, it should be assumed that the cryptanalyst has continuous access to the encryption algorithm, which allows him to select a pair of plain texts and analyse the generated ciphertexts. It is important that the tested pairs must differ from each other in a certain way. Most symmetric block ciphers determine this difference on the basis of a simple symmetric difference operation, which is written as $P' = P \oplus P^*$, where P and P^* are two crafted plain texts. Pairs may be generated in a pseudorandom way, although the most important condition is the difference P', which must follow the established process. Next, the cryptanalyst checks how the determined difference changes in the subsequent phases of the cipher. Using the difference between the texts in individual iterations of the cipher, for a sufficiently large number of pairs, it is possible to assign different probabilities, suggesting the correctness of some subkeys [3]. When analyzing subsequent pairs of plain texts and ciphertexts, it turns out that one key may be more probable than the others.

Every modern cipher is non-linear—it means that it is not possible to find any pattern or rule by which to determine the value of a function for the next argument [3]. This nonlinearity is obtained via the round f function. Each of all possible differences is characterized by a certain probability, which determines how often the f function returns the expected value [3]. These differences are called characteristics Ω. All possible characteristics can be determined by means of an additional matrix, where the rows correspond to all possible symmetric differences of the input blocks, and the columns to all possible symmetric differences of the output blocks [1]. Each of the elements will determine how many times the sum of the output bits occurs for the selected sum of the input bits.

By analysing the diagram shown in Figure 2, the input symmetric difference B' can be determined assuming that $E = E(R_{i-1})$:

$$B' = \overset{8}{\underset{j=1}{\|}} B_j \oplus B_j^* = \overset{8}{\underset{j=1}{\|}} (E_j(R_i) \oplus K_i) \oplus (E_j(R_i^*) \oplus K_i) = \overset{8}{\underset{j=1}{\|}} E_j \oplus E_j^*, \tag{1}$$

where symbol $\|$ stands for the concatenation of the successive data blocks. From the expression above, it can be seen that B' has nothing to do with the subkey. When the value of each B_j' is known, the set of all ordered pairs (B_j, B_j^*) can be determined for the input symmetric difference as suggested in [31]:

$$\Delta(B_j') = \{(B_j, B_j \oplus B_j') : B_j \in (\mathbb{Z}_2)^6\}. \tag{2}$$

Knowing the output difference $C_j' = S_j(B_j) \oplus S_j(B_j^*)$, it becomes possible to generate the distribution of all possible input differences to all output differences according to the theorem described in [31]:

$$IN_j(B_j', C_j') = \{B_j \in (\mathbb{Z}_2)^6 : S_j(B_j) \oplus S_j(B_j \oplus B_j') = C_j'\}. \tag{3}$$

Most often, this distribution will be steady. The cryptanalyst's task is to find distributions that are as unsteady as possible. Based on the expression (3), an additional test set can be determined using the following formula [31]:

$$test_j(E_j, E_j^*, C_j') = \{B_j \oplus E_j : B_j \in IN_j(E_j', C_j')\}. \tag{4}$$

If the number of elements in $test_j$ is equal to the power of IN_j set, then the set must contain bits of the K_{ij} subkey [31].

This method makes it possible to restore the correct decryption key using 2^{47} selected plain texts and the corresponding ciphertexts.

4. Metaheuristics Differential Cryptanalysis

From the point of view of the developed attack, the IP and IP^{-1} permutations may be omitted. The algorithm begins by selecting the two most probable 3-round characteristics Ω_P^1 and Ω_P^2 mentioned in [31,32], which are presented in Figure 3, where P denotes characteristics for plaintext and C for ciphertexts.

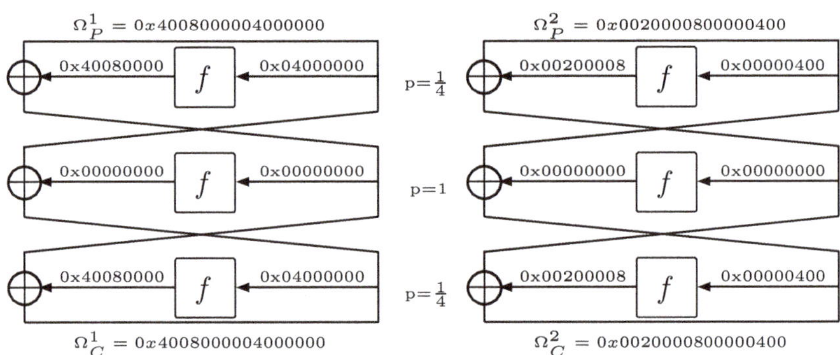

Figure 3. The two the most probable 3-round characteristics Ω_P^1 and Ω_P^2 for six rounded cipher DES [31,32].

The probability of each characteristic is exactly $P_\Omega = \frac{1}{16}$ in the fourth round of the encryption algorithm S-Blocks S_2, S_5, S_6, S_7, S_8 for Ω_P^1 and S_1, S_2, S_4, S_5, S_6 for Ω_P^2 for some input symmetric difference B_j' return an output symmetric difference C_j' equal to zero. Owing to this, it becomes possible to determine the sets $I_1 = \{2, 5, 6, 7, 8\}$ for Ω_P^1 and $I_2 = \{1, 2, 4, 5, 6\}$ for Ω_P^2. The further description of the attack is identical for each of the characteristics Ω so it was decided to generalize it by introducing one generic I set consisting of elements of sets I_1 and I_2.

The next step will be to generate a set of plain text pairs, along with a set of corresponding cryptograms, where the symmetrical difference will correspond to the characteristics Ω_1 and Ω_2. The number of pairs needed is calculated using the signal-to-noise ratio [3]:

$$S/N = \frac{m \cdot p}{m \cdot \alpha \cdot \beta / 2^k} = \frac{2^k \cdot p}{\alpha \cdot \beta} = \frac{2^{30} \cdot {}^1/_{16}}{4^5} = 2^{16}, \tag{5}$$

where:

- m—the number of pairs generated, having no effect on S/N;
- p—the probability of the selected characteristic Ω;
- k—number of bits of the subkey;
- α—the average number of subkeys, suggested by one pair;
- β—the ratio of the analysed pairs to all possible ones.

As suggested in [3], for $S/N = 2^{16}$, 7–8 correct pairs are needed for each of the characteristics. Due to the probability of P_Ω, a minimum of 150–200 pairs of plain text should be generated [3].

Additionally, the $test_j$ test set is determined, owing to which it will be possible to partially filter pairs from the set. If the power of the test set for at least one element from set I is equal to 0, the pair may be rejected:

$$\bigwedge_{j\in I} |test_j| > 0. \tag{6}$$

The aim of the suggested attack is to guess the last K_6 encryption subkey. If the difference of C' and part of R_5 is known, it becomes possible to analyze the various subkeys closely by comparing all bits of the output of the S-blocks with C'. A brute-force attack would need to check all 2^{30} solutions. MA can be used as an optimization tool that finds the correct solution in much shorter time.

Each individual is represented by a 30-bit K_j subkey. The fitness function is defined with the following formula:

$$F_f = \sum_{i=0}^{n} L - \sum_{j\in I} H((S_j(B_j) \oplus S_j(B_j^*)), P^{-1}(R_6' \oplus L_3')), \tag{7}$$

where:

- H—is the Hamming distance;
- L—the length of the subkey.

Owing to the knowledge of the probability of P_Ω, it is possible to estimate the value of L_3', while R_6' can be obtained by analyzing a pair of generated ciphertexts. F_f counts the number of overlapping bits between the difference obtained from the S-blocks and the C' difference.

The algorithm uses standard one-point crossover. The locus is selected pseudorandomly from 1 to 30. The newly created subkeys can be modified with the use of a mutation operator—which consists in replacing two pseudorandomly selected bits. The algorithm selects individuals using tournament selection. A leader is elected from the set of all subkeys and it is passed to the crossover operator.

There is an additional local search process in the algorithm—it is performed using the simulated annealing algorithm. The MASA attack pseudocode for the ΩP characteristic is shown below. Due to the complexity of this algorithm, it was decided to divide it into two parts:

- the first one, Algorithm 1—responsible for generating a set of filtered pairs of plain text, ciphertexts and determining the $test_j$ test set for each of the indexes;
- the second one, presented in Algorithm 2—describing the memetic algorithm, along with the processes of selection, crossing, mutation and exploitation, taking into account the pseudocode of the basic simulated annealing algorithm.

Algorithm 1: The pseudocode of the set of pairs preparation process for the MASA attack.

```
 1  Ωp := find_most_probabilistic_characteristic()
 2  I := determine_set_of_indexes()
 3  set_of_pairs := generate_set_of_plaintext_and_ciphertext_pairs()
 4  for i := 0 to size(set_of_pairs) do
 5      pair := set_of_pairs[i]
 6      foreach j ∈ I do
 7          testj := determine_test_set(pair)
 8          if |testj| == 0 then
 9              filter_invalid_pair(set_of_pairs, pair)
10              break
11          end
12      end
13  end
```

Running the MASA algorithm for Ω_P^1 will make it possible to guess 30 out of 48 bits of the K_6 subkey. Re-running the algorithm, this time for Ω_P^2, allows us to find an extra 12 bits. In order to obtain the remaining 6 bits of the last K_6 subkey—coming from the S-block S_3, we can use the brute-force method. Having the K_6 subkey, it is possible to recover 48 out of 56 bits of the decryption key by reversing the key decomposition process. The remaining 8 bits can be guessed using the brute-force method once again—for example, a brute force attack.

Algorithm 2: MASA attack pseudocode.

```
 1  P(0) := create_initial_population()
 2  for i := 0 to number_of_iterations do
 3      calculate_fitness_function_value_for_each_individual()
 4      for j := 0 to population_size do
 5          parent_A := tourney_selection()
 6          parent_B := tourney_selection()
 7          offspring := [parent_A, parent_B]
 8          if random(0, 1) ≥ crossover_probability then
 9              child_A, child_B := crossover(parent_A, parent_B)
10              if random(0, 1) ≥ mutation_probability then
11                  child_A := mutation(child_A)
12              end
13
14              if random(0, 1) ≥ mutation_probability then
15                  child_B := mutation(child_B)
16              end
17              offspring := [child_A, child_B]
18          end
19
20          foreach child ∈ offspring do
21              T = T_0
22              while T ≥ T_MIN do
23                  new_child := change_random_bit(child)
24                  difference := new_child.fitness - child.fitness
25                  if difference > 0 or
26                      probability_fun(difference, T) > random(0, 1) then
27                      child := new_child
28                  end
29                  T = T · α
30              end
31          end
32      end
33  end
```

---

5. Experimental Results

This chapter describes the analysis of the proposed memetic attack MASA and NGA in terms of the quality and number of solutions obtained [22]. It was important to check whether the suggested algorithms make it possible to improve the time of finding the correct subkey. Another important aspect was to check whether the MASA memetic algorithm enables a more effective, and therefore more successful, differential cryptanalysis.

5.1. Selecting Parameters

As part of the experiments, the impact of the parameters listed below for each of the attacks on the convergence of the algorithm and the quality of the obtained solutions was examined:

- number of iterations for the MASA and NGA algorithms;
- population size for the MASA i NGA algorithms;
- number of plaintext and ciphertext pairs γ for the MASA and NGA algorithms;
- probability of the heuristic negation P_n for the NGA algorithm.

In the conducted experiments, the parameter values were used in various combinations and for the subsequent experiments, potentially the best values in terms of the running time of the algorithm were established. For the MASA memetic algorithm, the parameters were set according to Table 2 below:

Table 2. Parameters of the MASA algorithm.

Id	Parameter	Symbol	Value
1	Maximum number of iterations	It_{MAX}	100
2	Population size	N	10
3	Number of plaintext pairs	γ	200
4	Tourney size	T_{SIZE}	10
5	Crossover probability	P_c	0.9
6	Mutation probability	P_m	0.02
7	Initial temperature	T_0	1
8	Minimal temperature	T_{MIN}	0.1
9	Cooling rate	α	0.9

The description of the NGA algorithm parameters has been described in detail in the publication [19]. Table 3 presents the most important parameters of the NGA algorithm:

Table 3. Parameters of the NGA algorithm.

Id	Nazwa	Symbol	Value
1	Maximum number of iterations	It_{MAX}	100
2	Population size	N	10
3	Number of plaintext pairs	γ	200
4	Tourney size	T_{SIZE}	10
5	Crossover probability	P_c	0.9
6	Mutation probability	P_m	0.02
7	Heuristic operator probability	P_n	0.25

As was mentioned before, for the purposes of the tests, a simplified version of the DES cipher was used, in which the number of rounds was limited from 16 to 6. All other processes in the encryption algorithm, such as subkey generation and S-block compression, remained unchanged.

5.2. Comparative Study

Each of the algorithms was tested 30 times for each of the characteristics Ω. Table 4 below shows the value of the F_f fitness function for the MASA and NGA algorithms for the first characteristic Ω_1. The remaining results—for the characteristic Ω_2 are given in Appendix A in the Table A1.

Table 4. Fitness function values for MASA and NGA algorithms for characteristic Ω_P^1.

ID	MASA					NGA				
	Min	**Med**	**Avg**	**Max**	**Std. Dev.**	**Min**	**Med**	**Avg**	**Max**	**Std. Dev.**
1	885	953	95.5	1014	39.9	982	993	994.7	1014	11.5
2	929	997	989.2	1014	30.8	**899**	**947**	**960.3**	**1012**	**40.3**
3	**888**	**935**	**948.7**	**1012**	**49.2**	916	992	977.4	1014	37.8
4	978	1010	1003.9	1014	12.3	**886**	**945**	**943.0**	**997**	**33.5**
5	910	950	960.9	1014	38.9	922	978	982.3	1014	25.5
6	915	971	971.6	1014	35.1	871	978	960.1	1014	53.2
7	877	925	953.7	1014	52.6	928	998	990.1	1014	30.3
8	920	982	983.6	1014	31.1	**900**	**960**	**958.9**	**1012**	**36.2**
9	895	997	978.6	1014	35.2	**943**	**980**	**981.3**	**1012**	**20.7**
10	949	957	981.0	1014	30.1	863	934	945.9	1014	50.5
11	938	1014	996.8	1014	25.5	**921**	**974**	**973.0**	**997**	**27.6**
12	947	995	988.6	1014	22.1	899	975	965.1	1014	42.1
13	903	936	952.0	1014	36.9	891	978	962.3	1014	48.7
14	886	997	975.5	1014	46.6	855	991	958.0	1014	55.8
15	960	990	992.3	1014	20.0	881	920	951.3	1014	52.3
16	892	996	970.6	1014	42.3	884	998	978.0	1014	45.4
17	880	984	960.8	1014	50.1	**911**	**954**	**962.5**	**1012**	**29.2**
18	983	1014	1008.8	1014	10.0	865	978	958.1	1014	54.5
19	893	992	975.8	1014	38.1	**878**	**951**	**951.3**	**998**	**44.0**
20	956	1010	1003.3	1014	17.7	922	990	977.3	1014	34.1
21	**892**	**979**	**965.9**	**998**	**38.3**	875	929	945.5	1010	45.3
22	962	1009	1003.6	1014	15.1	909	1014	990.2	1014	38.2
23	885	939	960.7	1014	41.2	**940**	**981**	**978.3**	**1010**	**23.2**
24	901	970	962.2	1014	40.6	**872**	**935**	**936.3**	**988**	**41.9**
25	864	949	949.9	1014	50.3	**890**	**954**	**944.9**	**988**	**34.9**
26	958	992	991.8	1014	15.9	931	955	972.2	1014	28.4
27	899	920	949.2	1014	44.8	888	965	964.5	1014	43.2
28	902	966	965.4	1014	40.2	893	957	959.4	1014	40.6
29	971	997	999.2	1014	13.0	912	980	973.1	1014	37.2
30	922	997	977.2	1014	36.0	953	976	987.7	1014	21.2

Experiments in which the correct decryption key could not be guessed were marked in bold in the table above.

The probability of each of the characteristics for this cipher is not 100%. It means that despite striving for the maximum value of the fitness function, it will never be achieved. The inability to obtain the maximum value means that we are not able to terminate the running of the algorithm earlier than after the completion of all predetermined iterations.

Figures 4 and 5 present a list of all correctly guessed bits of the K_6 subkey for the MASA and NGA algorithms for the Ω_1 characteristic. The remaining results—for the Ω_2 characteristic are present in Appendix A in the Figures A1 and A2.

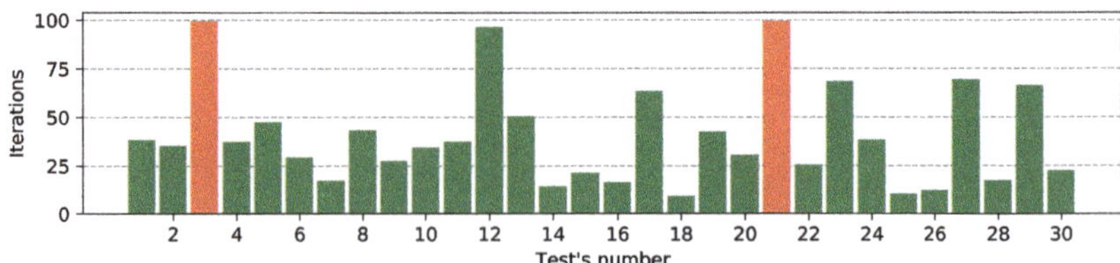

Figure 4. List of correctly guessed bits of MASA attack for the Ω_1 characteristic.

In a large number of cases, the MASA attack finds the correct subkey in the first 25 iterations. In approximately 6–7 cases, the algorithm found a solution using half of the available iterations, while in the other two cases (tests #3 and #21, marked as red on the figure) the attack failed to cope with the given ciphertext. The algorithm found the correct decryption subkey in 93% of the cases - markes as green on the figure.

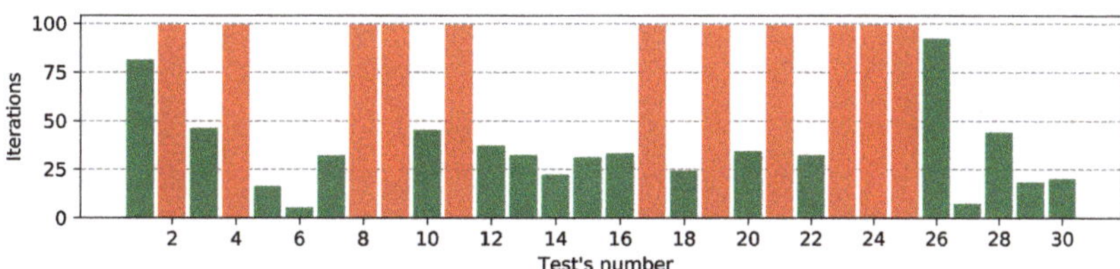

Figure 5. List of correctly guessed bits of NGA attack for the Ω_1 characteristic.

In the case of the NGA algorithm, the cipher was not cracked 11 times—which is over 37% of all possible approaches—red bars on the figure. During the remaining 63% of the tests, it was possible to crack the cipher with the decryption algorithm—green color. In most cases, it was possible to guess the correct subkey using only 30–40 iterations. The tests with identifiers #1 and #26 also deserve special attention. They show a very large number of iterations (over 80), which means that the NGA algorithm found the correct solution at the very end of its running.

On the presented bar plots we can notice the MASA algorithm is much effective because it successfully found the correct subkey in almost every test when NGA attack has worked in only 63% of experiments. Simulated annealing, used as an additional exploitation step of the MA, is more effective than the heuristic negation operator used in the NGA attack.

The next stage of the experiments was to analyze the course of the fitness function value using the convergence diagrams, which were presented successively, for the MASA attack and Ω_1 in Figures 6 and 7, for the NGA algorithm. Convergence diagrams for the Ω_2 were present in Appendix A in the Figure A3, for the MASA algorithm, and Figure A4 for the NGA attack.

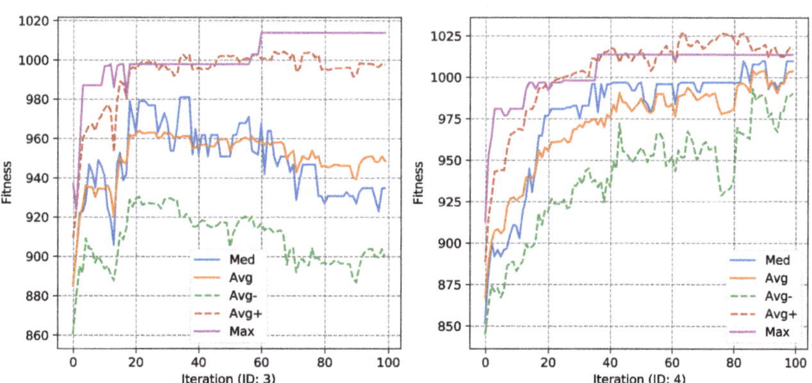

Figure 6. The MASA fitness function F_f convergence diagrams for Ω_1 (tests #3 and #4).

The above graph shows tests #3 and #4 with minimum, maximum, medians and averages—and average values increased and decreased by the standard deviation of the fitness function. The tests were selected in such a way as to visualize both a positive case—when it was possible to guess the correct subkey, and a negative one.

In the case of both tests of the MASA algorithm, a rapid increase in the maximum value of F_f can be noticed at the very beginning of the algorithm's running. In further iterations, there are single drops of this value, after which the maximum value is stabilized and then increased again. The median for 60% of the algorithm's running time remains similar, only at the very end of its running we can notice its decrease. When analyzing the case #4 diagram, already in the first iterations of the algorithm, a rapid increase in the median value can be observed—the majority of individuals in the population have a similar value of the fitness function. This may be related to the algorithm falling into the local extreme, which it has not managed to leave.

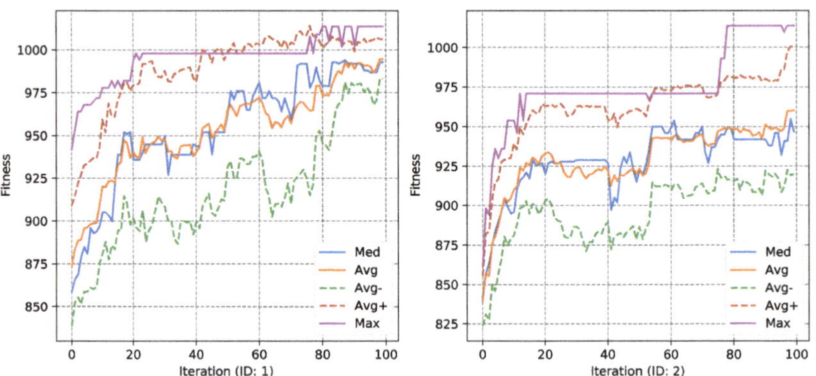

Figure 7. The NGA fitness function F_f convergence diagrams for Ω_1 (tests #1 and #2).

The next stage of the tests was to review the distribution of the fitness function values in the last iteration of each attack—the distribution is presented in Figure 8 for the MASA algorithm, and Figure 9 in the case of an NGA attack. Boxplots for the Ω_2 characteristic were present in Appendix A in the Figure A5, for the MASA algorithm, and Figure A6 for the NGA attack.

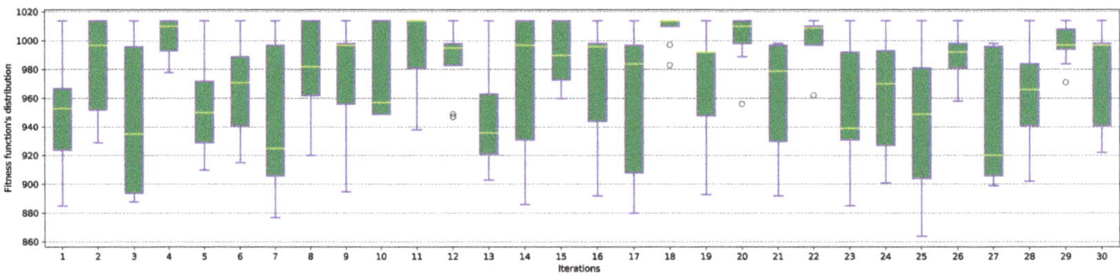

Figure 8. The distribution of the fitness function F_f values in the last iteration for the MASA algorithm and Ω_1 characteristic.

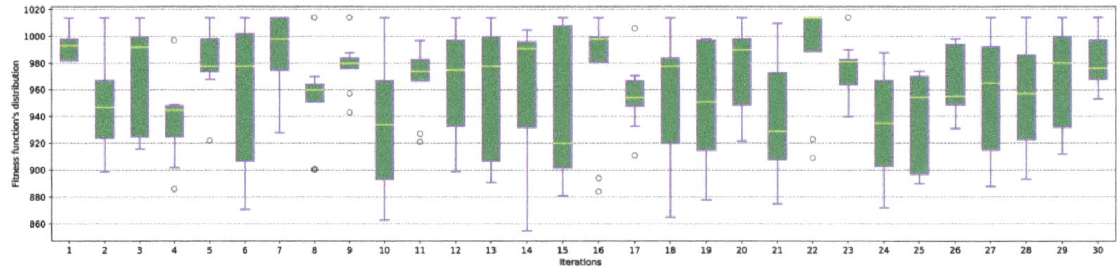

Figure 9. The distribution of the fitness function F_f values in the last iteration for the NGA algorithm and Ω_1 characteristic.

In the case of the MASA algorithm, some of the tests—for example, #18, #20 or #22—are characterized by a high degree of homogeneity, which means that the population is characterized by a low diversity of individuals. When analyzing each of the attacks, a large degree of variability between individuals can be observed, which is undoubtedly indicated by the median value, changing its position between the first and the third quartiles. In the case of the NGA algorithm, in some experimentes, an unexpected increase of the value of the fitness function can be observed at the very end of the algorithm's running—it is evidenced by the presence of the outlier of the maximum value.

The MASA and NGA attacks are characterized by a certain degree of pseudo-randomness. In order to perform statistical verification of the algorithms, a non-parametric Wilcoxon's test was used to compare the results. The hypothesis H_0, specifying no difference when comparing the samples, and the hypothesis H_1, assuming a difference between the two samples, were set. The following criteria were used to perform the test:

- value of the fitness function—performed for the best quality subkeys found for each run;
- number of subkeys checked.

The weight of each criterion was expressed at the same value, set to 0.5. For the analyses performed, hypothesis H_0 was rejected at $p < 0.05$—thus indicating the statistically important differences between the best results retrieved. The results obtained through the MASA algorithm are significantly better than the NGA attack.

5.3. Entropy Study

The possibility to maintain a highly diverse population may improve the algorithm's ability not to fall into local extremes. In order to estimate the size of the disorder in the system, the entropy was used:

$$H(X) = \sum_{i=1}^{n} p(x_i) log_2 \frac{1}{p(x_i)} = -\sum_{i=1}^{n} p(x_i) log_2 p(x_i). \tag{8}$$

The entropy was computed by comparing the respective bits of each subkey with the corresponding bits of the best-adapted individual. An example for the population $P = \{11101, 10101, 11011, 11110\}$, where the last individual 11110 is the leader, is presented below (Table 5):

Table 5. Example scenario of the entropy calculation.

Subkey	Bit 1	Bit 2	Bit 3	Bit 4	Bit 5
A	1	1	1	0	1
B	1	0	1	0	1
C	1	1	0	1	1
Leader	1	1	1	1	0
$p(x_1)$	1	0.75	0.75	0.50	0.25
$p(x_2)$	0	0.25	0.25	0.50	0.75
$H(x_1)$	$4 \cdot log_2(1)$	$3 \cdot \frac{3}{4} log_2(\frac{3}{4})$	$3 \cdot \frac{3}{4} log_2(\frac{3}{4})$	$2 \cdot \frac{1}{2} log_2(\frac{1}{2})$	$\frac{1}{4} log_2(\frac{1}{4})$
$H(x_2)$	0	$\frac{1}{4} log_2(\frac{1}{4})$	$\frac{1}{4} log_2(\frac{1}{4})$	$2 \cdot \frac{1}{2} log_2(\frac{1}{2})$	$3 \cdot \frac{3}{4} log_2(\frac{3}{4})$

where:

- $p(x_1)$—the probability of an identical bit occurring in a given position between individuals and the leader;
- $p(x_2)$—the probability of a different bit occurring in a given position between individuals and the leader;
- $H(x_1)$—entropy values for the probability $p(x_1)$, at a given position;
- $H(x_2)$—entropy values for the probability $p(x_2)$, at a given position.

Based on the example listed in Table 5, the entropy value of the entire system can be computed as follows:

$$H = -(0 + 0 - 0.93 - 0.5 - 0.93 - 0.5 - 1 - 1 - 0.5 - 0.93) = 6.29. \tag{9}$$

Entropy for the MASA and NGA algorithms was visualized respectively in Figures 10 and 11. The charts show the maximum, minimum and average values. Moreover, it was decided to visualize the average value of entropy for both attacks on one graph, which is presented in Figure 12. The remaining results—for the characteristic Ω_2 are given in Appendix A in Figures A7–A9.

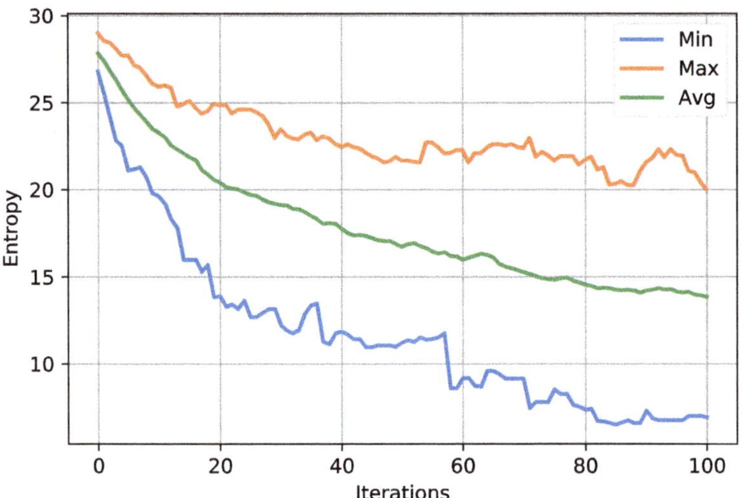

Figure 10. Minimum, maximum and average entropy, during all iterations, for MASA algorithm and Ω_1 characteristic.

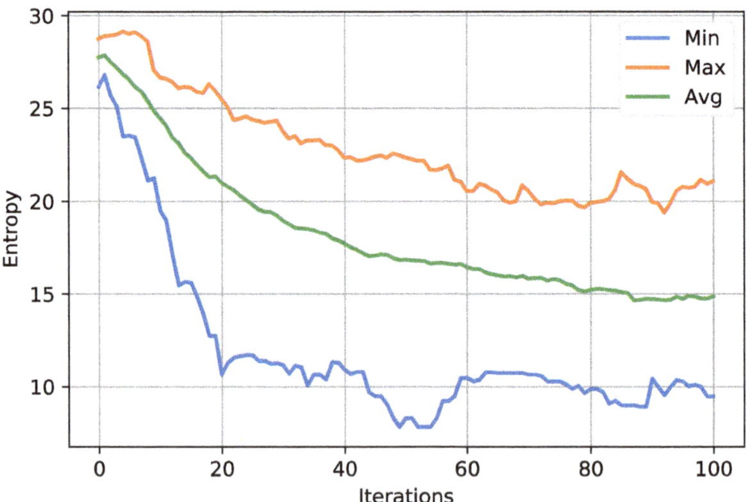

Figure 11. Minimum, maximum and average entropy, during all iterations, for NGA algorithm and Ω_1 characteristic.

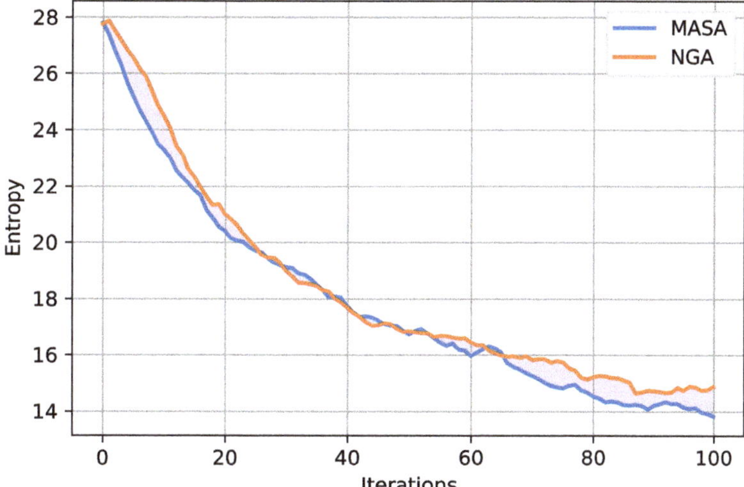

Figure 12. The comparison of the entropy of the MASA and NGA algorithms for the Ω_1 characteristic.

The entropy value was computed during each iteration and 30 launches of MASA and NGA attacks. During all the conducted tests, identical pairs of plain text and the corresponding ciphertexts were used, as well as the same encryption key—owing to which it was possible to make the most reliable comparison.

When analyzing the graphs presented above (Figures 10 and 11), a decrease in the entropy value can be noticed from the very beginning of the running of each of the algorithms. In the last iterations, a gradual stabilization of the system becomes visible, which would most probably be more noticeable after increasing the number of iterations. Comparing the average courses, it can be noticed in Figure 12 that the entropy value for the MASA attack is lower from the very beginning. Only from about the thirtieth iteration, the NGA algorithm obtains a similar value, and sometimes even lower, in relation to the MASA attack. Eventually, the entropy values for the NGA algorithm begin to stabilize at around the sixtieth iteration, while in the case of the MASA attack it continues to decrease. At the end of the algorithms' running, the difference in entropy value between attacks becomes visible.

The experiments carried out and described above clearly confirm the effectiveness of the suggested MASA attack, based on the use of memetic algorithms and simulated annealing. This information may be important during the running of the algorithm, since the probability of leaving the local extremum will be higher, and thus the quality of the final results will be better.

6. Conclusions

The article presents the results for the NGA genetic algorithm enriched with an additional heuristic negation operator and the MASA memetic algorithm that performs the local search process through simulated annealing. Both algorithms undoubtedly improve the process of an attack of differential cryptanalysis against the ciphertexts generated with the DES standard. An important aspect is the attempt to minimize the number of verified subkeys, which is presented in the table below:

The developed algorithms improve the effectiveness and efficiency of the attack, which is extremely important from the point of view of a cryptanalyst. Presented metaheuristics cryptanalysis, based on the differential cryptanalysis approach, can be helpful to raise the security level in already implemented IT systems. It can also be used to improve the complexity of ciphers at the design level. Proposed attacks, verified on the DES cipher, can be tested on more complicated modern encryption algorithms like AES or GOST ciphers.

Based on the tests presented in the previous section and Table 6, it is possible to clearly state the superiority of the MASA attack and the NGA algorithm over the classic differential cryptanalysis attack, due to the frequency of correctly guessed subkey and the number of proven solutions.

Table 6. Comparison of checked subkeys between MASA, NGA and differential cryptanalysis attacks.

Attack	Total Number of Checked Subkeys	Average Number of Checked Subkeys
MASA algorithm		
Ω_1	687,752	22,925.1
Ω_2	687,788	22,926.3
Σ	1,375,540	45,851.3
NGA algorithm		
Ω_1	252,456	8415.2
Ω_2	252,899	8430.0
Σ	505,355	16,845.2
Differential Cryptanalysis		
Ω_1	$30 \cdot (6 \cdot 2^{30} + 1024)$	$6 \cdot 2^{30} + 1024$
Ω_2	$30 \cdot (6 \cdot 2^{30} + 1024)$	$6 \cdot 2^{30} + 1024$
Σ	$30 \cdot (12 \cdot 2^{30} + 1024)$	$12 \cdot 2^{30} + 1024$

There are many parameters that influence the quality of offered solutions. Analyzing the importance of individual parameters, we intend in the future to conduct an analysis based on removing some of them or replacing them with a simplified version, without losing the quality of the offered solutions. Such approach (an ablation study) is very common when estimating costs of deep learning solutions and we hope that it will also be very effective here.

Work is currently underway on modifications of the developed attack, which would enable an even faster exploration of the solution space. In the future, an adaptive version of the memetic algorithm is expected to be developed to automatically adjust the attack parameters. A parallel implementation is also planned, which should be much more effective.

Simplified and the original DES encryption algorithms are commonly used by many cryptanalysts as a starting point to perform research and experimental studies in this discipline of science. It can be found in the literature review, presented in Table 1, in the introduction section. The authors of this article decided to use a reduced DES cipher for the purposes of developing new metaheuristic attacks described in the paper. Starting experiments from modern ciphers could be too complicated and significantly extend the research process. At the current state, we can test the proposed algorithms against more advanced symmetric block ciphers such as Twofish, AES, or GOST, which will definitely be the next step in future works.

Author Contributions: Conceptualization, K.D. and U.B.; formal analysis, K.D and U.B.; investigation, K.D. and U.B.; methodology, K.D and U.B.; project administration, K.D.; resources, K.D.; software, K.D.; supervision, K.D.; validation, K.D. and U.B.; visualization, K.D.; writing, original draft, K.D. and U.B.; writing, review and editing, K.D. and U.B. All authors have read and agreed to the published version of the manuscript.

Funding: This research received no external funding.

Institutional Review Board Statement: Not applicable.

Informed Consent Statement: Not applicable.

Data Availability Statement: Not applicable.

Conflicts of Interest: The authors declare no conflict of interest.

Abbreviations

The following abbreviations are used in this paper:

ACO	Ant Colony Optimization
BCSO	Binary Cat Swarm Optimization
DC	Differential Cryptanalysis
DES	Data Encryption Standard
DSA	Dolphin Swarm Algorithm
EA	Evolutionary Algorithms
FEAL	Fast Data Encipherment Algorithm
GSO	Genetic Swarm Optimization
MA	Memetic Algorithms
MASA	Memetic Algorithm Simmulated Annealing
NGA	Negation Genetic Algorithms
PSO	Particle Swarm Optimization
RC4	Rivest Cipher 4
SDES	Simplified Data Encryption Standard
TEA	Tiny Encryption Algorithm
VMPC	Variably Modified Permutation Composition

Appendix A. The Comparative and Entropy Studies for the Ω_2 Characteristic

Table A1. Fitness function values for MASA and NGA algorithms for characteristic Ω_P^2.

ID	MASA					NGA				
	Min	Med	Avg	Max	Std. Dev.	Min	Med	Avg	Max	Std. Dev.
1	906	984	1002.6	1095	57.1	987	1027	1041.5	1095	38.0
2	967	1059	1043.0	1095	42.6	**1008**	**1034**	**1043.3**	**1074**	**21.3**
3	946	973	997.8	1095	50.7	**993**	**1010**	**1025.2**	**1075**	**29.2**
4	1008	1041	1048.5	1095	29.2	**968**	**1047**	**1046.8**	**1074**	**31.5**
5	1008	1044	1045.6	1095	20.2	922	1021	1024.3	1095	52.7
6	943	989	1018.7	1095	60.7	**964**	**1044**	**1043.5**	**1074**	**35.4**
7	959	1041	1038.8	1095	40.0	888	957	994.0	1095	72.2
8	940	1011	1020.6	1095	44.0	953	1059	1036.3	1095	49.8
9	928	1028	1029.0	1095	56.6	958	1030	1031.6	1095	39.1
10	953	1011	1019.9	1095	47.3	**895**	**1059**	**1017.1**	**1074**	**64.7**
11	941	1041	1022.8	1095	50.8	958	1059	1043.7	1095	44.2
12	993	1033	1045.5	1095	30.0	**1006**	**1021**	**1036.7**	**1075**	**26.7**
13	955	1060	1039.2	1095	50.0	959	1053	1049.5	1095	38.1
14	946	1006	1012.2	1095	49.0	927	1027	1022.9	1095	59.4
15	**949**	**1021**	**1019.4**	**1053**	**27.0**	916	1034	1028.2	1095	56.3
16	**891**	**979**	**992.2**	**1075**	**58.3**	995	1068	1054.6	1095	32.5
17	897	1002	1004.4	1095	55.4	958	1054	1045.2	1095	46.4
18	**902**	**952**	**974.1**	**1075**	**53.2**	**939**	**963**	**989.1**	**1074**	**47.4**
19	969	1025	1033.3	1095	45.5	**884**	**989**	**990.6**	**1074**	**62.6**
20	950	1023	1037.6	1095	52.0	985	1027	1039.0	1095	35.3
21	899	1036	1026.9	1095	57.7	**940**	**977**	**992.1**	**1075**	**44.3**
22	1016	1032	1043.9	1095	27.2	957	1007	1021.2	1095	49.1
23	947	1036	1027.0	1095	56.6	**902**	**1021**	**1007.4**	**1039**	**38.0**
24	977	1068	1053.6	1095	39.7	**913**	**1013**	**1016.1**	**1074**	**48.6**
25	945	1021	1031.1	1095	42.9	**905**	**1028**	**1024.2**	**1075**	**53.2**
26	1011	1041	1043.7	1095	25.1	952	1031	1036.4	1095	50.1
27	937	1013	1010.0	1095	45.0	961	996	1013.6	1095	49.9
28	971	1002	1027.2	1095	52.1	**936**	**1017**	**1023.8**	**1074**	**46.6**
29	907	1038	1027.5	1095	64.6	**949**	**1018**	**1018.6**	**1075**	**32.0**
30	950	1018	1016.0	1095	50.4	949	1065	1054.8	1095	41.8

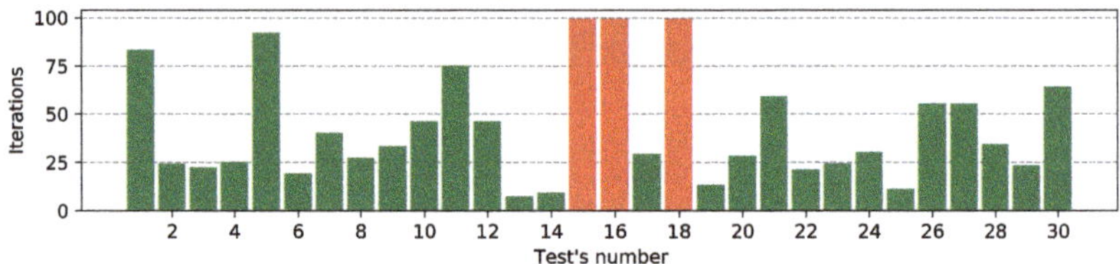

Figure A1. List of correctly guessed bits of MASA attack for the Ω_2.

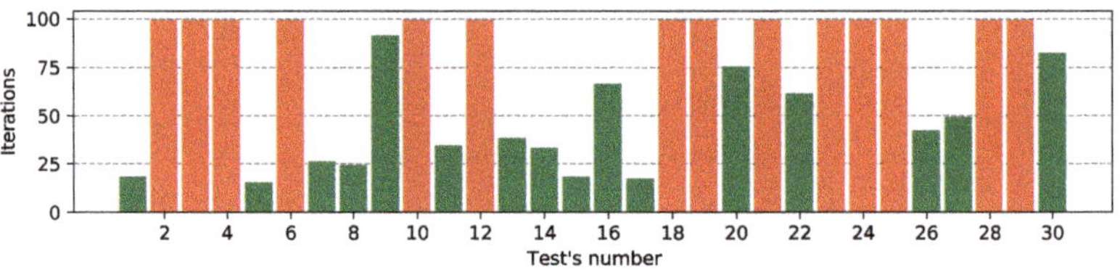

Figure A2. List of correctly guessed bits of NGA attack for the Ω_2.

Where the red color indicates experiments when the algorithm wasn't able to find the correct subkey and the green bars indicate are tests when the subkey was successfully guessed.

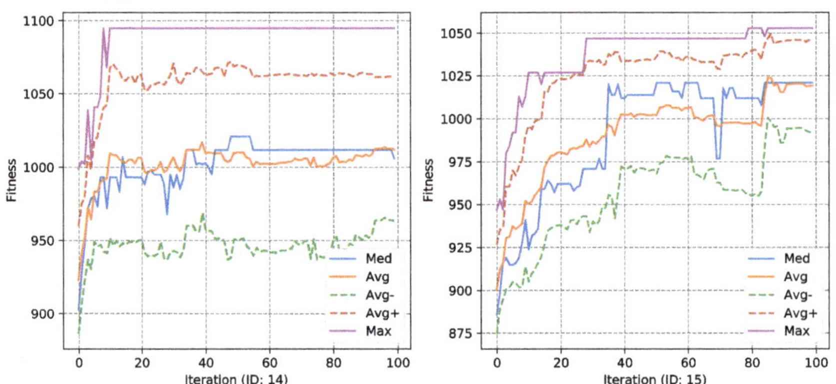

Figure A3. The MASA fitness function F_f convergence diagrams for Ω_2 (tests #14 and #15).

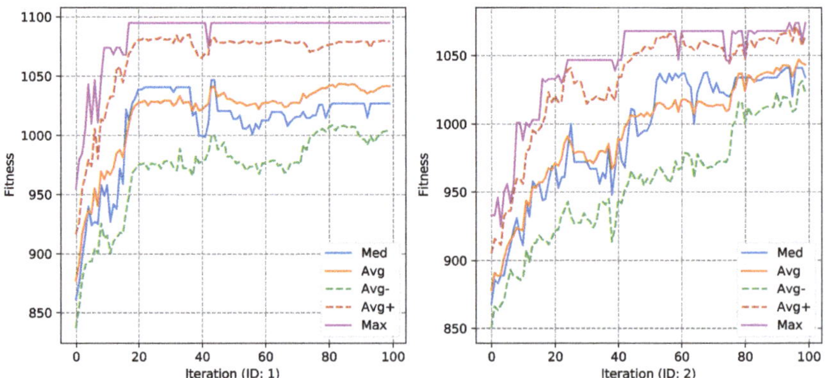

Figure A4. The NGA fitness function F_f convergence diagrams for Ω_2 (tests #1 and #2).

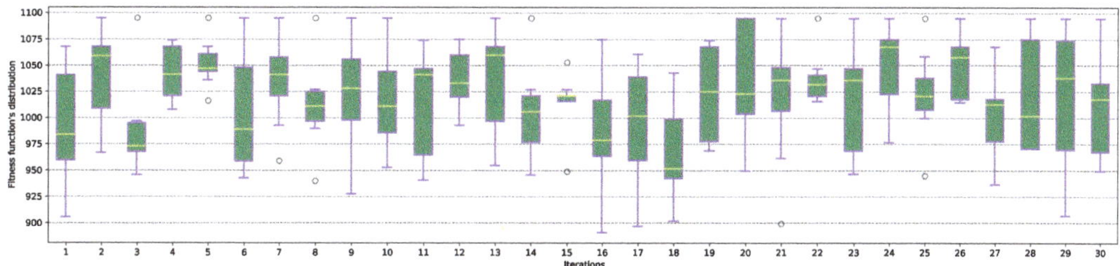

Figure A5. The distribution of the fitness function F_f values in the last iteration for the MASA algorithm and Ω_2 characteristic.

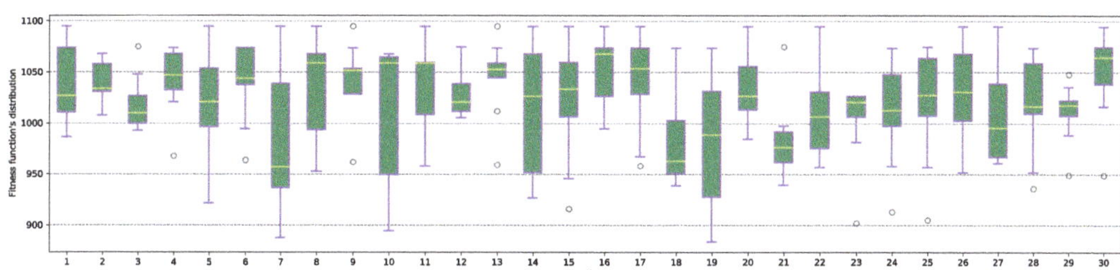

Figure A6. The distribution of the fitness function F_f values in the last iteration for the NGA algorithm and Ω_2 characteristic.

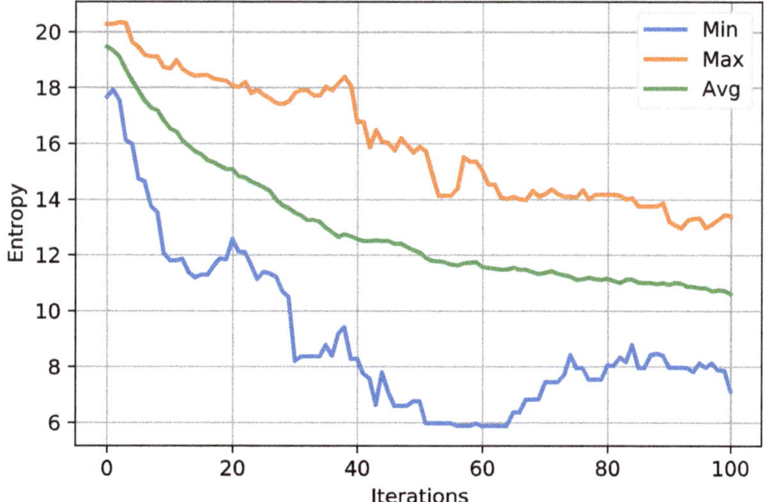

Figure A7. Minimum, maximum and average entropy, during all iterations, for MASA algorithm and Ω_2 characteristic.

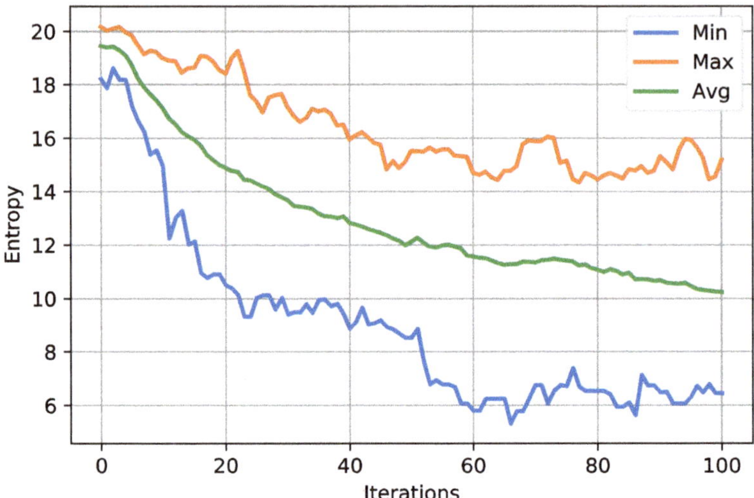

Figure A8. Minimum, maximum and average entropy, during all iterations, for NGA algorithm and Ω_2 characteristic.

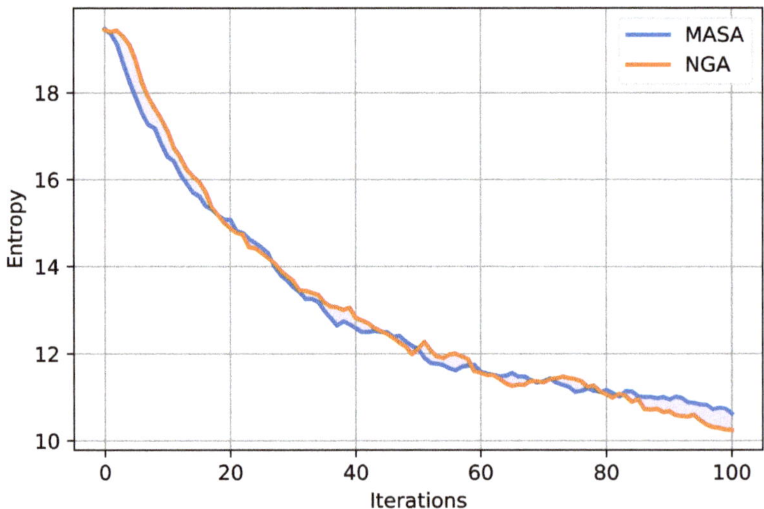

Figure A9. The comparsion of the entropy of the MASA and NGA algorithms for the Ω_2 characteristic.

References

1. Schneier, B. *Applied Cryptography: Protocols, Algorithms, and Source Code in C*; Wiley: Hoboken, NJ, USA, 1996.
2. Menezes, A.J.; Oorschot, P.C.; Vanstone, S.A. *Handbook of Applied Cryptography*; CRC Press: Boca Raton, FL, USA, 1997.
3. Biham, E.; Shamir, A. Differential cryptanalysis of DES-like cryptosystems. *J. Cryptol.* **1991**, *4*, 3–72. [CrossRef]
4. Song, J.; Zhang, H.; Meng, Q.; Zhangyi, W. Cryptanalysis of Four-Round DES Based on Genetic Algorithm. *Wirel. Commun. Netw. Mob. Comput. IEEE* **2007**, *10*, 2326–2329.
5. Tadros, T.; Hegazy, A.; Badr, A. Genetic Algorithm for DES Cryptanalysis. *Int. J. Comput. Sci. Netw. Secur.* **2007**, *10*, 5–11.
6. Garg, P. A Comparison between Memetic algorithm and Genetic algorithm for the cryptanalysis of Simplified Data Encryption Standard algorithm. *Int. J. Netw. Secur. Its Appl. (IJNSA)* **2009**, *1*, 34–42.
7. Hu, W. Cryptanalysis of TEA using quantum-inspired genetic algorithms. *J. Softw. Eng. Appl.* **2010**, *3*, 50–57. [CrossRef]

8. Abd-Elmonim, W.G.; Ghali, N.I.; Hassanien, A.E.; Abraham, A. Known-Plaintext Attack of DES16 Using Particle Swarm Optimization. In Proceedings of the Third IEEE World Congress on Nature and Biologically Inspired Computing, Salamanca, Spain, 19–21 October 2011; pp. 12–16.
9. Vimalathithan, R.; Valarmathi, M.L. Cryptanalysis of simplified-DES using computational intelligence. *WSEAS Trans. Comput.* **2011**, *10*, 210–219.
10. Jadon, S.S.; Sharma, H.; Kumar, E.; Bansal, J.C. Application of binary particle swarm optimization in cryptanalysis of DES. In *Proceedings of the International Conference on Soft Computing for Problem Solving*; Deep, K., Nagar, A., Pant, M., Bansal, J., Eds.; Advances in Intelligent and Soft Computing; Springer: New Delhi, India, 2012; Volume 130, pp. 1061–1071.
11. Pandey, S.; Mishra, M. Particle swarm optimization in cryptanalysis of DES. *Int. J. Adv. Res. Comput. Eng. Technol.* **2012**, *4*, 379–381.
12. Ali, I.K. Cryptanalysis of simple substitution ciphers using bees algorithm. *J. Baghdad Coll. Econ. Sci. Univ.* **2013**, *36*, 373–382.
13. Boryczka, U.; Dworak, K. Cryptanalysis of Transposition Cipher Using Evolutionary Algorithms. In *Computational Collective Intelligence. Technologies and Applications*; Hwang, D., Jung, J.J., Nguyen, N.T., Eds.; Lecture Notes in Computer Science; Springer: Cham, Switzerland, 2014; Volume 8733, pp. 623–632.
14. Mekhaznia, T.; Menai, M.E.B. Cryptanalysis of classical ciphers with ant algorithms. *Int. J. Metaheuristics* **2014**, *3*, 175–198. [CrossRef]
15. Bhateja, A.K.; Bhateja, A.; Chaudhury, S.; Saxena, P.K. Cryptanalysis of vigenere cipher using cuckoo search. *Appl. Soft Comput.* **2015**, *26*, 315–324. [CrossRef]
16. Jain, A.; Chaudhari, N.S. A New Heuristic Based on the Cuckoo Search for Cryptanalysis of Substitution Ciphers. In Proceedings of the International Conference on Neural Information Processing, Istanbul, Turkey, 9–12 November 2015; Sabri, A., Tingwen, H., Weng, K.L., Qingshan, L., Eds.; Volume 9490, pp. 206–215.
17. Amic, S.; Soyjaudah, K.S.; Mohabeer, H.; Ramsawock, G. Cryptanalysis of DES16 using binary firefly algorithm. In Proceedings of the 2016 IEEE International Conference on Emerging Technologies and Innovative Business Practices for the Transformation of Societies, Balaclava, Mauritius, 3–6 August 2016; IEEE: Balaclava, Mauritius, 2016; pp. 94–99.
18. Dworak, K.; Nalepa, J.; Boryczka, U.; Kawulok, M. Cryptanalysis of SDES using genetic and memetic algorithms. In *Recent Developments in Intelligent Information and Database Systems*; Król, D., Madcyski, L., Nguyen, N.T., Eds.; Springer International Publishing: Da Nang, Vietnam, 2016; pp. 3–14.
19. Dworak, K.; Boryczka, U. Differential Cryptanalysis of FEAL4 using Evolutionary Algorithm. In *Computational Collective Intelligence*; Nguyen, N.T., Iliadis, L., Manolopoulos, Y., Trawiński, B., Eds.; Springer International Publishing: Halkidiki, Greece, 2016; Volume 9876, pp. 102–112.
20. Amic, S.; Soyjaudah, K.S.; Ramsawock, G. Dolphin swarm algorithm for cryptanalysis. In *Information Systems Design and Intelligent Applications*; Satapathy, S., Bhateja, V., Somanah, R., Yang, X.S., Senkerik, R., Eds.; Advances in Intelligent Systems and Computing; Springer: Singapore, 2019; Volume 863, pp. 149–163.
21. Jain, A.; Chaudhari, N.S. A novel cuckoo search strategy for automated cryptanalysis: A case study on the reduced complex knapsack cryptosystem. *Int. J. Syst. Assur. Eng. Manag.* **2017**, *9*, 942–961. [CrossRef]
22. Dworak, K.; Boryczka, U. Genetic Algorithm as Optimization Tool for Differential Cryptanalysis of DES6. In *Computational Collective Intelligence*; Nguyen, N.T., Papadopoulos, G.A., Jędrzejowicz, P., Trawiński, B., Vossen, G., Eds.; Springer International Publishing: Nicosia, Cyprus, 2017; Volume 10449, pp. 107–116.
23. Polak, I.; Boryczka, M. Tabu search against permutation based stream ciphers. *Int. J. Electron. Telecommun.* **2018**, *64*, 137–145.
24. Kamal, R.; Bag, M.; Kule, M. On the cryptanalysis of SDES using binary cuckoo search algorithm. In *Computational Intelligence in Pattern Recognition*; Das, A., Nayak, J., Naik, B., Pati, S., Pelusi, D., Eds.; Advances in Intelligent Systems and Computing; Springer: Singapore, 2019; Volume 999, pp. 23–32.
25. Polak, I.; Boryczka, M. Tabu Search in revealing the internal state of RC4+ cipher. *Appl. Soft Comput.* **2019**, *77*, 509–519. [CrossRef]
26. Sabonchi, A.K.S.; Akay, B. Cryptanalysis of Polyalphabetic Cipher Using Differential Evolution Algorithm. *Tehnički Vjesnik* **2020**, *27*, 1101–1107.
27. Grari, H.; Lamzabi, S.; Azouaoui, A.; Zine-Dine, K. Cryptanalysis of Merkle-Hellman cipher using ant colony optimization. *IAES Int. J. Artif. Intell.* **2021**, *10*, 490–500. [CrossRef]
28. Amic, S.; Soyjaudah, K.S.; Ramsawock, G. Binary cat swarm optimization for cryptanalysis. In Proceedings of the 2017 IEEE International Conference on Advanced Networks and Telecommunications Systems (ANTS), Bhubaneswar, India, 17–20 December 2017; IEEE: Bhubaneswar, India, 2017; pp. 1–6.
29. Pieprzyk, J.; Hardjono, T.; Seberry, J. *Fundamentals of Computer Security*; CRC Press: Boca Raton, FL, USA, 2003.
30. Stallings, W. *Cryptography and Network Security: Principles and Practice*; Pearson: London, UK, 2011.
31. Stinson, D.R. *Cryptography: Theory and Practice*; CRC Press: Boca Raton, FL, USA, 1995.
32. Stamp, M.; Low, R.M. *Applied Cryptanalysis. Breaking Ciphers in the Real World*; Wiley-Interscience: Hoboken, NJ, USA, 2007.

MDPI

Article

Immunity in the ABM-DSGE Framework for Preventing and Controlling Epidemics—Validation of Results

Jagoda Kaszowska-Mojsa [1,2,3,*,†], Przemysław Włodarczyk [4,†] and Agata Szymańska [4,†]

[1] Institute for New Economic Thinking at the Oxford Martin School, University of Oxford, Manor Road, Oxford OX1 3UQ, UK

[2] Institute of Economics, Polish Academy of Sciences, Nowy Świat St. 72, 00-330 Warsaw, Poland

[3] Department of Macroeconomics, Institute of Economics, Cracow University of Economics, Rakowicka St. 27, 31-510 Cracow, Poland

[4] Department of Macroeconomics, Faculty of Economics and Sociology, University of Łódź, 90-136 Łódź, Poland; przemyslaw.wlodarczyk@uni.lodz.pl (P.W.); agata.szymanska@uni.lodz.pl (A.S.)

* Correspondence: jagoda.kaszowska-mojsa@maths.ox.ac.uk or jagoda.kaszowska@inepan.waw.pl

† These authors contributed equally to this work.

Abstract: The COVID-19 pandemic has raised many questions on how to manage an epidemiological and economic crisis around the world. Since the beginning of the COVID-19 pandemic, scientists and policy makers have been asking how effective lockdowns are in preventing and controlling the spread of the virus. In the absence of vaccines, the regulators lacked any plausible alternatives. Nevertheless, after the introduction of vaccinations, to what extent the conclusions of these analyses are still valid should be considered. In this paper, we present a study on the effect of vaccinations within the dynamic stochastic general equilibrium model with an agent-based epidemic component. Thus, we validated the results regarding the need to use lockdowns as an efficient tool for preventing and controlling epidemics that were obtained in November 2020.

Keywords: COVID-19; vaccination; agent-based modelling; dynamic stochastic general equilibrium models; scenario analyses; validation of results

Citation: Kaszowska-Mojsa, J.; Włodarczyk, P.; Szymańska, A. Immunity in the ABM-DSGE Framework for Preventing and Controlling Epidemics— Validation of Results. *Entropy* **2022**, *24*, 126. https://doi.org/10.3390/e24010126

Academic Editors: Jan Kozak and Przemysław Juszczuk

Received: 30 November 2021
Accepted: 10 January 2022
Published: 14 January 2022

Publisher's Note: MDPI stays neutral with regard to jurisdictional claims in published maps and institutional affiliations.

Copyright: © 2022 by the authors. Licensee MDPI, Basel, Switzerland. This article is an open access article distributed under the terms and conditions of the Creative Commons Attribution (CC BY) license (https://creativecommons.org/licenses/by/4.0/).

1. Introduction

Last year was dominated by discussions on how to contain the spread of the Sars-CoV-2 virus and the economic impact of the prevention and control measures. The need to effectively introduce lockdowns has been discussed in the literature, in particular, when there had not yet been widespread vaccinations and there was no consensus on how to treat patients who had contracted COVID-19. It was also unclear how contagious the virus was and whether it was possible to get re-infected or whether the body developed an immunity against the virus. While many open questions have already been answered since the outbreak of the pandemic, some crucial questions are still unanswered. It is especially worth considering how justified and effective lockdowns are in the era of widespread vaccinations against COVID-19.

In our article [1], which was published in November 2020, we studied the shape and range of state interventions whose goal was to limit the negative effects of the pandemic, in particular, by reducing the number of infections and deaths that were caused by the pandemic. As it was emphasised, lockdowns have been introduced in many countries around the world in order to limit the spread of the virus and to prevent the collapse of the health systems. In some countries, a deep lockdown strategy was abruptly adopted, while in others, the focus was put on gradually closing certain sectors of the economy. The effectiveness of a lockdown in a given country is influenced by many factors, the most important of which are legal, behavioural, cultural and social factors. Hence, the impact of a lockdown on the course of the epidemic and its impact on the given economy could be different than in other countries. In our previous paper, we advised on how the lockdown policy should be implemented. In this article, we would like to validate the results taking into account that the COVID-19 vaccinations started in most developed countries in 2021. Therefore, we address the same two major questions:

- Should we freeze an economy in order to decrease the pace of SARS-CoV-2 transmission?
- What should the scale and composition of an efficient lockdown policy look like?

However, this time, we did take immunity into account in our analysis.

2. Literature Review

The efficiency of the COVID-19 vaccination process in the context of the potential to achieve herd immunity by a society is one of the most frequently discussed topics in the literature today. In simple terms, "herd immunity works through achieving a threshold immunity at the population level that is able to theoretically cut the transmission chain of a given infectious disease, be it obtained through natural infection or vaccination" [2]; for general studies on the topic, see also [3–6]. In the most general terms, two main sources of achieving herd immunity were discussed—through a natural infection and recovery or by a vaccination [3].

In the literature, the results of many studies whose goal was to compute the thresholds for infectious diseases have been presented, including recent works that have focused on the threshold for the COVID-19 disease. Fontanet and Cauchemez [6] suggested that under the assumption about the absence of control measures $(p_c = 0)$, i.e., without pharmacological interventions, among others, "the condition for herd immunity ($R < 1$, where $R = (1 - p_i)R_0$) is attained when the proportion of immune individuals reaches $p_i = 1 - \frac{1}{R_0}$, where R_0 denoted the reproduction number in the absence of control measures in a fully susceptible population" and that it can vary across populations and over time. In the formula above, which was presented by [6], R denotes the effective reproduction number that is explained as "the average number of persons infected by a case". As is clarified, in the absence of interventions, the number R is lower than 1 and this case denotes the possibility of the occurrence of herd immunity, i.e., a situation in which one infected person is responsible for inducing "less than one secondary case on average", see [6]. The rest of the abbreviations that are used by the authors in the formula explaining herd immunity are as follows: p_i is the proportion of the society that is immune and p_c denotes the relative reduction in the transmission rates that is achieved by using non-pharmaceutical interventions.

The estimates for the COVID-19 pandemic differ. As was mentioned in [6], R_0 varies across populations and over time. The literature review emphasises the differences across countries and regions. For example, Kwok et al. [7], in a sample of 32 countries, estimated that the effective reproduction number ranged from 1.06 in Kuwait to 6.64 in Bahrain as of 13 March 2020. As a result, the minimum proportion of the total population, in percentage terms, that would be required to recover from COVID-19 was between 5.66% in Kuwait to 85% in Bahrain, see [8]. As was emphasised by Fontanet and Cauchemez [6], in the case of SARS-CoV-2, most estimates of R_0 are in the range of 2.5 and 4. Moreover, Kwok et al. [7] obtained the R_0 between 2 and 4 for the largest group of countries. For $R_0 = 2$, which they estimated for Iran, the herd immunity threshold for SARS-CoV-2 was expected to require 50% of the population to have immunity, for $R_0 = 2.09$, while the threshold increased to 65.5% for the UK. The estimates for Israel (a country with one of the widest distributions of vaccines against COVID-19 among its population, [9]), the R_0 was 3.02 and the estimated threshold was 66.9%. As has been argued, the low figures for Kuwait reflected the fact that the country had strong lockdowns and put many measures in place to control the SARS-CoV-2 virus and an escalation of the COVID-19 pandemic [10]. Aschwanden in her article [10] emphasised that the threshold for herd immunity, taking into account the literature overview, ranged from 10% to more than 70%. However, in the most compelling research, it has ranged between 60 and 70% [10]. Gomes et al. [11] indicated that the initial and simple estimates of a COVID-19 threshold that were based on relying on homogeneity assumptions ranged between 60 and 80% of a population to become immune. In their study, which was based on the earlier models that had been explored in the literature, but with the individual variations in susceptibility or in exposure to infection incorporated into the data, they obtained a lower herd immunity threshold, for example, for SARS-CoV-2, they calculated the threshold that was associated with a natural infection to be in the range of 10–30%. As has been presented, in order to obtain the required threshold for herd immunity, the specified size of the proportion of the population to have immunity is required.

The thresholds of herd immunity that have been presented in the literature are influenced by the vaccination process and vaccine coverage. Using the SMEIHRDV model, Dashtbali and Mirzaie [12] predicted that the number of infected cases at the height of the COVID-19 pandemic was significantly reduced by the increasing vaccine coverage of between 0.2 and 0.6. Their results were predicted for Egypt and Germany. In the case of Egypt, the number of infected cases at the height of the epidemic was estimated to decrease from around 540 cases with a vaccine coverage of 0.2 to around 200 cases when the vaccine coverage increased to 0.6. In the case of Germany, the obtained predictions suggested that increasing the vaccine coverage from 0.2 to 0.6 affected the pandemic peak, in which the cases decreased from approximately 320 to 120. Moreover, the German case enabled it to be predicted that, both investing in a strategy of social distancing and increasing the vaccine coverage, the length of the epidemic peak of infected cases might shorten from approximately 200 to 55 by increasing vaccine coverage from 0.2 to 0.6.

The results obtained by Makhoul et al. [13] indicated that even a partially efficient vaccine was able to affect the spread of SARS-CoV-2 virus and the COVID-19 pandemic. The authors calibrated and estimated their model for the Chinese case and assumed a long duration of vaccine protection that lasted ten years. As was argued, the simulated scenarios emphasised that the three vaccines do not need to provide complete immunity to be able to completely control the infection. As it was also emphasised, a vaccine with $VE_s \geq 70\%$ (i.e., a vaccine efficacy in reducing susceptibility of greater than 70%) would have enabled us to control the pandemic at $\geq 80\%$ coverage before its onset. However, as was estimated, when the reproduction number R_0 is assumed to be three, then the minimum VE_s that is required to eliminate infections is about 90%.

The simulations of Charumilind et al. [14] for the US and the UK showed that the use of a vaccine that is 95% effective at preventing transmission and with a natural immunity between 5–20%, the required vaccine coverage had to increase when more than a 40–80% transmissible COVID-19 strain was predominant, and more stringent non-pharmaceutical interventions should be used to manage the pandemic. For example, under the assumption that a COVID-19 strain is 40% more transmissible, then the required coverage for the two countries would be 65–72% (or 78–86% if limited only to those more than 12 years old). However, when a new COVID-19 strain was assumed to be 80% more transmissible, then the required coverage should be increased and it would be 75–80% (or 89–95% if limited to only those more than 12 years old). As was investigated, if the transmissibility of a new variant increased by 40% or 80%, then COVID-19 herd immunity can only be achieved once the total immune population reaches 70% (under assumption $R_0 = 3.4$) or 77% (with $R_0 = 4.3$).

Using a deterministic model, Han et al. [15] analysed "the connection between the daily vaccination capacity (rollout speed) and transmissibility in determining the optimal (vaccine prioritisation) strategies" based on the case of China. They argued that introducing a high vaccination capacity in the early phase of a vaccination campaign is crucial for achieving large increases in strategic prioritisations. The simulations were based on a time-varying optimisation of the COVID-19 vaccine prioritisation. The obtained findings enabled them to conclude that increasing the vaccination capacity to 2.5 million first doses per day (0.17% roll-out speed) or higher could considerably reduce the COVID-19 burden, when the assumed reproduction number was equal to 1.5 ($R_0 = 1.5$).

The Chen et al. [16] study was based on eight selected countries: Chile, Hungary, Israel, Serbia, Qatar, the UAE, the UK and the USA. These countries were selected due to their high degree of effectiveness in mitigating the COVID-19 pandemic in a situation in which the rate of vaccination achieved the level of criticality, even if it was lower than the herd immunity threshold. The results of the research suggested that a value of the vaccination rate of 50.91 doses per 100 people could be perceived as the minimum condition for avoiding an exacerbation of the pandemic in a society.

Coccia's [8] study, which was based on 192 countries with data from March to May 2021, indicated that the optimal levels of vaccination in the global context for decreasing the number of infected cases and deaths would require about "80 doses of vaccines per 100 inhabitants in order to sustain a decrease in confirmed cases and the number of deaths". As was shown, approximately 47 doses of vaccines have to be administered in order to reduce infected cases when an intensive vaccination campaign was introduced at the beginning of the pandemic wave. However, the findings also indicate a need to increase the number of doses when the pandemic grew—data retrieved from May 2021 enabled them to conclude that an increase of COVID-19 wave required a higher optimal level of vaccines to be administered—it was estimated to be about 90 doses.

Finally, in the literature, there are also many interesting studies related to analyses of the impact of vaccinations and their combination with other non-pharmaceutical measures were presented. Among others, Viana et al. [17] studied the case of Portugal, while Maghadas et al. [18] and Coccia [19] provided evidence for the US.

The short literature review, presented above, strongly emphasises the fact that the assumptions concerning the size of the proportion of the immune population, the coverage of vaccinations or the reproduction number all affect herd immunity.

In our new study, we would like to present the results of virus spread simulations in three scenarios with immunity. We will also refer to the scenarios described in [1], which did not include immunity. The new scenarios took into account the process of vaccinating the population over time and its impact on the course of the pandemic. In the first scenario, approximately 50% of the population was vaccinated. In the second one, approximately 80% of the population was immune. In the last scenario, we present the conditions under which herd immunity can be expected in a relatively small economy, and therefore, we offer evidence about why lockdowns are still an important tool in the fight against a pandemic.

3. Updated ABM Component for Studying the Dynamics of the COVID-19 Pandemic

We updated the agent-based model that was presented in [1] in order to introduce immunity into the system (we present the details of the ABM model in Section 3). Using this ABM component, we simulated the spread of the COVID-19 virus and analysed the impact of the COVID-19 pandemic on society's overall labour productivity within three scenarios that took into account both vaccinations and natural immunity. Those scenarios will be described in Section 4. In this section, the results that were obtained in 2020 in connection with the introduction of vaccination will be validated. Then, we will estimate the economic impact of the COVID-19 pandemic using the dynamic stochastic general equilibrium (DSGE) model (see Section 5).

We describe the functioning of the model in the following six modules. The way in which the program works is analogous to the one described in [1], but this time, the program and the analyses included the vaccinations and immunity of individuals who had previously contracted COVID-19. We listed the most important elements of the program and we emphasised the changes that had been made recently.

In the first module, the initial conditions were defined. The variables and parameters that had to be specified in order to run the simulations are presented in Tables 1–3.

We estimated the values of the parameters and the transition probabilities that would be assumed in a specific scenario using the empirical data. We present the calibration for a given scenario in Tables 4 and 5.

The health status of an agent, their age and location were randomly assigned at the beginning of the simulation (the number of infected agents were set in the initial conditions). In the second module, the characteristics of the agents were recorded in the matrices after each time step. The simulation was conducted for the values of the parameters that had previously been defined. Among the most important characteristics that were recorded after each time step (weekly) were: the health status of each individual in a society (an $M \times T$ matrix H^*), the productivity of each individual in a society (an $M \times T$ matrix W^*), the age of each individual in a society (an $M \times T$ matrix A^*) and the location of each individual on the map after each iteration (x- and y-coordinates) (an $M \times 2T$ matrix X^*). The full dataset was also recorded in the matrix (an $M \times 4$ matrix F^*).

Table 1. The list of initial conditions to be set.

Initial Conditions	Explanation	Restrictions
T	Number of time steps (weeks)	≥ 0
s_t^{Ind}	Health status of the individual at time $t = 0$ (1—healthy, 2—infected, 3—treated, 4—healthy individual in preventive quarantine, 5—deceased; 6—recovered, 7—vaccinated)	Int $\in \{1, 2, 3, 4, 5, 6, 7\}$
$(Age)_t^{Ind}$	Age of an individual at time $t = 0$	
N^{Ind}	Number of individuals at time $t = 0$	Int ≥ 0
K^{Ind}	Number of infected individuals at time $t = 0$ (including asymptomatically infected)	Int ≥ 0
λ^{max}	The parameter corresponding to the maximum number of vaccinated persons in the iteration (week)	Int ≥ 0
$S_t \times S_t$	Dimensions of the grid at time t *	Int ≥ 0
$(Ag)_t^1$	Share of citizens of pre-working age at time t	$\in \langle 0,1 \rangle$
$(Ag)_t^2$	Share of citizens of working age at time t	$\in \langle 0,1 \rangle$
$(Ag)_t^3$	Share of retired individuals at time t	$\in \langle 0,1 \rangle$
$(Wp)_t^{av\_h}$	The productivity of an individual when healthy at time t (it was assumed to be equal to one)	$\in \langle 0,1 \rangle$
$(Wp)_t^{av\_inf}$	The productivity of an individual when infected at time t (the decline in productivity was estimated based on empirical data)	$\in \langle 0,1 \rangle$
$(Wp)_t^{av\_r}$	The productivity of an individual after recovery at time t (the decline in productivity was estimated based on empirical data)	$\in \langle 0,1 \rangle$
$(Wp)_t^{av\_t}$	The productivity of an individual when treated or who is infected and in quarantine at time t (the decline in productivity was estimated based on empirical data)	$\in \langle 0,1 \rangle$
$(Wp)_t^{av\_q}$	The productivity of an individual who is healthy and in quarantine at time t (the decline in productivity was estimated based on empirical data)	$\in \langle 0,1 \rangle$
$(Wp)_t^{av\_v}$	The productivity of an individual who has been vaccinated at time t (it was assumed to be equal to one)	$\in \langle 0,1 \rangle$

* The dimensions do not have to be constant in all scenarios for all t. We assumed that in baseline scenario and in scenarios with immunity $S_t = S$.

Note that this approach is analogous to the one that we adopted in our study in November 2020. However, because we added two variables (recovered and vaccinated agents), this also affected the way the transition probabilities had to be defined. Therefore, the matrices H^*, W^*, A^* and X^* are different from the H, W, A, X that were presented in November 2020. Moreover, because the simulations were stochastic, the results that were recorded for each simulation also differed. In the article, we present the results that were averaged for 100 simulations.

The movements of the agents were described in the third module. The grid represented a closed economy. Although this simplification is easily modifiable and there is the possibility to introduce new infections from outside the economy, the aim of this study was to show the validity of lockdowns in the simplest way. The research results would be similar, even if this assumption was lifted. The logic that is known from the cellular automata models was adopted in the study, see [20,21]. We tested several neighbourhoods of a cell in which a healthy agent could move. As the adoption of a specific neighbourhood did not significantly affect the results, we present the conclusions for a simulation in which the agents can move around in the Moore neighbourhood of a cell, which was defined as a two-dimensional square lattice and was composed of a central cell and the eight cells that surround it. An infected agent (symptomatically or asymptomatically), while moving on the grid, encounters other agents and thus spreads the virus. Agents that are receiving treatment, in quarantine or are deceased stop moving on the grid. The size of the grid was carefully selected in order to represent the actual scaled empirical population density of the selected country.

Table 2. Probabilities that are set as parameters *.

Parameter	Explanation	Restrictions
$(Pr)_t^{12}$	The probability that a healthy agent (1) will become infected (2) at time t	$\in (0, 1)$
$(Pr)_t^{14}$	The probability that a healthy agent (1) will be in quarantine (although she is healthy) (4) at time t	$\in (0, 1)$
$(Pr)_t^{15}$	The probability that a healthy agent (1) will become infected and will die almost instantly (within week) (5)	$\in (0, 1)$
$(Pr)_t^{17}$	The probability that the healthy agent (1) will be vaccinated (7)	$\in (0, 1)$
$(Pr)_t^{26}$	The probability that an infected agent (2) will become healthy (will recover) (6)	$\in (0, 1)$
$(Pr)_t^{23}$	The probability that an infected agent (2) will be treated in a hospital or will stay in quarantine (3)	$\in (0, 1)$
$(Pr)_t^{25}$	The probability that an infected agent (2) dies (5)	$\in (0, 1)$
$(Pr)_t^{35}$	The probability that an infected agent in a hospital or quarantine (3) dies (5)	$\in (0, 1)$
$(Pr)_t^{36}$	The probability that an infected agent in a hospital or quarantine (3) gets better (6) (recovers)	$\in (0, 1)$
$(Pr)_t^{41}$	The probability that a healthy agent in quarantine (4) will end the quarantine, that is, is healthy (1)	$\in (0, 1)$
$(Pr)_t^{43}$	The probability that a healthy agent in quarantine (4) will become infected during the quarantine and she is still in quarantine (but now is already infected) (3) at time t	$\in (0, 1)$
$(Pr)_t^{45}$	The probability that a healthy agent in quarantine (4) dies (5)	$\in (0, 1)$
$(Pr)_t^{46}$	The probability that a healthy agent in quarantine (4) was not infected and returned to the state "recovered" (6)	$\in (0, 1)$
$(Pr)_t^{47}$	The probability that a healthy agent in quarantine (4) was not infected and returned to the state "vaccinated" (7)	$\in (0, 1)$
$(Pr)_t^{61}$	The probability that the recovered agent (6) will get infected (1)	$\in (0, 1)$
$(Pr)_t^{64}$	The probability that the recovered agent (6) will go to the quarantine (4)	$\in (0, 1)$
$(Pr)_t^{65}$	The probability that the recovered agent (6) will die (5)	$\in (0, 1)$
$(Pr)_t^{67}$	The probability that the recovered agent (6) will get vaccinated (7)	$\in (0, 1)$
$(Pr)_t^{72}$	The probability that the vaccinated agent (7) will get infected (2)	$\in (0, 1)$
$(Pr)_t^{74}$	The probability that the vaccinated agent (7) will go to the quarantine (4)	$\in (0, 1)$
$(Pr)_t^{75}$	The probability that the vaccinated agent (7) will die (5)	$\in (0, 1)$

* Estimated on empirical data.

In the fourth module, we defined how the virus can be spread in the society. The program analyses the neighbourhood of each individual and determines whether there is someone who might infect other agents.

Cases for Healthy Individuals

The program determines whether there were any infected ($s_t^{Ind} = 2$) or treated individuals ($s_t^{Ind} = 3$) in the neighbourhood of a healthy agent ($s_t^{Ind} = 1$). If there were, they could have been infected ($s_t^{Ind} = 2$) with a certain probability. With a given probability, they could also have been treated in hospital (or put in isolation) ($s_t^{Ind} = 3$). Infection was not equivalent to a diagnosis of sickness. This part of the program is based on two probabilistic tests. The first probabilistic test determined whether an individual had been infected. However, only the second one determined whether the individual had been diagnosed and had been receiving treatment. If an individual was not infected, they could still be directed into preventive quarantine ($s_t^{Ind} = 4$) with a certain probability. There are also non-negative chances that a healthy agent might die within one week ($s_t^{Ind} = 5$). If the system determined that no prior changes in status could be applied, then with a certain probability, a person could be vaccinated and remain healthy ($s_t^{Ind} = 7$). The state transition probabilities in the agent-based epidemic component that included immunity are presented in Figure 1.

Table 3. Variables and parameters that were computed by the program after each iteration.

Variable	Explanation	Restr.
$(Pr)_t^{13}$	The probability that a healthy agent (1) will become treated in the hospital (or isolation) after becoming infected (3) at time t	$\in (0, 1)$
$(Pr)_t^{42}$	The probability that a healthy agent in quarantine (4) will become infected at the end of her quarantine at time t	$\in (0, 1)$
$(Pr)_t^{63}$	The probability that a recovered agent (6) will be hospitalised (3) at time t	$\in (0, 1)$
$(Pr)_t^{73}$	The probability that a vaccinated agent (7) will be hospitalised (3) at time t	$\in (0, 1)$
p	Temporal variable that defines a threshold probability 1	$\in (0, 1)$
q	Temporal variable that defines a threshold probability 2	$\in (0, 1)$
r	Temporal variable that defines a threshold probability 3	$\in (0, 1)$
z	New temporal variable that defines a threshold probability 4	$\in (0, 1)$
s_t^{Ind}	Health status of the agent at time $t > 0$ (1—healthy, 2—infected, 3—treated, 4—healthy individual in preventive quarantine, 5—deceased, 6—recovered, 7—vaccinated)	Int $\in \{1, 2, 3, 4, 5, 6, 7\}$
$(Age)_t^{Ind}$	Age of an agent at time $t > 0$	≥ 0
$(Wp)_t^{Ind}$	Productivity of an agent at time $t > 0$	$\in \langle 0, 1 \rangle$

Cases for Infected Individuals

For all of the agents that were already infected ($s_t^{Ind} = 2$), the program performed probabilistic tests that determined whether an agent should be referred for treatment ($s_t^{Ind} = 3$) or whether they had managed to overcome the virus ($s_t^{Ind} = 6$) ('recoveries') or had died ($s_t^{Ind} = 5$).

Cases for Treated or Infected Individuals in Isolation

With certain probabilities, an agent that was also receiving treatment ($s_t^{Ind} = 3$) could change their state to recovered ($s_t^{Ind} = 6$) or deceased ($s_t^{Ind} = 5$). They could also remain in hospital or in isolation ($s_t^{Ind} = 3$).

Cases for Healthy Individuals in Preventive Quarantine

For the agents in preventive quarantine ($s_t^{Ind} = 4$), the program determined the length of time that an individual had remained in quarantine. There were two alternatives based on a probabilistic test. The individual could be released after two time steps (weeks) or an agent would have to remain in quarantine. If an agent was healthy after quarantine, the program would assign his prior state (healthy, recovered, vaccinated) with a certain probability: ($s_t^{Ind} = 1$, $s_t^{Ind} = 6$ or $s_t^{Ind} = 7$). Moreover, a quarantined agent could have contracted the virus as a result of contacts during or at the end of the quarantine (respectively, states $s_t^{Ind} = 3$ and $s_t^{Ind} = 2$) with a certain probability. In the worst-case scenario, an agent could have died in isolation with a very low probability ($s_t^{Ind} = 5$).

Cases for Recovered Individuals

These agents were treated in the same way as healthy ones. However, in their case, we assumed a decreased probability of them becoming infected or being hospitalised. In addition, recovered agents would become immune to COVID-19 for several weeks in all three scenarios.

Cases for Vaccinated Individuals

These agents were treated in the same way as recovered ones. However, we assumed a lower probability of them becoming infected or being hospitalised. In addition, vaccinated agents would become immune for a longer period of time than recovered patients.

In our stylised simulation, we attempted to take into account the most important characteristics that could affect the dynamics of the spread of the virus and the impact of the pandemic on an economy. Therefore, all of the probability tests considered the age of an individual. This is important because, according to the empirical data, in the first waves of the coronavirus, the elderly were more likely to suffer with a severe disease or die from the coronavirus. Apart from changes in the health status, one of the most important characteristics of the agents was productivity. When their health status changed, an agent's productivity was updated accordingly. Any decrease in an agent's productivity was extensively consulted with both medical specialists and economists. The calibration was also consistent with the conclusions that had been extracted from the literature.

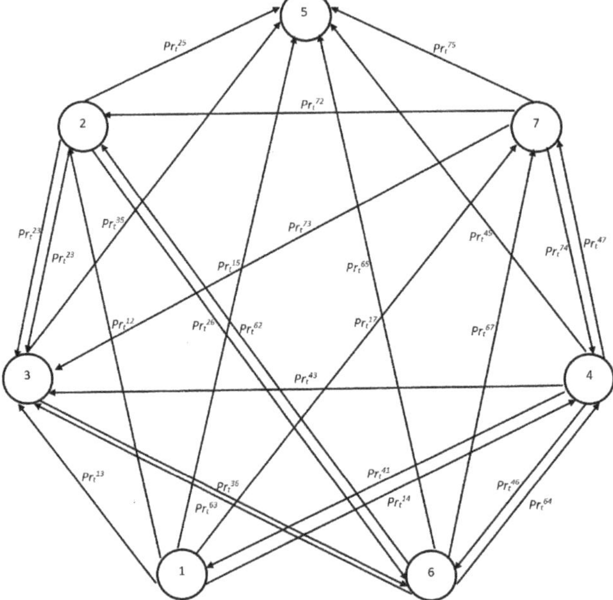

Figure 1. State transition probabilities in the agent-based epidemic component. Health status: 1—healthy (*h*), 2—infected (*i*), 3—treated (*t*), 4—healthy individuals in preventive quarantine (*q*), 5—deceased (*d*), 6—recovered (*r*), 7—vaccinated (*v*) P^{ij}—transition probability between states *i* and *j*, see Tables 2 and 3.

In the fifth module, the aggregation for each iteration is performed. As a result, we obtained the overall number of:

- Healthy individuals by age for each iteration;
- Infected agents by age for each iteration;
- Recovered agents by age for each iteration;
- Vaccinated agents by age for each iteration;
- Individuals receiving treatment by age for each iteration;
- Agents in preventive quarantine by age for each iteration;
- Agents deceased by age for each iteration.

These data were used as the input data for the dynamic stochastic general equilibrium model, which will be described in the following sections.

The last part of the code helps us visualise the results of the simulations. It also permits the results to be systematised in csv tables for further analysis using the DSGE model. The most important input data were the productivity shock.

4. Validation of Scenarios in Connection with the Introduction of Vaccination

In Kaszowska-Mojsa and Włodarczyk (2020) [1], four scenarios were analysed. However, because it was the early stage of the COVID-19 pandemic, we did not assume immunity or the effects of vaccinations in our study.

In the first scenario in [1], we studied the spread of the coronavirus in a country that was under mild restrictions. Home isolation was compulsory in this scenario. We also assumed that in more severe cases, people would be hospitalised. In both cases, the agents spent at least three weeks there. However, agents who had contact with an infected individual were quarantined only with a given probability. The quarantine period was a minimum of two weeks. At the same time, no additional restrictions were assumed by the regulator. In 2020, we treated this scenario as a baseline scenario (1). In 2021, we updated this scenario in order to introduce different levels of population immunity into the model (see: scenarios 1.1., 1.2. and 1.3) and hence validated the results. We managed to prove that the conclusions that had been presented in [1] were still valid when the vaccination process and natural immunity after recovery were taken into account.

The results for scenarios 1.1, 1.2 and 1.3 were presented in relation to scenarios 1–4 from the research conducted in 2020 [1]. For ease of reading, we briefly described below the assumptions of scenarios 2–4 and the main conclusions. For specific calibration of scenarios 1–4, see Table 4. For more information, see [1].

Table 4. Comparison of the calibration of scenarios 1–4.

Notation	Scenario 1	Scenario 2	Scenario 3	Scenario 4
T	104	104	104	104
N^{Ind}	10,000	10,000	10,000	10,000
K^{Ind}	150	150	150	150
$S_t \times S_t$	100×100 for all t	Dynamic adjustment *	Dynamic adjustment *	100×100 for all t
$(Ag)_t^1$	0.181	0.181	0.181	0.181
$(Ag)_t^2$	0.219	0.219	0.219	0.219
$(Ag)_t^3$	0.6	0.6	0.6	0.6
$(Wp)_t^{av\_h}$	1 for all t	Dynamic adjustment *	Dynamic adjustment *	1 for all t
$(Wp)_t^{av\_inf}$	0.9	0.9	0.9	0.9
$(Wp)_t^{av\_q}$	0.8	0.8	0.8	–
$(Wp)_t^{av\_t}$	0.3	0.3	0.3	0.3
$(Pr)_t^{12}$	0.03	0.03	Dynamic adjustment *	0.2
$(Pr)_t^{13}$	0.1	0.1	Dynamic adjustment *	0
$(Pr)_t^{15}$	0.00002	0.00002	Dynamic adjustment *	0.00002
$(Pr)_t^{21}$	0.6998	0.6998	Dynamic adjustment *	0.6998
$(Pr)_t^{24}$	0.2	0.2	Dynamic adjustment *	0.2
$(Pr)_t^{25}$	0.0002	0.0002	Dynamic adjustment *	0.005
$(Pr)_t^{41}$	0.6	0.6	Dynamic adjustment *	–
$(Pr)_t^{43}$	0.1	0.1	Dynamic adjustment *	–
$(Pr)_t^{45}$	0.0002	0.0002	Dynamic adjustment *	–
$(Pr)_t^{31}$	0.7	0.7	Dynamic adjustment *	0.7
$(Pr)_t^{35}$	0.0002	0.0002	Dynamic adjustment *	0.002

* The details of dynamic adjustment were described in [1].

In the second scenario, we simulated the spread of the COVID-19 pandemic under mobility restrictions, i.e., we focused on the impact of a lockdown on the spread of the SARS-CoV-2 virus and on the economy. We assumed that the duration of a lockdown would be at least two months (the lockdown was relatively long). The main observation was that "an extreme lockdown resulted in the long-term decrease in productivity in the economy" [1]. However, the pre-crisis level of productivity was achieved within two years after the outbreak of the COVID-19 pandemic. There was no permanent loss of productivity due to "an increase in the number of deaths and the permanent destruction of jobs".

In the third scenario, we studied the effects of gradually introducing preventive restrictions on a society, such as mobility restrictions, restrictions that could affect the probability of infection and a lockdown. Then, we analysed the impact of these restrictions on the spread of the virus and the economy.

In the fourth scenario, we described the situation in which the coronavirus spread in a society in a much more aggressive manner. At the same time, the death rate was also higher. In this scenario, no restrictions were imposed on a society by the government and the spread of the virus was unrestricted. No large-scale testing was performed. Quarantine or home isolation was not mandatory.

For each of the four scenarios, we generated the labour productivity paths using the agent-based epidemic component. We then used these to obtain conditional forecasts of the main macroeconomic indicators, i.e., output, capital and investments as well as the unemployment rate, see Sections 5 and 6 (using the DSGE model).

To validate the results after introduction of COVID-19 vaccine, we developed three new scenarios. In the first one (1.1), we updated the baseline scenario (1) with the immunity periods for vaccinated and recovered agents. Like in baseline scenario (1), we assumed the existence of only mild restrictions. In this scenario, the agents were vaccinated under certain probability (representing their willingness to get vaccinated) or under the condition that they were healthy or had recovered. In addition, there was a restriction on how many agents could be vaccinated in one period of time. We assumed that vaccinated agents would be immune for 20 weeks and that they would have a lower probability of being infected afterwards. Similarly, recovered people would be protected for ten weeks and would have lower a probability of being reinfected. Furthermore, our code enabled the period of immunity and transition probabilities to be modified. In this scenario, on average, we achieved up to 50% of immune agents in the population. In addition, we assumed that the effects of the vaccine would decrease over time. However, when the positive effects of vaccinations would be fading away, new individuals would be getting vaccinated. We also included the possibility of receiving additional dose of vaccine. For those reasons, we observed fluctuations in the number of infected people (as well as of the other states, see Figure 2) that translated into fluctuations in productivity.

In the second scenario (1.2), we tested whether a higher percentage of immune agents in a society (80%) would enable herd immunity to be achieved and thus could avoid introducing further prevention and control measures. In order to obtain this higher percentage of vaccinated agents, we assumed that the vaccination process would be more effective (i.e., a larger number of agents could be vaccinated each day and we also increased probability of getting vaccinated. This probability of getting vaccinated would be higher, e.g., if there had been efficiently conducted pro-vaccination campaigns that increased public willingness to get vaccinated). In Figure 3, a gradual decrease in the number of agents in quarantine can be observed, however, a relatively high percentage of individuals were still being infected and hospitalised. Unfortunately, a large percentage of people also died. After recovery, immunity was gradually built up in a society. Building herd immunity is supported by the vaccination process, although it should be emphasised that herd immunity was not achieved. A higher vaccination coverage led to a lower number of infected individuals and a correspondingly fewer number of recoveries over time. Vaccinating agents at a higher level also reduced the burden on the health care system. Changes between the states would also update the states of the agents in terms of their productivity. Taken together, in the aggregate, we observed changes in the productivity shock, which then fed the DSGE model.

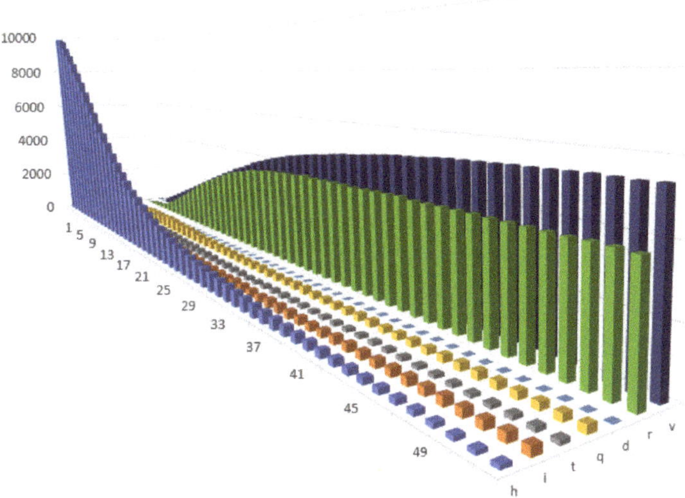

Figure 2. Changes in the health states in Scenario 1.1 (with immunity).

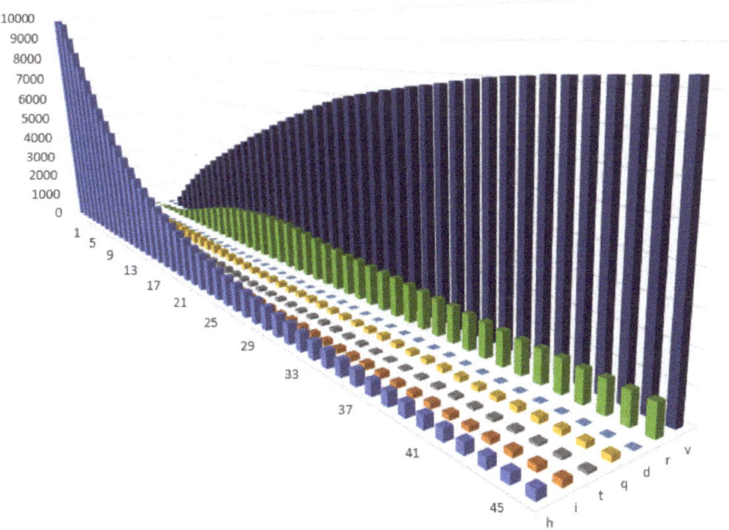

Figure 3. Changes in the health states in Scenario 1.2 (with immunity).

In the third scenario (1.3), we further increased the effectiveness of the vaccination process as well as the probability of getting vaccinated. We did this until we achieved herd immunity. This was achieved when approximately 90% of the population was vaccinated or had already recovered, see Figure 4. In all three scenarios, we observed a moderate decrease in productivity during the second year. This can be easily explained by the existence of a group of individuals who were against vaccinations and the other new forms of COVID-19 treatment. This trend also reflected the fact that part of population had died as a result of being infected with COVID-19 or from natural causes (for now, we did not allow new agents to be created in the model). In Table 5, the calibration for all three scenarios with vaccinations and immunity after recovery is compared.

Table 5. Comparison of the calibration of scenarios 1.1–1.3 (with immunity).

Notation	Scenario 1.1	Scenario 1.2	Scenario 1.3
T	104	104	104
N^{Ind}	10,000	10,000	10,000
K^{Ind}	150	150	150
$S_t \times S_t$	100×100 for all t	100×100 for all t	100×100 for all t
$(Ag)_t^1$	0.181	0.181	0.181
$(Ag)_t^2$	0.219	0.219	0.219
$(Ag)_t^3$	0.6	0.6	0.6
$(Wp)_t^{av\_h}$	1 for all t	1 for all t	1 for all t
$(Wp)_t^{av\_inf}$	0.9	0.9	0.9
$(Wp)_t^{av\_q}$	0.8	0.8	0.8
$(Wp)_t^{av\_t}$	0.3	0.3	0.3
$(Pr)_t^{12}$	0.03	0.03	0.03
$(Pr)_t^{13}$	0.1	0.1	0.1
$(Pr)_t^{14}$	0.1	0.1	0.1
$(Pr)_t^{15}$	0.00002	0.00002	0.00002
$(Pr)_t^{17}$	0.05	0.05	0.3
$(Pr)_t^{23}$	0.2	0.2	0.2
$(Pr)_t^{25}$	0.0002	0.0002	0.0002
$(Pr)_t^{26}$	0.6998	0.6998	0.6998
$(Pr)_t^{35}$	0.0002	0.0002	0.0002
$(Pr)_t^{36}$	0.7	0.7	0.7
$(Pr)_t^{41}$	0.6	0.6	0.6
$(Pr)_t^{43}$	0.1	0.1	0.1
$(Pr)_t^{45}$	0.0002	0.0002	0.0002
$(Pr)_t^{46}$	0.06	0.06	0.06
$(Pr)_t^{47}$	0.06	0.06	0.06
$(Pr)_t^{62}$	0.01	0.01	0.01
$(Pr)_t^{63}$	0.0005	0.005	0.005
$(Pr)_t^{64}$	0.05	0.05	0.05
$(Pr)_t^{65}$	0.00001	0.00001	0.00001
$(Pr)_t^{67}$	0.009	0.1	0.2
$(Pr)_t^{72}$	0.009	0.005	0.005
$(Pr)_t^{73}$	0.00045	0.00025	0.00025
$(Pr)_t^{74}$	0.05	0.05	0.05
$(Pr)_t^{75}$	0.00001	0.00001	0.00001

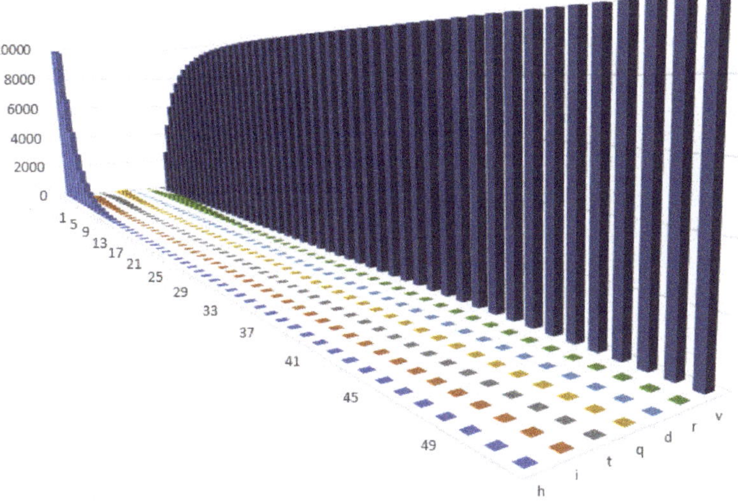

Figure 4. Changes in the health states in Scenario 1.3 (with immunity).

In Figure 5, the labour productivity paths for four scenarios without immunity are presented.

In Figure 6, the productivity paths for all three scenarios with immunity are presented. It is easy to observe that all of the scenarios that included the vaccination process achieved much better results than the baseline scenario that had only standard preventive measures (face masks and quarantine). However, because approximately 90% of population needs to be vaccinated or has to recover in order to obtain herd immunity, the use of lockdowns seems to be indispensable. In the following sections, we used those productivity shocks as an input into the DSGE model in order to obtain a conditional forecast of the main macroeconomic variables, which would indicate the impact of vaccinations on the economy (i.e., output, investment, capital and unemployment rate).

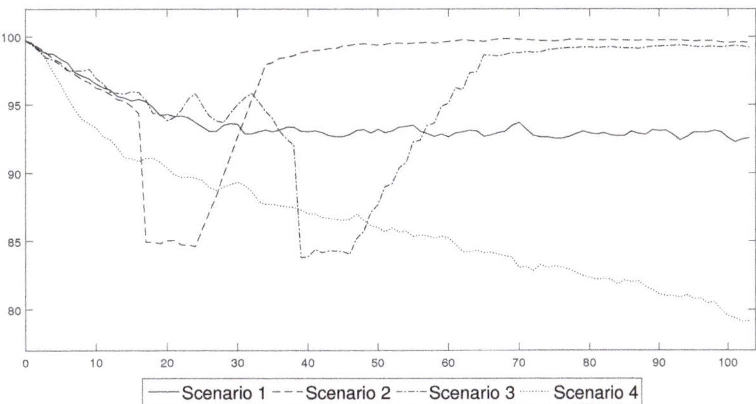

Figure 5. Aggregate labour productivity under the different COVID-19 prevention and control schemes. Please note that this figure is similar to the one that was published in [1] in November 2020. This figure enables the results for the scenarios that were analysed in 2021 to be compared with those from 2020.

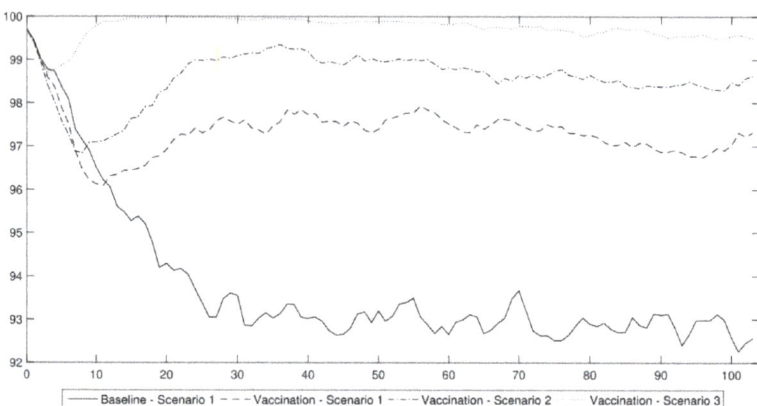

Figure 6. Aggregate labour productivity under the different COVID-19 vaccination schemes. Vaccination Scenario 1 is (1.1); Vaccination Scenario 2 is (1.2) and Vaccination Scenario 3 is (1.3).

5. Macroeconomic Consequences of Pandemics—The DSGE Approach

Like in [1], we also used the DSGE model to assess the macroeconomic consequences of the COVID-19 pandemic under the different prevention and control schemes, with a special emphasis on vaccinations. We used the approach that enabled the business cycles of modern economies to be replicated. The model was based on the model that was elaborated by Gali [22]. However, our aim was to extend it in a such way that the capital accumulation and labor market components could be introduced. We developed our model in line with the works of Christiano et al. [23], Gali [24,25], Gali et al. [26]. A description of the equations that were used in the modelling process can be found in our previous paper [1]. The changes in the calibration are explained in this section.

In order to study the impact of the COVID-19 pandemic on the economy using this framework, we introduced an additional labour productivity shock into the DSGE model. This shock was obtained from the agent-based component that was previously described. This approach enabled us to assess the consequences of a change in the availability of employees because they were infected, hospitalized, quarantined and because of the introduction of remote work. A change in their health status or working remotely made employees less effective or prevented them from working at all. It should be noted that these employees continued to work for the company in question and received either wages or sickness benefits for this work. Therefore, the COVID-19 shock should not be considered to be a labour supply shock, which pushes part of the labour force into inactivity as was the case in [27]. In our view, illness may have caused employees to be unproductive or not fully productive, but in many cases, it did not have any negative impact on the formal employment relationship. Such an approach located the first impact of a COVID-19 pandemic on the supply side of an analysed economy, which led to better reproducing the character of the pandemic disturbances. The demand-side effects were a second-order phenomenon. Such an approach is also in line with the results of the research on the nature of pandemic shock that assesses the supply-side effects as being the major factor that is responsible for the economic disturbances that have been caused by the SARS-CoV-2 pandemic [28].

The model assumed that "an economy was populated by a unit mass continuum of households that maximised their utility levels by solving the optimisation problem" as was described in [1]. The model is expressed in weekly terms in order to be able to study the dynamics of the COVID-19 pandemic. This approach was also used in [27,29]. In Table 6, the calibration of the model is presented. It was performed in such a way so that it matches the standard stylised facts associated with the business cycles of developed economies. Our model successfully reproduced the results of the empirical studies such as, for example, the estimated model of Christiano et al. [30].

Table 6. Proposed calibration of the parameters of the model.

Variable	Description	Calibrated Values
$\mathcal{A}$	Elasticity of output towards the changes of labour	0.25
φ	Reverse of the labour supply elasticity	5
ϵ_w	Elasticity of substitution between types of labour	4.52
ϵ_p	Elasticity of substitution between types of goods	9
θ_w	Calvo index of wage rigidity	0.9807
θ_p	Calvo index of price rigidity	0.9807
β	Discount factor	0.9996
δ	Capital depreciation rate	0.0175
ϕ_k	Capital adjustment costs' scaling parameter	12
h	Habit persistence parameter	0.9
ρ_a	Autoregressive parameter of the technological shock	0.99
ρ_χ	Autoregressive parameter of the labour supply shock	0.99
ρ_a	Autoregressive parameter of the technological shock	0.99
ρ_χ	Autoregressive parameter of the labour productivity shock	0.99
ρ_M	Autoregressive parameter of the monetary policy shock	0.965
ϕ_π	Central bank's reaction to the deviation of inflation from its steady state value	0.115
ϕ_y	Central bank's reaction to the deviation of output gap from its steady state value	0.0096

The model was slightly recalibrated with respect to [1] based on the fact that new information about the COVID-19 pandemic was provided. We assumed a discount factor $\beta = 0.9996$, which resulted in a steady-state interest rate of 2.1% in annual terms. Following the approach adopted by Christiano et al. [30] and Gali [25], we set the expected duration of prices and wages to 52 weeks, which makes $\theta_p = \theta_w = 0.9807$. As in the study of in Gali [25], we assumed that $\epsilon_w = 4.52$ and $\varphi = 5$. As a result of the adopted calibration, the steady-state unemployment rate was approximately 4.8%. In our model, the unemployment rate could be identified with the natural unemployment rate under certain restrictions. The habit persistence parameter, h, was set at a relatively high level of 0.9. Nonetheless, this value is acceptable if we consider the fact that we adopted the calibration in weekly terms. As was expected, consumption was characterised by a relatively high week-to-week inertia. We calibrated the capital share in production to 0.25 ($\alpha = 0.25$). Following the analysis of the empirical data concerning the behaviour of capital during the pandemic, we decided to slightly recalibrate the parameters concerning the capital accumulation $\phi_k = 12$ and $\delta = 0.0175$ (compared to $\phi_k = 8$ and $\delta = 0.05$ in [1]). Both parameters enabled us to obtain the reactions of capital and investment that were better fitted to the actual tendencies that were observed in the data. These values also permitted the model to be identified.

Because the model was calibrated in weekly terms, the parameters of the Taylor rule also had to be adjusted. We assumed $\phi_\pi = 0.115$ and $\phi_y = 0.0096$. This calibration is consistent with the values of 1.5 and 0.125 in quarterly terms, respectively. Finally, we had to recalibrate the autoregressive parameters of the shocks in order to obtain the duration of the shocks in weekly terms. As a result, the values of $\rho_a = \rho_\chi = \rho_N = 0.99$ and $\rho_M = 0.965$ were assumed. The main advantage of the proposed calibration is that the Blanchard–Kahn conditions were fulfilled and the model could be identified. In both this article and in the research that was conducted in November 2020, the model was solved in nonlinear terms (no log-linearisation around the steady state was required).

6. COVID-19 Prevention and Control Schemes—What Does the Vaccination Change?

Using the DSGE model and the labour productivity shocks that had been obtained from the agent-based epidemic component, we generated conditional forecasts of the standard macroeconomic indicators: output, capital, investments and the unemployment rate. In Figure 7, we present the results of the four scenarios without immunity (for a description of the scenarios, see Section 4). The analyses were performed for 104 weeks (two years). The results are expressed as the relative difference from the steady-state value. A mean of 10,000 simulations of the model is reported.

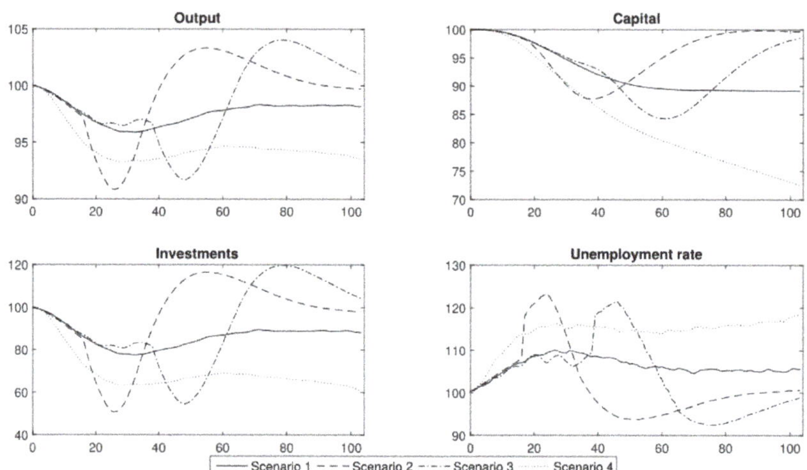

Figure 7. The major macroeconomic indicators under the different COVID-19 prevention and control schemes (conditional forecasts using the DSGE model). Please note that this figure is similar to the one that was published in [1] in November 2020. However, the capital accumulation process was recalibrated in the DSGE model as is explained in Section 5. This figure enables the results for scenarios analysed in 2021 to be compared with those from 2020.

Our analysis in November 2020 showed that the scenarios could easily be divided into two groups, that produced similar economic trends. The first group consisted of scenarios 1 and 4, which "resulted in the occurrence of negative economic trends that persisted in an economy in the medium or even long term". The other group was composed of scenarios with lockdowns (2 and 3). The use of lockdowns led to a deeper response of macroeconomic variables, but the negative effects were observable over a shorter period of time.

The first group consisted of "the scenarios that assumed that the government permitted the persistent spread of the disease by introducing only general sanitary restrictions (scenario 1) or by not introducing any restrictions at all and hoping that the propagation of the virus would finally cease at some point (scenario 4). Both of these approaches resulted in a relatively high share of people who were either infected or were placed in quarantine, which translated into a persistent decrease in the productivity of labour" [1].

In the case of the first scenario, the labour productivity stabilised at a level of approximately 92% of full capacity. In this scenario, we observed that the output initially decreased to approximately 97.5% of the steady value. Nonetheless, then it stabilised at 98% of its steady-state value. At the same time, it should be emphasised that there was a decrease in capital and investment by approximately 10% during the first year after the outbreak of the COVID-19 pandemic. During the second year after the onset of the pandemic, the unemployment rate stabilised at 5 pp. above the steady state, which translated into an actual unemployment rate of approximately 9%.

After the introduction of vaccinations, we modified this scenario to include immunity. We will present the results after commenting on the main results that were obtained from the scenarios 2–4 in 2020 in order to facilitate the reference of the results to the original study that we wanted to validate.

The strategy that enabled the virus to spread without restriction led to a permanent decrease in productivity to a level of 80% within two years after the outbreak of the COVID-19 pandemic that was described in scenario 4. In this scenario, output decreased by approximately 4% in the first half of the year. Then, it stabilised for another six months. During the second year, the output continued in a downward trend and reached approximately 94% of the steady-state value. While the output decreased, firms stopped making further investments. As a result, the level of investment was lower and the capital decreased as well. The unemployment rate increased by approximately 15 pp. within the next two years. We estimated a vast social cost as the actual unemployment rate reached 20%. These estimations did not include the long-term effects, i.e., due to the loss of human capital. The results of our analysis clearly show that the policy makers should not have followed the strategy of no reaction.

In scenarios 2 and 3, we compared the strategy of a strict lockdown with a gradual one. In both cases, the macroeconomic variables (output, capital, investment and unemployment rate) decreased by almost the same amount, see Figure 7. A lockdown that lasted for two months caused a contraction of output and economic activity that disappeared within six months. The main difference between scenarios 2 and 3 was the duration of the economic downturn that was caused by a lockdown. Gradual lockdowns unnecessarily prolonged the duration of restrictions as well as the adverse effects that were caused by the introduction of the prevention and control schemes, and thus seems to be suboptimal compared to an immediate action strategy.

To summarise, the main conclusion was that "if we decide to shape our policy according to scenarios 2 or 3, the changes in economic activity might be abrupt but short-lived. In the case of scenarios 1 or 4, the decrease in economic activity might not be as deep but would be rather permanent" [1]. However, the question of whether in the era of widespread vaccination, it would not make sense to follow the first baseline scenario instead of continuing to implement costly lockdowns remains? As we show, our results are still valid despite the introduction of a vaccine against COVID-19.

In Figure 8, the results of three scenarios with immunity (1.1–1.3) compared to the baseline scenario (1) are presented. In the first scenario with immunity (1.1), the output initially dropped by at least 2%. For the following 1.5 year, the output stabilised at 98.5% of its steady-state value. In this scenario, the contraction of capital and investment (of approx. 4%) was permanent. In the first 25 weeks, the unemployment rate increased by 5 pp. Then, it stabilised at 2.5 pp. above the steady state.

In the second scenario with immunity (1.2), we observe that the output decreased by at least 1.5% in the first 20 weeks. In the next few weeks, it returned to a level of approximately 99% of its steady-state value. There was a permanent decrease in capital and investment of approximately 2%. The situation on the labour market deteriorated in this first period (the unemployment rate was greater by 3 pp. during that period). However, in the second half of the year of the COVID-19 pandemic, the unemployment rate stabilised at 2 pp. above the steady state.

Only in the third scenario with immunity (1.3) was the decrease in output temporary and negligible in the long term. Capital and investment decreased but the contraction was not permanent. The social costs were also low. The unemployment rate increased in the first weeks of the pandemic but decreased after a society had been vaccinated (after obtaining immunity at a level of 90%).

To conclude, the use of lockdowns is still an effective strategy that we should use because a vaccination coverage of 0.5 (50% of the population) does not contain the spread of the virus enough. Herd immunity was only achieved when the vaccination coverage was very high (approximately 90% of the society needs to be immune). Such a high vaccination coverage of the population is virtually impossible when people's skepticism towards vaccinations is considered and the presence of contraindications to vaccinations (i.e., children under five years of age cannot be vaccinated and certain diseases are also contraindications).

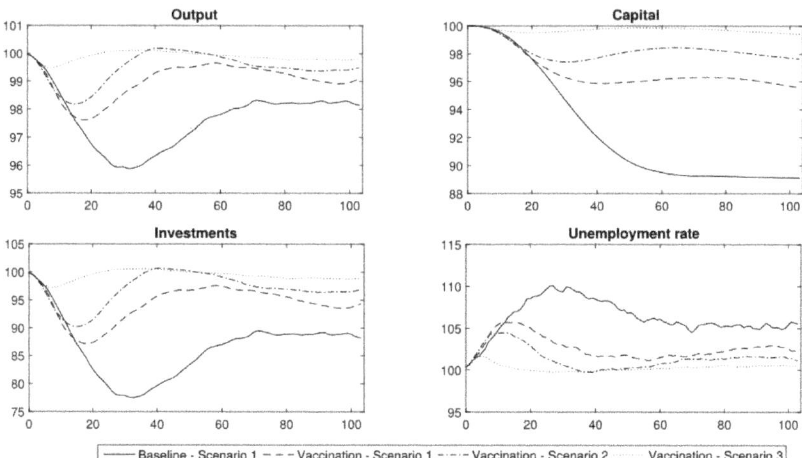

Figure 8. The major macroeconomic indicators under the different COVID-19 vaccination schemes (conditional forecasts using the DSGE model). Vaccination Scenario 1 is (1.1); Vaccination Scenario 2 is (1.2) and Vaccination Scenario 3 is (1.3).

7. Conclusions

Our analysis showed that vaccines and after-recovery immunity do change the dynamics of a contagion or reduce the adverse effects of pandemics on an economy. Although the consecutive pandemic waves have developed some self-limiting characteristics, we proved that it is hard to develop "herd immunity". Even when vaccines are available, the disease remains a constant feature of the economic landscape, which causes losses in economic welfare. The outcomes that resemble "herd immunity" might only be generated for very high vaccination rates (i.e., approximately 90% of the population should be vaccinated or should obtain natural immunity).

The introduction of vaccines in the analysis did not change the main conclusions of the research that was conducted in November 2020. In the article, we showed that the changes in labour productivity that were caused by the spread of disease still lead to negative changes of the macroeconomic aggregates. Despite the fact that the magnitude of these changes is much smaller than in the case in which people did not vaccinate at all, the pandemic still depresses economic activity over a relatively long period of time. As the resulting economic fluctuations are not very abrupt, strategies that promote vaccination do permit the economic costs of pandemic to be reduced in terms of output losses, capital depreciation and unemployment. On the other hand, they prevent the occurrence of an economic recovery after the peak of pandemic wave as the labour productivity does not reach the steady state level.

A lockdown strategy causes bigger falls of productivity, output and capital, which are accompanied by increases in unemployment, in the initial phase, when harsh constraints on personal and economic activity are introduced. On the other hand, as the constraints are lifted, an economic recovery is observed. That recovery is a representation of the "creative destruction" phenomenon, which leads to the occurrence of microcycles of capital working towards an improvement of the economic perspectives after the lockdown. As a result, lockdowns do not extend the duration of a recession, which confirms that they are still a viable alternative in the fight against an epidemic.

Author Contributions: Conceptualization, J.K.-M. and P.W.; methodology, J.K.-M. and P.W.; software, J.K.-M. and P.W.; validation, J.K.-M. and P.W. and A.S.; formal analysis, J.K.-M., P.W. and A.S.; investigation, J.K.-M., P.W. and A.S.; resources, J.K.-M., P.W. and A.S.; data curation, J.K.-M., P.W. and A.S.; writing—original draft preparation, J.K.-M., P.W. and A.S.; writing—review and editing, J.K.-M., P.W. and A.S.; visualization, J.K.-M. and P.W.; project administration, J.K.-M., P.W. and A.S.; funding acquisition, J.K.-M., P.W. and A.S. All authors have read and agreed to the published version of the manuscript.

Funding: The article was funded within the MACROPRU project. This project has received funding from the European Union's Horizon 2020 research and innovation programme under the Marie Skłodowska–Curie grant agreement No 101023445. The research of P. Włodarczyk and A. Szymańska was funded by the University of Łódź.

Institutional Review Board Statement: Not applicable.

Informed Consent Statement: Not applicable.

Data Availability Statement: Not applicable.

Acknowledgments: We would like to thank the reviewers for their valuable comments.

Conflicts of Interest: The authors declare no conflict of interest.

Abbreviations

The following abbreviations are used in this manuscript:

ABM	Agent-Based Modelling
COVID-19	Coronavirus Disease 2019
DSGE	Dynamic Stochastic General Equilibrium
pp.	percentage points
SARS-CoV-2	Severe acute respiratory syndrome coronavirus 2

References

1. Kaszowska-Mojsa, J.; Włodarczyk, P. To Freeze or Not to Freeze. Epidemic prevention and control in the DSGE model with agent-based epidemic component. *Entropy* **2020**, *22*, 1345. [CrossRef]
2. Kadkhoda, K. Herd Immunity to COVID-19: Alluring and Elusive. *Am. J. Clin. Pathol.* **2021**, *155*, 471–472. [CrossRef]
3. Desai, A.; Majumder, M. What Is Herd Immunity? *JAMA* **2020**, *324*, 2113. [CrossRef] [PubMed]
4. Fine, P.; Eames, K.; Heymann, D. "Herd Immunity": A Rough Guide. *Clin. Infect. Dis.* **2011**, *52*, 911–916. doi: 10.1093/cid/cir007. [CrossRef] [PubMed]
5. Ashby, B.; Best, A. Herd immunity. *Curr. Biol.* **2021**, *31*, R174–R177. [CrossRef]
6. Fontanet, A.; Cauchemez, S. COVID-19 herd immunity: Where are we? Nature reviews. *Nat. Rev. Immunol.* **2020**, *20*, 583–584. [CrossRef] [PubMed]
7. Kwok, K.; Lai, F.; Wei, W.; Wong, S.; Tang, J. Herd immunity—Estimating the level required to halt the COVID-19 epidemics in affected countries. *J. Infect.* **2020**, *80*, e32–e33. [CrossRef]
8. Coccia, M. Optimal levels of vaccination to reduce COVID-19 infected individuals and deaths: A global analysis. *Environ. Res.* **2021**, *204*, 112314. [CrossRef]
9. Aschwanden, C. Five reasons why COVID herd immunity is probably impossible. *Nature* **2021**, *591*, 520–522. [CrossRef]
10. Aschwanden, C. The false promise of herd immunity for COVID-19. *Nature* **2020**, *587*, 26–28. [CrossRef] [PubMed]
11. Gomes, M.G.; Aguas, R.; King, J.; Langwig, K.; Souto-Maior, C.; Carneiro, J.; Penha-Gonçalves, C.; Gonçalves, G.; Ferreira, M. Individual variation in susceptibility or exposure to SARS-CoV-2 lowers the herd immunity threshold. *Prepr. Health Sci. (medRxiv)* **2020**. [CrossRef]
12. Dashtbali, M.; Mirzaie, M. A compartmental model that predicts the effect of social distancing and vaccination on controlling COVID-19. *Sci. Rep.* **2021**, *11*, 8191. [CrossRef] [PubMed]
13. Makhoul, M.; Ayoub, H.H.; Chemaitelly, H.; Seedat, S.; Mumtaz, G.R.; Al-Omari, S.; Abu-Raddad, L.J. Epidemiological Impact of SARS-CoV-2 Vaccination: Mathematical Modeling Analyses. *Vaccines* **2020**, *8*, 668. [CrossRef] [PubMed]
14. Charumilind, S.; Craven, M.; Lamb, J.; Sabow, A.; Wilson, M. *When Will the COVID-19 Pandemic End? An Update. Healthcare Systems & Services Practice*; McKinsey & Company: New York, NY, USA, 2021.
15. Han, S.; Cai, J.; Yang, J.; Zhang, J.; Wu, Q.; Zheng, W.; Shi, H.; Ajelli, M.; Zhou, X.-H.; Yu, H. Time-varying optimization of COVID-19 vaccine prioritization in the context of limited vaccination capacity. *Nat. Commun.* **2021**, *12*, 4673. [CrossRef]
16. Chen, Y.-T. The Effect of Vaccination Rates on the Infection of COVID-19 under the Vaccination Rate below the Herd Immunity Threshold. *Int. J. Environ. Res. Public Health* **2021**, *18*, 7491. [CrossRef] [PubMed]

17. Viana, J.; van Dorp, C.; Nunes, A.; Gomes, M.; van Boven, M.; Kretzschmar, M.; Veldhoen, M.; Rozhnova, G. Controlling the pandemic during the SARS-CoV-2 vaccination rollout. *Nat. Commun.* **2021**, *12*, 3674. [CrossRef]
18. Moghadas, S.; Vilches, T.; Zhang, K.; Wells, C.; Shoukat, A.; Singer, B.; Meyers, L.; Neuzil, K.; Langley, J.; Fitzpatrick M.; et al. The impact of vaccination on coronavirus disease 2019 (COVID-19) outbreaks in the United States. *Clin. Infect. Dis.* **2021**, *73*, 2257–2264. [CrossRef]
19. Coccia, M. Preparedness of countries to face COVID-19 pandemic crisis: Strategic positioning and factors supporting effective strategies of prevention of pandemic threats. *Environ. Res.* **2022**, *203*, 111678. [CrossRef]
20. Wolfram, S. Cellular automata as models of complexity. *Nature* **1984**, *311*, 419–424. [CrossRef]
21. Ilachinski, A. *Cellular Automata. A Discrete Universe*, 1st ed.; World Scientific Publishing Co. Pte. Ltd.: London, UK, 2001; pp. 1–462.
22. Galí, J. *Monetary Policy, Inflation, and the Business Cycle. An Introduction to the New Keynesian Framework*, 1st ed.; Princeton University Press: Princeton, NJ, USA, 2008; pp. 3–224.
23. Christiano, L.; Eichenbaum, M.; Evans, C. Nominal Rigidities and the Dynamic Effects of a Shock to Monetary Policy. *J. Political Econ.* **2005**, *113*, 1–45. [CrossRef]
24. Galí, J. *Unemployment Fluctuations and Stabilization Policies. A New Keynesian Perspective*, 1st ed.; MIT Press: Cambridge, MA, USA, 2011; pp. 2–120.
25. Galí, J. *Monetary Policy, Inflation, and the Business Cycle. An Introduction to the New Keynesian Framework and Its Applications*, 1st ed.; Princeton University Press: Princeton, NJ, USA, 2015; pp. 2–224.
26. Galí, J.; Smets, F.; Wouters, R. Unemployment in an Estimated New Keynesian Model. In *NBER Macroeconomics Annual 2011*, 1st ed.; Acemoglu, D., Woodford, M., Eds.; Chicago University Press: Chicago, IL, USA, 2012; Volume 26, pp. 329–360.
27. Eichenbaum, M.; Rebelo, S.; Trabandt, M. *The Macroeconomics of Epidemics*; NBER Working Paper No. 26882; NBER: New York, NY, USA, 2020; pp. 1–35. [CrossRef]
28. del Rio-Chanona, R.M.; Mealy, P.; Pichler, A.; Lafond, F.; Farmer, J.D. Supply and demand shocks in the COVID-19 pandemic: An industry and occupation perspective. *Oxf. Rev. Econ. Policy* **2020**, *36*, S94–S137. [CrossRef]
29. Brzoza-Brzezina, M.; Kolasa, M.; Makarski, K. *Monetary Policy and COVID-19*; IMF Working Papers WP/21/274; IMF: Washington, DC, USA, 2021; pp. 1–42.
30. Christiano, L.; Trabandt, M.; Walentin, K. DSGE Models for Monetary Policy Analysis. In *Handbooks in Economics. Monetary Economics*, 1st ed.; Friedman, B., Woodford, M., Eds.; North-Holland: Amsterdam, The Netherlands, 2011; Volume 3A, pp. 285–367.

Article

Stock Index Prediction Based on Time Series Decomposition and Hybrid Model

Pin Lv, Qinjuan Wu, Jia Xu * and Yating Shu

School of Computer, Electronics and Information, Guangxi University, Nanning 530004, China; lvpin@gxu.edu.cn (P.L.); 1913392058@st.gxu.edu.cn (Q.W.); 2013391057@st.gxu.edu.cn (Y.S.)
* Correspondence: xujia@gxu.edu.cn

Abstract: The stock index is an important indicator to measure stock market fluctuation, with a guiding role for investors' decision-making, thus being the object of much research. However, the stock market is affected by uncertainty and volatility, making accurate prediction a challenging task. We propose a new stock index forecasting model based on time series decomposition and a hybrid model. Complete Ensemble Empirical Mode Decomposition with Adaptive Noise (CEEMDAN) decomposes the stock index into a series of Intrinsic Mode Functions (IMFs) with different feature scales and trend term. The Augmented Dickey Fuller (ADF) method judges the stability of each IMFs and trend term. The Autoregressive Moving Average (ARMA) model is used on stationary time series, and a Long Short-Term Memory (LSTM) model extracts abstract features of unstable time series. The predicted results of each time sequence are reconstructed to obtain the final predicted value. Experiments are conducted on four stock index time series, and the results show that the prediction of the proposed model is closer to the real value than that of seven reference models, and has a good quantitative investment reference value.

Keywords: stock index forecasting; CEEMDAN; ADF; ARMA; LSTM; hybrid model

Citation: Lv, P.; Wu, Q.; Xu, J.; Shu Y. Stock Index Prediction Based on Time Series Decomposition and Hybrid Model. *Entropy* **2022**, *24*, 146. https://doi.org/10.3390/e24020146

Academic Editors: Jan Kozak and Przemysław Juszczuk

Received: 18 November 2021
Accepted: 12 January 2022
Published: 19 January 2022

Publisher's Note: MDPI stays neutral with regard to jurisdictional claims in published maps and institutional affiliations.

Copyright: © 2022 by the authors. Licensee MDPI, Basel, Switzerland. This article is an open access article distributed under the terms and conditions of the Creative Commons Attribution (CC BY) license (https://creativecommons.org/licenses/by/4.0/).

1. Introduction

The stock index is calculated based on some representative listed stocks. To some extent, it can reflect price changes of the whole financial market, hence its use as an important indicator of the country's future macroeconomic performance. Forecasting the stock index accurately is of paramount importance for reducing risks in decision-making, by providing some important reference information [1]. However, owing to the complexity of the internal structure and the variability of external factors, changes of the stock market are dynamic and uncertain, and forecasting the stock index has always been a challenge. Many stock forecasting models are mostly classified as either statistical or machine learning models [2]. Statistical models were first used to predict the stock market in finance, and have made some achievements. However, they assume a linear and stationary time series, which is inconsistent with the dynamic, non-linear characteristics of the real stock market, so they have great limitations. A deep learning model can overcome the defects of traditional statistical models in time series prediction but is easily affected by noise in some complex and dynamic financial systems, making it difficult to mine the hidden features of time series, resulting in poor learning ability and limited prediction accuracy.

Therefore, a single statistical or machine learning model cannot well predict the stock index. To overcome these limitations, we propose a hybrid stock index forecasting model based on Complete Ensemble Empirical Mode Decomposition with Adaptive Noise (CEEMDAN) [3]. In this model, CEEMDAN is first used to decompose the original financial time series into a series of Intrinsic Mode Functions (IMFs) and a residual term. Then, the stability of the IMFs and the residual term is characterized using the Augmented Dickey Fuller (ADF) method, the low-volatility time series are classified as linear components, and high-volatility time series are classified as non-linear components. In the final step, the Autoregressive Moving Average (ARMA) model is applied to the linear component, and Long Short-Term Memory (LSTM) is applied to the non-linear component. The final prediction result is obtained by reconstructing each prediction series. This method makes full use of ARMA in linear problems and uses LSTM to identify and abstract non-linear features, mining the movement rules of hidden components in time series and improving prediction accuracy. Hence, our proposed method is referred to as CAL (CEEMDAN-ARMA-LSTM). In the CAL model, CEEMDAN sequence decomposition can reduce the complexity of time series, and the sequences that pass the ADF stationarity test have significant linear trends. Therefore, we employ ARMA to predict the data of the linear part, avoiding the waste of effective information caused by differential operation.

The hybrid model combining linear and non-linear methods has great advantages in time series prediction [4]. Ref. [5] proposed a hybrid time-series prediction model taking the residual generated by Autoregressive Integrated Moving Average (ARIMA), combining the differences in a non-stationary time series with ARMA, as the input of LSTM for fitting. The ARIMA-LSTM model has achieved more accurate forecasting results than the individual LSTM and ARIMA models. A moving average filter was used to decompose a time series into linear and non-linear components [6]. ARIMA and Artificial Neural Network (ANN) were used to model low- and high-volatility data, respectively. This hybrid ARIMA-ANN model can achieve good prediction results. Each hybrid model in the literature combined linear and non-linear models in different ways, providing different perspectives for time series data prediction. However, these methods have the limitations that the error sequence generated by a linear model is assumed to be non-linear [5], and the original sequence is decomposed into single linear and non-linear components, which cannot mine the internal features of an overly complicated time series [6].

Our proposed model can properly decompose the original time series, and the ARMA and LSTM models are applied, which overcomes the defects of strong assumptions [5] and insufficient decomposition [6]. We validate our model's effectiveness on four stock market indices. The experimental results show that the proposed model has higher prediction accuracy than seven reference models on these indices. The main contributions of this study are summarized as follows:

1. The advantages of CEEMDAN are used to decompose the original complex sequential data into trends of different scales. This reduces the complexity of the original time series to extract abstract and deep features.
2. The ADF test method effectively combines the linear and non-linear models. This method can judge the stationarity of data. The linear prediction method of ARMA is used for the stationary time series, and the non-linear prediction method of LSTM for unstable time series.
3. The proposed CAL model is compared with the individual LSTM, Gated Recurrent Units (GRU), Bi-directional LSTM (Bi-LSTM), ARIMA models and the hybrid EMD-ARMA-LSTM, CEEMDAN-LSTM [7], and ARIMA-ANN [6] models. Experiments on different datasets show that the CAL model outperforms traditional hybrid models, improved deep learning model, and their separate component models.

The remainder of this article is organized as follows. Section 2 summarizes related work. Section 3 introduces the proposed CAL model. Section 4 experimentally evaluates the proposed method on real stock index datasets. Section 5 summarizes the paper and points out future research directions.

2. Related Work

Time series analysis is an important tool in many stock market prediction methods, and it makes predictions by analyzing observed points in the series. As one of the most widely used linear time series forecasting methods, the ARIMA model [8] integrates the Autoregressive (AR) and Moving Average (MA) models. It assumes that future predictions have a linear dependence on the current and past data values. Therefore, ARIMA can only fit linear stationary time series data; the non-stationary time series might not be modeled effectively.

Deep learning can overcome the limitations of traditional linear models, such as weak fitting ability and weak feature extraction ability with non-linear data, and has gradually become a key research method in stock prediction. Some deep learning models, such as Convolutional Neural Networks (CNNs), can identify non-linear relationships and extract hidden information from data. LSTM can retain long historical information and achieve high prediction accuracy in sequential pattern learning problems. It does not require selecting features manually [9] and the performance to be superior to that of Feedforward Neural Network (FNN) [10], a Deep Neural Networks (DNN) [11], and Support Vector Machines (SVM) [12]. Although deep learning well models some complex problems, the traditional linear model still has some advantages. For example, the regression method sometimes has better prediction performance than deep learning in power system prediction [13,14].

Based on the above analysis, no individual model can be applied well in all circumstances. In a practical problem, the appropriate model depends on the characteristics of the dataset. However, in time series prediction, it is sometimes difficult to define whether the data are linear or non-linear, especially when there are multiple linear or non-linear components, making it difficult to choose an appropriate prediction model.

Various hybrid techniques exploit the unique strengths of both types of model to effectively improve prediction performance [4–6]. Ref. [15] combined ARIMA and SVM, which showed that the combined model was better than either of its components at stock price prediction. LSTM and an Autoregressive Conditional Heteroscedasticity (GARCH) model were combined to predict stock price volatility, with relatively accurate results [16]. Ref. [17] proposed an ARIMA-ANN hybrid model to improve time series predictions when a time series has both linear and non-linear components. Ref. [18] developed three different hybrid models combining linear ARIMA and non-linear models, such as SVM, ANN, and random forest (RF) models, to predict stock index returns. Experimental results showed that the hybrid model ARIMA-SVM achieved the highest accuracy and the best return.

3. Stock Index Forecasting Model

3.1. Related Models

3.1.1. CEEMDAN

Empirical mode decomposition (EMD) [19] can decompose time series data into subseries according to their own time scales without setting a basis function, for effective treatment of non-linear and unstable data. However, mode aliasing can occur during EMD data decomposition. Ensemble Empirical Mode Decomposition (EEMD) addresses this problem but cannot completely eliminate reconstruction error after the introduction of Gaussian white noise [7]. In the process of decomposition, CEEMDAN adaptively adds white noise to avoid mode mixing of EMD, and addresses reconstruction error due to noise. The prediction of stock prices is affected by multiple factors and is a non-linear complex model. The components of CEEMDAN are relatively simple; hence, more accurate predictions can be obtained.

3.1.2. LSTM

As a special recurrent neural network, LSTM solves the problem of gradient disappearance and explosion in the training process of long sequences, and it has a more complex network structure. LSTM introduces a cellular state and combines forgetting, input, and output gates to discard, maintain, and update information. The output of the model is calculated by multiple functions involving some summation operations, so it is not easy to produce the problems of gradient disappearance and explosion in the process of backpropagation. LSTM has advantages in some problems related to time series, such as industrial time series prediction [20] and text translation [21]. We take this model as the non-linear part of time series prediction.

3.1.3. ARMA

ARMA is a linear sequential method that predicts a future according to historical and current data. ARMA data prediction must meet the requirements of stationarity. In practice, trends and periodicity often exist in many datasets, so there is a need to remove these effects before applying such models. Removal is typically carried out by including an initial differencing stage in the model, and the model is transformed into an ARIMA model. Therefore, ARIAM can be seen as an enhanced version of ARMA. It has a wider range of applications but a certain amount of information loss.

3.2. Proposed Model

It is widely accepted that the financial market is complex and dynamic, which calls for a noise elimination or time series decomposition. For this purpose, a multi-scale decomposition method called CEEMDAN is used in our model. The decomposed components have different scales; ARMA and LSTM are used as linear and non-linear prediction modules to exploit their respective advantages. Thus, a hybrid ARMA-LSTM model for time series forecasting based on CEEMDAN is proposed, which is called CAL (CEEMDAN-ARMA-LSTM). CEEMDAN can adaptively decompose a time series, yielding a series of IMFs and residue with different characteristic scales. The decomposition principle is given by

$$s(t) = \sum_{i=1}^{n} imf_i(t) + res(t),\tag{1}$$

where $s(t)$ represents given time series data; $imf_i(t)$ ($i = 1, 2,\ldots,n$) represents the different IMFs; and $res(t)$ is the residue. Each IMF and residue has its own local characteristic time scale. A low-volatility sequence contains more linear features, and ARMA is more suitable for processing. A high-volatility sequence can be considered non-linear, which better suits LSTM. We require a method to separate the linear and non-linear components and feed them into ARMA and LSTM.

Each hybrid model brings its own perspectives to time series decomposition. We use a statistical ADF method to separate linear and non-linear components. The ADF test can identify whether a time series is stationary. The existence of a unit root in a sequence indicates that a series is unstable. A more negative ADF test result indicates more stable data, and 0.05 is an accepted threshold to judge the stability of a dataset, which can used to separate linear and non-linear sequences [4].

$$s(t) = \sum_{i=1}^{m} l_i + \sum_{i=m+1}^{n+1} n_i.\tag{2}$$

An ADF stationary test separates time series decomposed by CEEMDAN in Equation (2), where l_i and n_i, respectively, denote linear and non-linear components.

$$L_t = g(l_{t-1}, l_{t-2}, \ldots, l_{t-p}, \varepsilon_{t-1}, \varepsilon_{t-2}, \ldots, \varepsilon_{t-q}).\tag{3}$$

After the linear and non-linear components, respectively. The modeling process of ARMA is described by Equation (3), where l_{t-1} to l_{t-p} are time sequence values of the past p days, ε_{t-1} to ε_{t-q} denote corresponding random error, and g is the linear function of ARMA. It can be seen from Equation (3) that the results are related to the sequential values and random errors in a past period of time, so it can be concluded that its prediction process can reflect the continuity of the original sequence in time.

LSTM can mine the characteristics of non-linear time series, which we use to fit non-stationary sequences.The LSTM modeling process is described by Equation (4), where f is the non-linear function of LSTM, and a is the number of days observed by the model, i.e., how far we will go back in time. The prediction results of the linear and non-linear parts are obtained by the corresponding models, and the final prediction is the integration of the linear and non-linear parts in Equation (5), where $y(t)$ denotes the final predictions.

$$N_t = f(n_{t-1}, n_{t-2}, \ldots, n_{t-a}),$$ (4)

$$y_t = \sum_{i=1}^{m} L_i + \sum_{i=m+1}^{n+1} N_i.$$ (5)

To sum up, the CAL model prediction consists of time series decomposition, an ADF stationary test, model fitting, and integration of results. Figure 1 shows the prediction model, where IMF_1-IMF_n are IMF components after time series decomposition, and *res* is the residue. $ARMA_1$-$ARMA_m$ denote that the m sequences pass the ADF test and are fitted using ARMA, and $LSTM_{(m+1)}$-$LSTM_{(n+1)}$ denote the $n - m + 1$ sequences that fail the ADF test and are modeled by LSTM. The steps of the proposed hybrid model are as follows.

1. Given time series decomposition, using a CEEMDAN method (Equation (1)), time series data are decomposed into finite IMFs and residue. Components can be more or less volatile.
2. Sequences with different stability are separated by an ADF stationary test (Equation (2)).
3. Low- and high-volatility components are fitted by ARMA (Equation (3)) and LSTM (Equation (4)), respectively.
4. The final result is the sum of the predictions of each component (Equation (5)).

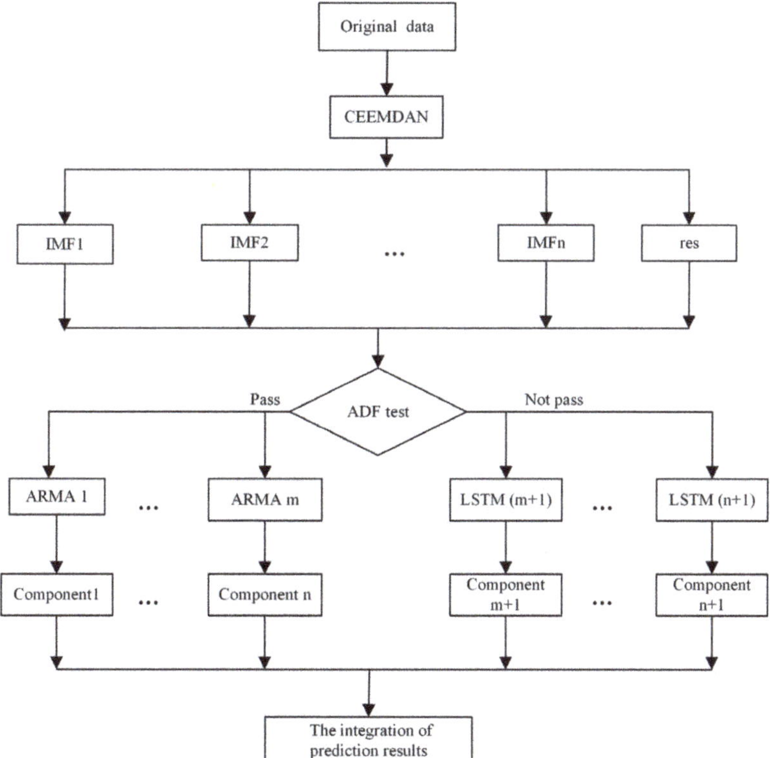

Figure 1. Stock market index forecasting model.

4. Experimental Results and Discussions

In this section, we experimentally present the predictive ability of the CAL model. In Section 4.1, datasets used in experiments are introduced. In Sections 4.2 and 4.3, the evaluation metrics and parameter settings in the experiment are discussed, respectively. The decomposition results of EMD and CEEMDAN are compared in Section 4.4. The models for comparison are listed in Section 4.5. The predicted effects of the CAL model and other comparative methods are evaluated in Section 4.6.

4.1. Datasets

We use one-step-ahead prediction to verify the prediction accuracy of the proposed CAL model on four major global stock indices: Deutscher Aktien (DAX), Hang Seng (HSI), Standard and Poor's 500 (S&P500), and Shanghai Stock Exchange Composite (SSE). These have strong representation in the global financial market and can reflect stock market changes, which has much research value. Stock market indices are affected by national policies, market environments, and other factors presenting different characteristics. Research on stock market indices in different financial markets can examine the prediction accuracy of the model.

The dataset comes from Yahoo! Finance. The range of each stock index is from 13 December 2007, to 12 December 2020, and the daily closing price is selected as the research object. The first 90% of the dataset in the time order of each stock index is used as the training set, and the last 10% is used as the test set. Only the data of trading days are used for research.

The statistical analysis of each stock index is shown in Table 1, where we determine the amount of data contained in each stock market index, as well as the average, maximum, minimum, standard deviation, and ADF test results of the closing index. As can be seen from Table 1, there is a large gap between the maximum and minimum values, and a large standard deviation, indicating that these closing indices have great volatility within the research range. Moreover, the ADF test results of the DAX and S&P500 are greater than the threshold 0.05, indicating that the dataset is highly volatile and non-stationary. SSE is somewhat more stable than the other three datasets. Figure 2 shows the sequential change of the closing index within the study range, from which it can be seen that the four indices all have great volatility and instability in the short term.

Table 1. Descriptive statistics of closing indices.

Index	Count	Mean	Max	Min	Standard Deviation	ADF Test
DAX	3300	9118.21	13,789.00	3666.41	2722.52	0.79
HSI	3219	23,206.70	33,154.12	11,015.84	3660.60	0.11
S&P500	3273	1915.40	3702.25	676.53	713.03	0.99
SSE	3163	2846.43	5497.90	1706.70	586.51	0.01

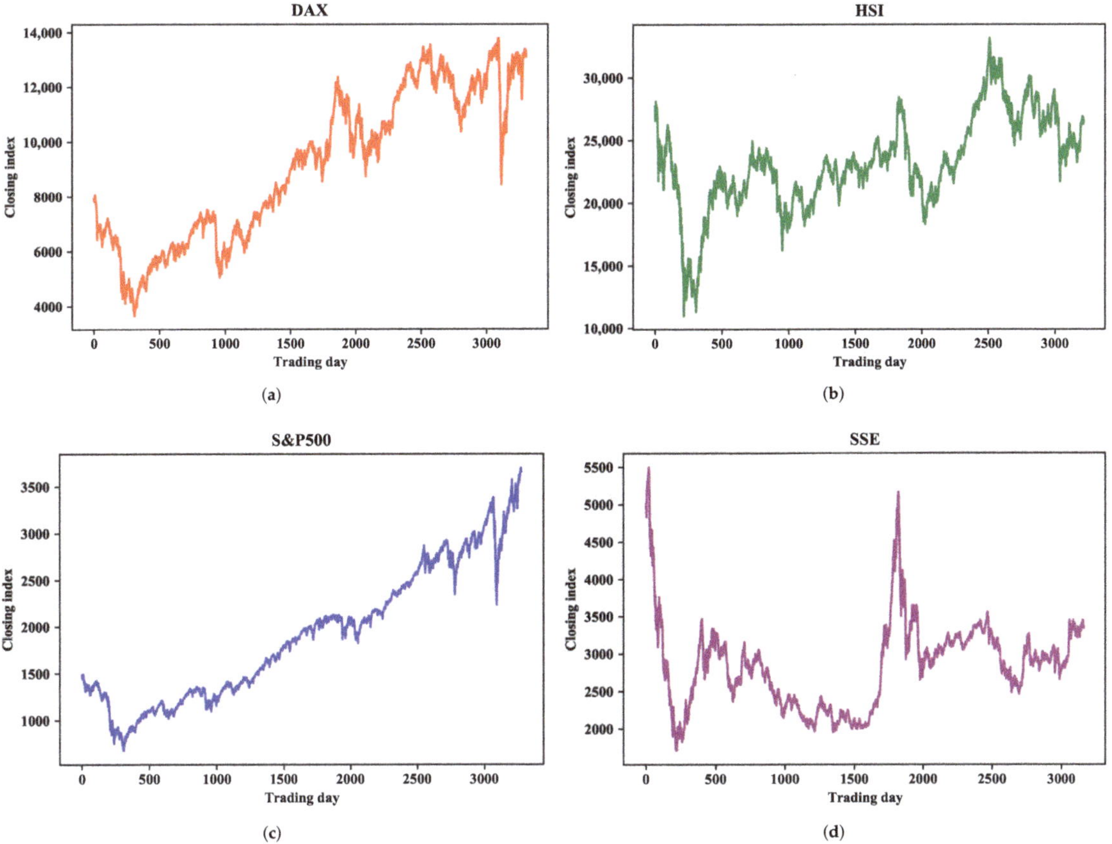

Figure 2. Daily closing index series of four financial markets. (**a**) DAX. (**b**) HSI. (**c**) S&P500. (**d**) SSE.

4.2. Evaluation Metrics

We evaluate the proposed CAL model by the Mean Absolute Error (*MAE*), Root Mean Square Error (*RMSE*), Mean Absolute Percentage Error (*MAPE*), and R-squared (R^2), defined as Equation (6) to Equation (9).

$$MAE = \frac{1}{n} \sum_{i=1}^{n} |p_t - y_t| \tag{6}$$

$$RMSE = \sqrt{\frac{1}{n} \sum_{t=1}^{n} (p_t - y_t)^2}, \tag{7}$$

$$MAPE = \frac{1}{n} \sum_{t=1}^{n} |\frac{p_t - y_t}{y_t}| \times 100, \tag{8}$$

$$R^2 = 1 - \frac{\sum_{i=1}^{n} (p_t - y_t)^2}{\sum_{i=1}^{n} (p_t - \bar{y}_t)^2}. \tag{9}$$

Here, p_t, y_t, and $\bar{y}_t$ are the predicted, actual, and average of actual values, respectively, and n is the prediction horizon. *MAE* measures the average magnitude of the errors in a set of predictions, without considering their direction. *RMSE* is a quadratic scoring rule that also measures the average magnitude of the error. It is the square root of the average of squared differences between prediction and actual observation. MAPE measures the percentage error of the forecast in relation to the actual values. R^2 is a statistical measure in a regression model that determines the proportion of variance in the dependent variable that can be explained by the independent variable. It corresponds to the squared correlation between the observed values and the predicted values by the model. A higher value of R^2 means a better prediction accuracy.

4.3. Parameter Settings

The sequential model structure in Keras is used to build the LSTM network. The batch size of the model is 128. Two layers of LSTM are employed to build the sequential model, and the output of the second layer of the last LSTM unit is connected to a fully connected layer. Then, the fully connected layer is connected to another fully connected layer for the final output. Figure 3 shows the LSTM network structure, where x_i ($i = 1, 2,\ldots, n$) is the input to the model. The numbers of units in each LSTM in the first and second layers are 128, 64, respectively. The third fully connected layer has 16 neurons, and the last layer has only one unit, which will provide a predicted value. Fully connected units and LSTM units use the ReLU and tanh activation function, respectively. We use MSE as a loss function, and use Adam as an optimization algorithm. Adam is an adaptive learning rate optimization algorithm that utilizes both momentum and scaling, and it has two decay parameters that control the decay rates and adjust the learning rate adaptively [22]. We explore the influence of different training epochs on the experimental results, and the results suggest that more training epochs result in a more skillful model, but it may lead to the problem of overfitting. Therefore, it is suitable to set the epoch to 200. The time steps works best at 10. The detailed parameter settings are shown in Table 2.

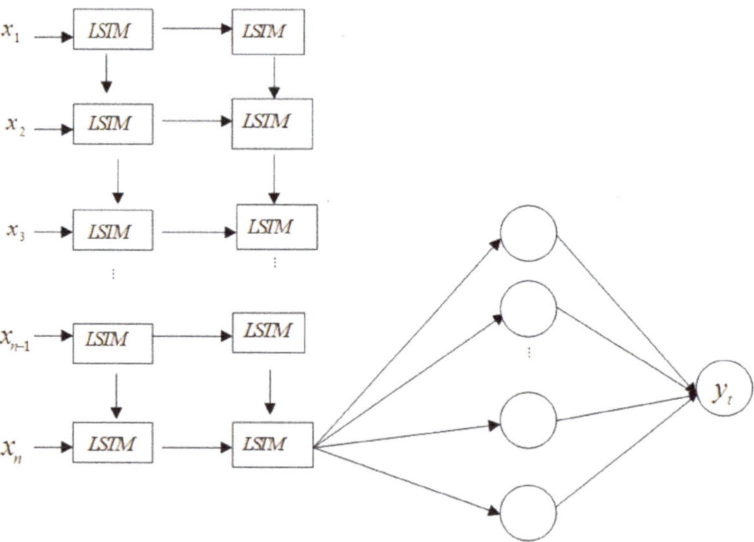

Figure 3. LSTM network architecture.

Table 2. Details of the parameters of the CAL model.

Parameter	Meaning	Value
Input layer	Number of input layer nodes	128
Hidden layer 1	Number of first hidden layer nodes	64
Hidden layer 2	Number of second hidden layer nodes	16
Output layer	Number of output layer nodes	1
Batch size	Pass through to the network at one time	128
Optimization algorithm	Select the training mode	Adam
Loss function	With the goal of minimizing the loss	MSE
Epochs	Number of training	200
Timesteps	Input time steps	10

The best fitted ANN of ARIMA-ANN model in comparison has a layered architecture of $17 \times 17 \times 1$ [4]. The parameters of CEEMDAN-LSTM refer to Ref. [7]. The parameters of LSTM, GRU, and Bi-LSTM, in comparison, are similar to that of LSTM in the CAL model.

Grid search is used to determine the optimal parameters p and q of the ARMA model. The range of the grid search is [0, 5], and the group with the smallest Akaike Information Criterion (AIC) value is selected.

4.4. Decomposition Results of EMD and CEEMDAN

Stock indices, which contain many influencing factors, can be decomposed used EMD or CEEMDAN. We take the SSE stock index as an example to decompose the original time series, so as to compare the two decomposition methods. To intuitively compare the results, we limit CEEMDAN and EMD to generate the same number of IMFs.

In Figure 4, the decomposing results of the original SSE index series are demonstrated. The results of sequence decomposition range from high to low frequency. The first few IMFs, with more noise, represent the high-frequency components in the original data; the middle IMFs, with reduced frequency, represent middle-frequency components; and the last few IMFs, with less volatility, which is similar to the long-term movement trend of a stock, represent the low-frequency components. The left and right sides of Figure 4 show the results of CEEMDAN and EMD data decomposition, respectively. It can be found that IMF5 and IMF6 on the right of Figure 4 have similar scales and are not easily distinguished. This is because the mode aliasing of EMD leads to the distribution of some similar time scales in different intrinsic mode functions, resulting in waveform aliasing and mutual influence. As a result, the features of a single sequence are not obvious, and feature extraction of later prediction models is more difficult. CEEMDAN data decomposition effectively solves this problem. As can be seen from the decomposition results on the left side of Figure 4, CEEMDAN decomposed the stock index into several components, from high- to low-frequency, whose characteristics are obvious, and there is no waveform aliasing.

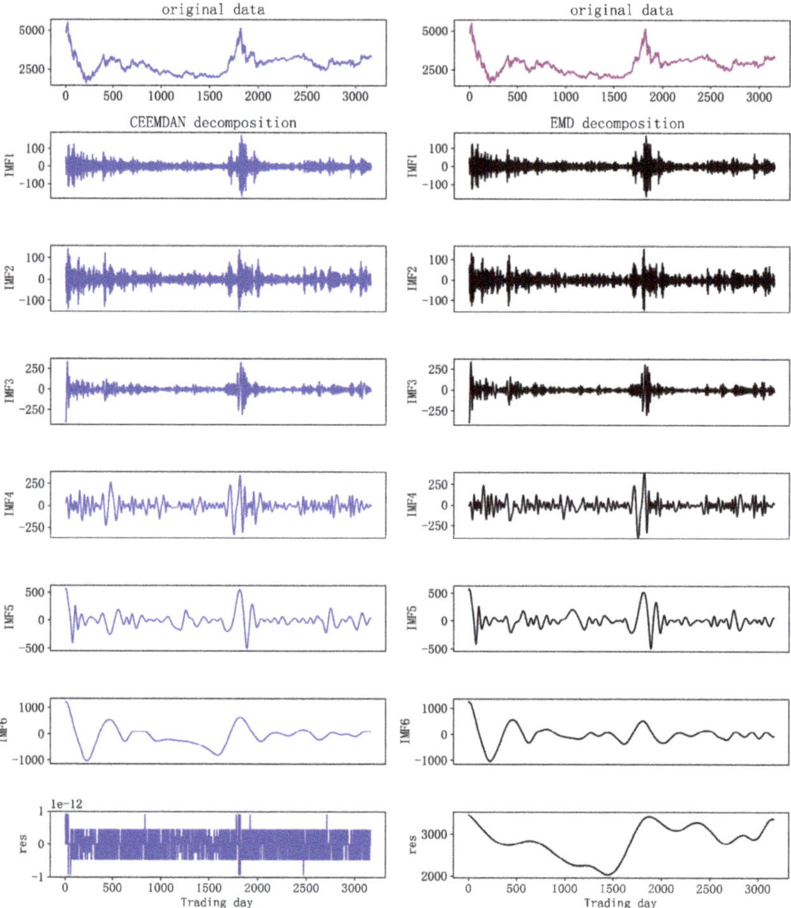

Figure 4. SSE decomposition results.

4.5. Comparative Models

To verify the effectiveness of the proposed CAL model for stock market prediction, we experimentally compare seven models. Table 3 lists the models and reference purposes of these seven controlled experiments, which verify the proposed model from different perspectives.

1. LSTM deep learning model: LSTM networks can automatically detect the best patterns suitable for raw data, and are widely utilized in financial time series modeling [23–25]. However, LSTM methods are susceptible to noise. The comparison result of CAL and LSTM can evaluate whether the proposed model can effectively improve the results of LSTM in complex time series modeling.

2. Linear ARIMA model: ARIMA can better predict linear time series, but is not suitable for complex non-linear time series [4]. We combine ARMA and LSTM to extend the application range of the ARIMA time series model. In addition, the prediction effects of the ARIMA and CAL models are compared, which verifies the effectiveness of the proposed model compared with a single linear model.

3. GRU: GRU is a simplified version of the LSTM. It uses only one state vector and two gate vectors, i.e., reset gate and update gate. The comparison result can evaluate whether the CAL model is better than other deep learning model.

4. Bi-LSTM: To preserve the future and the past information, Bi-LSTM makes the neural network have the sequence information in both directions, i.e., backwards (future to past) and forward (past to future). The aim of the experiments is to show whether Bi-LSTM improves the prediction accuracy of LSTM. The experiments also verify the effectiveness of the proposed model compared with a single improved model.

5. EMD-ARMA-LSTM model: EMD can generate more predictable components when fed into the decomposing module. CEEMDAN is designed to solve the problem of EMD mode mixing. To compare the prediction effects of EMD-ARMA-LSTM, and CAL, we verify the influence of different decomposition methods on model prediction.

6. Hybrid ARIMA-ANN model [6]: ARIMA and ANN are adopted to model the linear and non-linear data [6], and empirical results demonstrate that ensemble models can effectively improve performance. We use the ARIMA-ANN model for comparison. The results can demonstrate the advantages of CAL over ARIMA-ANN when combining linear and non-linear models. The advantages of LSTM over an ANN in abstract feature extraction and prediction ability could also be verified.

7. CEEMDAN-LSTM model [7]: The CEEMDAN-LSTM model integrates the advantages of CEEMDAN and LSTM but does not consider that the original time series may contain linearly correlated components, and the non-linear prediction of all decomposed sequences will affect the prediction performance of the model. The empirical results demonstrate the validity of the CAL model in comparison to the CEEMDAN-LSTM model.

Table 3. Contrastive experiments.

Model	Comparison Purpose of Model Settings
LSTM	Comparison to single deep learning model
ARIMA	Comparison to single linear model
GRU	Comparison to other single non-linear model
Bi-LSTM	Comparison to improved deep learning model
EMD-ARMA-LSTM	Evaluation of CEEMDAN and EMD
ARIAM-ANN	Comparison of CAL to hybrid models [6]
CEEMDAN-LSTM	Comparison of CAL to stock forecasting model [7]

4.6. Experiments and Discussions

We verify the effectiveness and superiority of the proposed model from three aspects:

1. Statistics of MAE, RMSE, MAPE, and R^2 are chosen to assess the consistency between predicted and observed terms. These indicators measure the deviation between forecast and reality from different aspects.

2. The deviation between real and predicted values can be observed from Figure 5, and the variation of the error can be utilized to observe the stability of the CAL model from Figure 6.

3. A linear regression model is then used to further observe the performance of the CAL model; then, a series of technical diagnostics are leveraged to check the regression models.

4.6.1. Observation of the Statistical Data

It can be observed from Table 4 that the CAL model has obvious advantages in stock index DAX series prediction, which decreases by 56.71% when compared to LSTM, and by 46.83% when compared to ARIMA in MAE. This indicates that a single model cannot effectively capture data patterns and make excellent predictions. Although GRU and Bi-LSTM improve the prediction accuracy of LSTM, their prediction accuracies are still lower than CAL.

Table 4. Prediction results of different models in DAX.

Model	MAE	RMSE	MAPE (%)	R^2
LSTM	167.0816	224.5003	1.4006	0.9570
ARIMA	136.0422	206.5253	1.1633	0.9650
GRU	153.5215	216.7465	1.2982	0.9608
Bi-LSTM LSTM	138.0041	209.2315	1.1768	0.9641
ARIMA-ANN	140.4099	211.9800	1.1966	0.9630
CEEMDAN-LSTM	97.2277	128.2331	0.8106	0.9866
EMD-ARMA-LSTM	127.1255	191.0622	1.0771	0.9687
CAL	72.3340	101.8321	0.6099	0.9915

Methods with EMD achieve remarkably less error in their forecasts than CEEMDAN-LSTM and CAL, which shows that experimental results vary with data decomposition, and CEEMDAN-based methods can achieve better predictive performance. The ARIMA-ANN model is inferior to EMD- and CEEMDAN-based methods, perhaps because it has limited decomposition ability to extract hidden features. CEEMDAN properly decomposes time series, reduces their complexity, and improves LSTM information extraction, so the hybrid CEEMDAN-LSTM model can achieve a better prediction effect than just LSTM. However, CEEMDAN-LSTM is not as good as CAL because it does not consider linear factors that may exist in the original sequence in time series prediction.

Table 5 lists the prediction performance of different models on the HSI stock index, where we find a large error between the real and predicted values. This is mainly because the data of the HSI stock index are more volatile and difficult to predict. The CAL model achieves the best prediction accuracy, followed by CEEMDAN-LSTM, EMD-ARMA-LSTM, and ARIMA-ANN. ARIMA-ANN achieve higher prediction accuracy than the individual ARIMA and LSTM models, and ARIMA obtains better results than LSTM. As deep learning is easily affected by noise, it is difficult to learn effective data patterns in complex dynamic time series. Deep learning methods, such as LSTM, GRU, and Bi-LSTM, have the largest prediction error on the HSI stock index. Although ARIAM has a higher prediction accuracy than them, the gap between predicted and actual values of ARIAM is still large. This indicates the predictive performance of a single model is very limited. The hybrid model performs better than the single ARIMA and LSTM models. The experimental results show that ARIAM-ANN gives poorer results than CEEMDAN-LSTM, EMD-ARMA-LSTM, and CAL, perhaps due to an insufficient scale of decomposition. CEEMDAN-LSTM and EMD-ARMA-LSTM effectively improve prediction accuracy, but the effect is still inferior to the proposed CAL model, which has advantages and good potential in high-volatility time series data.

Table 5. Prediction results of different models in HSI.

Model	MAE	RMSE	MAPE (%)	R^2
LSTM	257.7703	347.1944	1.0197	0.9454
ARIMA	250.9188	345.3399	0.995	0.9470
GRU	256.1635	345.9382	1.0134	0.9451
Bi-LSTM	258.2292	353.4523	1.0249	0.9450
ARIMA-ANN	249.1046	344.5775	0.9882	0.9469
CEEMDAN-LSTM	127.0750	168.3214	0.5023	0.9879
EMD-ARMA-LSTM	181.7516	235.1773	0.7187	0.9751
CAL	120.8184	159.8226	0.4789	0.9885

Figure 2c shows that the movement trend of S&P500 is relatively stable, with little fluctuation in the research interval, and an overall upward trend. Hence, the predicted results are closer to the observed values of stock indices. Table 6 shows the experimental results of S&P500. The data show that the CAL model yields the smallest prediction error, with MAE 48.84% less than LSTM and 49.75% less than ARIMA. This shows that the single model has better prediction performance in some stable time series sets, but there is still room for improvement. However, GRU and Bi-LSTM cannot effectively improve the prediction accuracy. The prediction effect of EMD-ARMA-LSTM is still inferior to that of CAL, which further demonstrates the superiority of CEEMDAN over EMD data decomposition. CEEMDAN-LSTM achieves better prediction performance than the single LSTM model, and ARIMA-ANN yields higher prediction accuracy than ARIAM, showing that sequence decomposition and model combination can improve the prediction accuracy of financial series.

Table 6. Prediction results of different models in S&P500.

Model	MAE	RMSE	MAPE (%)	R^2
LSTM	33.4958	53.4345	1.1207	0.9595
ARIMA	34.1031	54.8336	1.1411	0.9598
GRU	43.3137	63.2251	1.4416	0.9469
Bi-LSTM	33.5198	53.4177	1.1262	0.9610
ARIMA-ANN	33.7170	53.6489	1.125	0.9608
CEEMDAN-LSTM	21.1496	30.1187	0.6964	0.9878
EMD-ARMA-LSTM	22.1886	33.4485	0.7334	0.9843
CAL	17.1362	26.1373	0.5645	0.9910

Table 7 shows the prediction performance results for SSE datasets. From Table 7, we can see that CAL has better predictive accuracy than the other seven models, with MAE up to 14.0294, followed by CEEMDAN-LSTM and EMD-ARMA-LSTM. ARIMA can achieve higher prediction accuracy than ARIMA-ANN and EMD-ARMA-LSTM. GRU and Bi-LSTM achieve higher prediction accuracy than LSTM.

Table 7. Prediction results of different models in SSE.

Model	MAE	RMSE	MAPE (%)	R^2
LSTM	38.3486	47.9563	1.2468	0.9475
ARIMA	25.1019	36.9815	0.819	0.9690
GRU	31.8217	43.1568	1.0355	0.9599
Bi-LSTM	31.8026	42.7439	1.0382	0.9596
ARIMA-ANN	25.6976	37.4014	0.8383	0.9686
CEEMDAN-LSTM	14.3562	19.6741	0.4681	0.9913
EMD-ARMA-LSTM	19.5074	28.5532	0.6382	0.9814
CAL	14.0294	19.9246	0.459	0.9911

Several important results are obtained on the SSE dataset. GRU and Bi-LSTM outperforms LSTM, but their prediction results are lower than ARIMA, which shows that a linear model can sometimes achieve a better prediction effect than a deep learning model. The prediction accuracy of EMD-ARMA-LSTM is relatively low, perhaps because the mode mixing of EMD leads to the inclusion of other scales of data in an IMF, and these abnormal data interfere with information extraction.

4.6.2. Prediction Results and Errors

As demonstrated in Figure 5, we zoom in a part of the prediction interval to observe the consistency between the real and predicted values of different models. It can be seen that the CAL model yields the closest prediction results, and CEEMDAN-LSTM is closer to the observed values in comparison with EMD-ARMA-LSTM and ARIAM. LSTM and GRU have larger volatility and prediction error than the other models. The stem diagram oscillates up and down around the zero axis in Figure 6 and is locally symmetrical concerning the zero axis, indicating that the prediction results of the CAL model are relatively stable within the prediction interval.

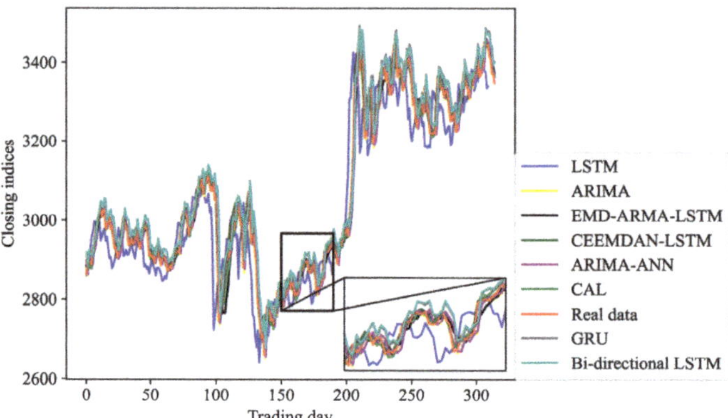

Figure 5. SSE comparison of sequence prediction results.

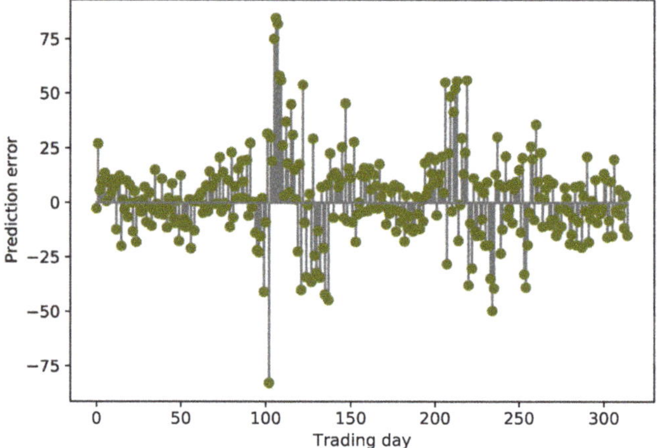

Figure 6. SSE error changes between real and predicted values.

4.6.3. Regression Analysis

We conduct a linear regression to assess the correlation between the real data and the predicted values. The predicted value is denoted as x, and the real value is y, respectively. The regression equation is $y = ax + b$. The metrics, including standard error (SE), p-value (p) and t-value (t), are used to test the results of regression analysis. The definitions of SE and t are as follows, and p is derived from the t distribution.

$$SE = \frac{\sigma}{\sqrt{n}}, \tag{10}$$

$$t = \frac{\bar{x} - \mu}{\frac{\sigma}{\sqrt{n}}}. \tag{11}$$

Here, σ is the standard deviation of the predicted values, n is the number of the predicted (or real) values, $\bar{x}$ is the mean of the predicted values, and μ is the mean of real values. Table 8 lists the regression parameters and diagnostics results. It is observed that the slope a of each stock index is close to 1, the SE for a is relatively small, which means that the predicted values are very close to the real values. Furthermore, for each linear regression model, *p* for *a* is below the standard cutoff of 0.05, and *t* for *a* is high, suggesting it is a good model. In addition, Figure 7 shows the linear regression results of each stock index. The scattered points are evenly distributed near the fitting line, which indicates that the predicted and real values are highly correlated.

Table 8. The regression parameters and diagnostics results.

Model	Parameter	Estimation	SE	*t*	*p*
DAX	*a*	0.9909	0.005	196.519	0.000
	b	104.2845	62.836	1.660	0.098
HSI	*a*	1.0012	0.006	167.616	0.000
	b	−12.2083	153.286	−0.080	0.937
S&P500	*a*	0.9844	0.005	192.819	0.000
	b	44.7296	16.195	2.762	0.006
SSE	*a*	0.9913	0.005	187.342	0.000
	b	28.5226	16.263	1.754	0.080

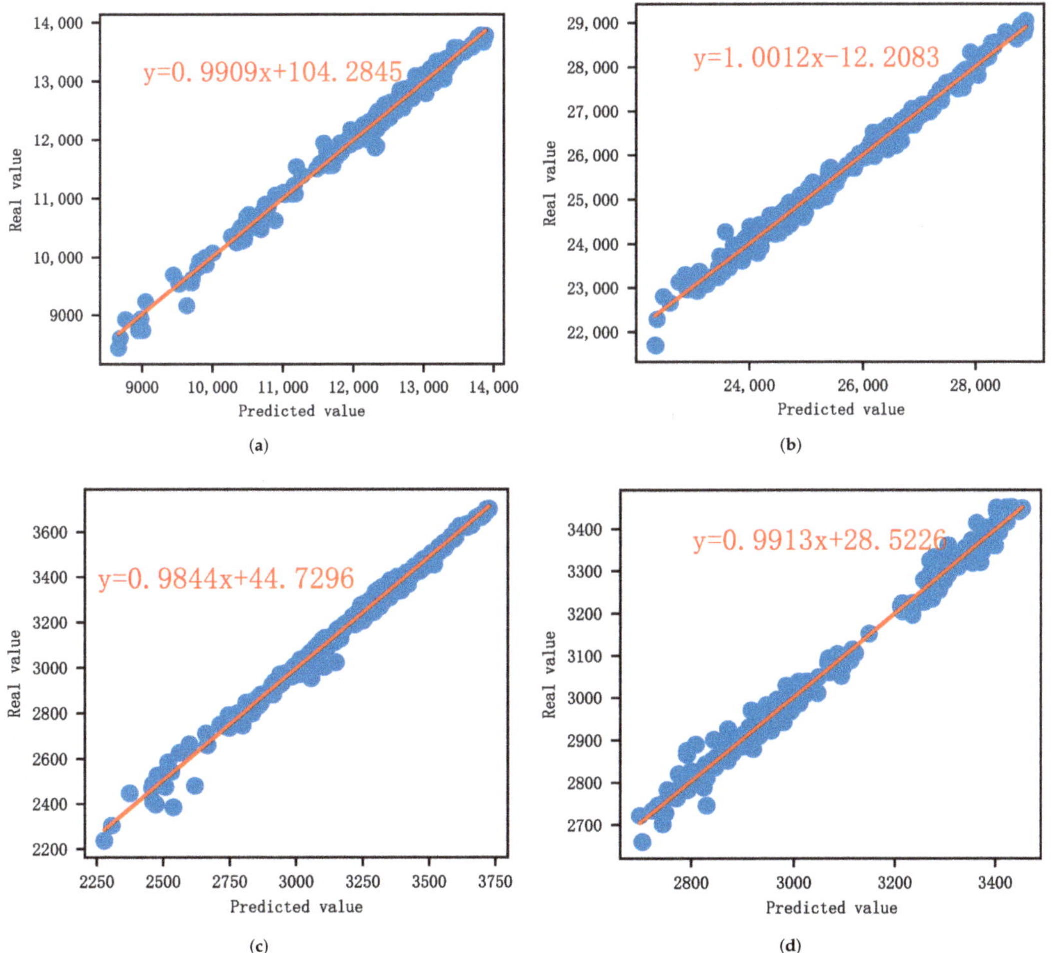

Figure 7. Linear regression analysis. (**a**) DAX. (**b**) HSI. (**c**) S&P500. (**d**) SSE.

4.6.4. Summary

Based on the above experiment results, the observations are summarized as follows.

1. Our proposed CAL model, with CEEMDAN-based methods, outperforms seven benchmark models in predictive accuracy on four stock indices from different developed stock markets, which indicates that methods with multi-scale decomposition can reduce the complexity of sequences, extract hidden features, and improve prediction accuracy.
2. CAL can obtain predictions closer to real values than CEEMDAN-LSTM, which indicates that components after decomposition may have both linear and non-linear characteristics. Therefore, models combining ARMA and LSTM can obtain more accurate predictions than individual LSTM models.
3. CAL can yield the closest prediction results in comparison to ARIMA-ANN. This indicates that the CAL model has advantages over some traditional hybrid models.
4. The prediction results show that CAL has a smaller prediction error than EMD-ARMA-LSTM does, and this indicates that the CEEMDAN method is superior to EMD in data decomposition.
5. In some volatile financial markets, a single prediction model, even improved deep learning model, has limited prediction ability because they cannot excavate internal movement rules of time series and reflect the multi-scale characteristics of financial time series.
6. The linear regression analysis shows the strong correlation between the predicted values and the real values, and the proposed prediction model is effective.

5. Conclusions and Discussion

Stock market index prediction plays an important role in reflecting overall stock market trends and has strong practical investment value. We proposed a hybrid stock index prediction model based on CEEMDAN and ARMA-LSTM. It takes the strengths of CEEMDAN in data decomposition, combines linear and non-linear models, and can well model complex time series. To verify the effectiveness of the prediction model, CAL was used to forecast the closing index of four stock markets, and seven control experiments were conducted for comparison. The results show that CAL can achieve the highest prediction accuracy. To optimize the model, future research can be conducted from the following aspects.

1. Single data source analysis has certain limitations. Combined analysis with different data sources, such as text information [26], can improve prediction to a certain extent.
2. Stock market data contain noise that affects forecast results. Methods, such as wavelet denoising [27] and principal component analysis [28], can eliminate the influence of irrelevant factors and improve the prediction effect to a certain extent.
3. Time series analysis has been applied in fields, such as natural science [29] and industrial time series prediction [30]. The application scope of the temporal sequence model in this paper can be extended, especially in some complicated temporal sequence scenes.

Author Contributions: Conceptualization, P.L.; Data curation, Q.W.; Formal analysis, Q.W.; Funding acquisition, J.X.; Investigation, Q.W. and Y.S.; Methodology, Q.W. and J.X.; Project administration, J.X.; Resources, Y.S.; Software, Q.W.; Supervision, P.L. and J.X.; Validation, P.L.; Visualization, Q.W. and Y.S.; Writing—original draft, P.L., Q.W. and J.X. All authors have read and agreed to the published version of the manuscript.

Funding: This research was funded by National Natural Science Foundation of China, grant number 62067001, Guangxi Natural Science Foundation, grant number 2019JJA170045, Special Funds for Guangxi BaGui Scholars.

Institutional Review Board Statement: Not applicable.

Informed Consent Statement: Not applicable.

Data Availability Statement: Publicly available datasets were analyzed in this study. DAX data can be found here: https://cn.investing.com/indices/germany-30-historical-data. HSI data can be found here: https://cn.investing.com/indices/hang-sen-40-historical-data. S&P500 data can be found here: https://cn.investing.com/indices/us-spx-500-historical-data. SSE data can be found here: https://cn.investing.com/indices/shanghai-composite-historical-data.

Conflicts of Interest: The authors declare no conflict of interest.

References

1. Yan, B.; Aasma, M. A novel deep learning framework: Prediction and analysis of financial time series using CEEMD and LSTM. *Expert Syst. Appl.* **2020**, *159*, 113609.
2. Zhou, F.; Zhou, H.M.; Yang, Z.; Yang, L. EMD2FNN: A strategy combining empirical mode decomposition and factorization machine based neural network for stock market trend prediction. *Expert Syst. Appl.* **2019**, *115*, 136–151. [CrossRef]
3. Torres, M.E.; Colominas, M.A.; Schlotthauer, G.; Flandrin, P. A complete ensemble empirical mode decomposition with adaptive noise. In Proceedings of the 2011 IEEE International Conference on Acoustics, Speech and Signal Processing (ICASSP), Prague, Czech Republic, 22–27 May 2011; pp. 4144–4147.
4. Büyükşahin, Ü.Ç.; Ertekin, Ş. Improving forecasting accuracy of time series data using a new ARIMA-ANN hybrid method and empirical mode decomposition. *Neurocomputing* **2019**, *361*, 151–163. [CrossRef]
5. Wang, Z.; Lou, Y. Hydrological time series forecast model based on wavelet de-noising and ARIMA-LSTM. In Proceedings of the 2019 IEEE 3rd Information Technology, Networking, Electronic and Automation Control Conference (ITNEC), Chengdu, China, 15–17 March 2019; pp. 1697–1701.
6. Babu, C.N.; Reddy, B.E. A moving-average filter based hybrid ARIMA–ANN model for forecasting time series data. *Appl. Soft Comput.* **2014**, *23*, 27–38. [CrossRef]
7. Cao, J.; Li, Z.; Li, J. Financial time series forecasting model based on CEEMDAN and LSTM. *Phys. Stat. Mech. Its Appl.* **2019**, *519*, 127–139. [CrossRef]
8. Khashei, M.; Bijari, M. A novel hybridization of artificial neural networks and ARIMA models for time series forecasting. *Appl. Soft Comput.* **2011**, *11*, 2664–2675. [CrossRef]
9. Goodfellow, I.; Bengio, Y.; Courville, A. *Deep Learning*; MIT Press: Cambridge, MA, USA, 2016.
10. Gers, F.A.; Eck, D.; Schmidhuber, J. Applying LSTM to time series predictable through time-window approaches. In *Neural Nets WIRN Vietri-01*; Springer: Berlin/Heidelberg, Germany, 2002; pp. 193–200.
11. Bao, W.; Yue, J.; Rao, Y. A deep learning framework for financial time series using stacked autoencoders and long-short term memory. *PLoS ONE* **2017**, *12*, e0180944. [CrossRef]
12. Chung, H.; Shin, K.S. Genetic algorithm-optimized long short-term memory network for stock market prediction. *Sustainability* **2018**, *10*, 3765. [CrossRef]
13. Foster, W.; Collopy, F.; Ungar, L. Neural network forecasting of short, noisy time series. *Comput. Chem. Eng.* **1992**, *16*, 293–297. [CrossRef]
14. Brace, M.C.; Schmidt, J.; Hadlin, M. Comparison of the forecasting accuracy of neural networks with other established techniques. In Proceedings of the First International Forum on Applications of Neural Networks to Power Systems, Seattle, WA, USA, 23–26 July 1991; pp. 31–35.
15. Pai, P.F.; Lin, C.S. A hybrid ARIMA and support vector machines model in stock price forecasting. *Omega* **2005**, *33*, 497–505. [CrossRef]
16. Kim, H.Y.; Won, C.H. Forecasting the volatility of stock price index: A hybrid model integrating LSTM with multiple GARCH-type models. *Expert Syst. Appl.* **2018**, *103*, 25–37. [CrossRef]
17. Zhang, G.P. Time series forecasting using a hybrid ARIMA and neural network model. *Neurocomputing* **2003**, *50*, 159–175. [CrossRef]
18. Kumar, M.; Thenmozhi, M. Forecasting stock index returns using ARIMA-SVM, ARIMA-ANN, and ARIMA-random forest hybrid models. *Int. J. Bank. Account. Financ.* **2014**, *5*, 284–308. [CrossRef]
19. Huang, N.E.; Shen, Z.; Long, S.R.; Wu, M.C.; Shih, H.H.; Zheng, Q.; Yen, N.C.; Tung, C.C.; Liu, H.H. The empirical mode decomposition and the Hilbert spectrum for nonlinear and non-stationary time series analysis. *Proc. R. Soc. London. Ser. Math. Phys. Eng. Sci.* **1998**, *454*, 903–995. [CrossRef]
20. Song, H.; Dai, J.; Luo, L.; Sheng, G.; Jiang, X. Power transformer operating state prediction method based on an LSTM network. *Energies* **2018**, *11*, 914. [CrossRef]
21. Ren, B. The use of machine translation algorithm based on residual and LSTM neural network in translation teaching. *PLoS ONE* **2020**, *15*, e0240663. [CrossRef]
22. Liu, M.D.; Ding, L.; Bai, Y.L. Application of hybrid model based on empirical mode decomposition, novel recurrent neural networks and the ARIMA to wind speed prediction. *Energy Convers. Manag.* **2021**, *233*, 113917. [CrossRef]
23. Fischer, T.; Krauss, C. Deep learning with long short-term memory networks for financial market predictions. *Eur. J. Oper. Res.* **2018**, *270*, 654–669. [CrossRef]

24. Petersen, N.C.; Rodrigues, F.; Pereira, F.C. Multi-output bus travel time prediction with convolutional LSTM neural network. *Expert Syst. Appl.* **2019**, *120*, 426–435. [CrossRef]
25. Kim, T.Y.; Cho, S.B. Web traffic anomaly detection using C-LSTM neural networks. *Expert Syst. Appl.* **2018**, *106*, 66–76. [CrossRef]
26. Hao, P.Y.; Kung, C.F.; Chang, C.Y.; Ou, J.B. Predicting stock price trends based on financial news articles and using a novel twin support vector machine with fuzzy hyperplane. *Appl. Soft Comput.* **2021**, *98*, 106806. [CrossRef]
27. Wu, D.; Wang, X.; Wu, S. A Hybrid Method Based on Extreme Learning Machine and Wavelet Transform Denoising for Stock Prediction. *Entropy* **2021**, *23*, 440. [CrossRef] [PubMed]
28. Yang, K.; Liu, Y.L.; Yao, Y.N.; Fan, S.D.; Mosleh, A. Operational time-series data modeling via LSTM network integrating principal component analysis based on human experience. *J. Manuf. Syst.* **2020**, *61*, 746–756. [CrossRef]
29. Coyle, D.; Prasad, G.; McGinnity, T.M. Extracting features for a brain-computer interface by self-organising fuzzy neural network-based time series prediction. In Proceedings of the 26th Annual International Conference of the IEEE Engineering in Medicine and Biology Society, San Francisco, CA, USA, 1–5 September 2004; Volume 2, pp. 4371–4374.
30. Wang, J.; Zhang, W.; Li, Y.; Wang, J.; Dang, Z. Forecasting wind speed using empirical mode decomposition and Elman neural network. *Appl. Soft Comput.* **2014**, *23*, 452–459. [CrossRef]

Article

Preference-Driven Classification Measure

Jan Kozak [1,*], **Barbara Probierz** [1], **Krzysztof Kania** [2] and **Przemysław Juszczuk** [1]

[1] Department of Machine Learning, University of Economics in Katowice, 1 Maja 50, 40-287 Katowice, Poland; barbara.probierz@ue.katowice.pl (B.P.); przemyslaw.juszczuk@ue.katowice.pl (P.J.)

[2] Department of Knowledge Engineering, University of Economics in Katowice, 1 Maja 50, 40-287 Katowice, Poland; krzysztof.kania@ue.katowice.pl

[*] Correspondence: jan.kozak@ue.katowice.pl

Abstract: Classification is one of the main problems of machine learning, and assessing the quality of classification is one of the most topical tasks, all the more difficult as it depends on many factors. Many different measures have been proposed to assess the quality of the classification, often depending on the application of a specific classifier. However, in most cases, these measures are focused on binary classification, and for the problem of many decision classes, they are significantly simplified. Due to the increasing scope of classification applications, there is a growing need to select a classifier appropriate to the situation, including more complex data sets with multiple decision classes. This paper aims to propose a new measure of classifier quality assessment (called the preference-driven measure, abbreviated p-d), regardless of the number of classes, with the possibility of establishing the relative importance of each class. Furthermore, we propose a solution in which the classifier's assessment can be adapted to the analyzed problem using a vector of preferences. To visualize the operation of the proposed measure, we present it first on an example involving two decision classes and then test its operation on real, multi-class data sets. Additionally, in this case, we demonstrate how to adjust the assessment to the user's preferences. The results obtained allow us to confirm that the use of a preference-driven measure indicates that other classifiers are better to use according to preferences, particularly as opposed to the classical measures of classification quality assessment.

Keywords: classification measure; quality of classification; quality measure; preference-driven classification; machine learning

Citation: Kozak, J.; Probierz, B.; Kania, K.; Juszczuk, P. Preference-Driven Classification Measure. *Entropy* **2022**, 24, 531. https://doi.org/10.3390/e24040531

Academic Editor: Sotiris Kotsiantis

Received: 7 February 2022
Accepted: 8 April 2022
Published: 10 April 2022

Publisher's Note: MDPI stays neutral with regard to jurisdictional claims in published maps and institutional affiliations.

Copyright: © 2022 by the authors. Licensee MDPI, Basel, Switzerland. This article is an open access article distributed under the terms and conditions of the Creative Commons Attribution (CC BY) license (https://creativecommons.org/licenses/by/4.0/).

1. Introduction

Classification continues to be one of the most important subjects in machine learning. Despite this, we still lack a general measure of quality independent of the specific characteristics of the data set. Moreover, in the situations where there is a necessity to involve the human decision maker in the classification process, we are forced to switch between different measures. Among the general ones, there are accuracy, precision or recall, and others that are data-dependent. Choosing the right (optimal) one is especially important because choosing a particular classification method depends heavily on the calculated quality measures.

Moreover, there is no single best classification measure that effectively identifies the method suitable for every task. Classification algorithms/methods have many characteristics. Consequently, there are many measures of classification because there is no single measure covering all the characteristics simultaneously [1]. Thus, finding an appropriate classification measure for a specific task is difficult and requires answering the question of what conditions, in specific circumstances, must be met by the measure.

One of the ways to bypass the problem of unambiguous assessment of the classifier's quality and selecting one best-suited classifier is ensemble and hybrid methods, which simultaneously use many different algorithms to perform a specific task. Within this approach, we can point out the homogeneous and heterogeneous solutions [2]. In the first group, we can find methods allowing us to create large groups of classifiers belonging to the same category (or even classifiers generated with the same method, but with different starting parameters), which allow the classification process simultaneously. The heterogeneous approach involves using a large variety of classifiers, for which the main advantage is the diversity of obtained results. Thus, the basic idea of these methods is the idea of collective decision making [3].

One should note that the above approach based on the ensemble methods allows for more robust classifier selection. However, it still leaves the decision maker with the problem of estimating the classification quality. In medicine, military or finance, the well-known accuracy measure seems to derive unsatisfactory results and present limited usability [4,5]. On the other hand, measures such as recall or precision are directed towards the binary classification problem. Most of the proposed measures are directly connected to the confusion matrix and related absolutely to the numerical outcome of the classification [6]. In most cases, it is strongly needed, but it makes the whole process independent of the user's preferences. In the decision-making context, taking them into account may be vital to make the process effective and at the same time maintain the sovereignty of the decision maker.

Incorporating users in the process of preparing a machine learning solution is an essential element of the entire procedure, the subject of many studies, and can take various forms [7,8]. One of the goals of the actions taken is to help the user to choose a suitable classifier. Most often, this task comes down to comparing simple measures of classification quality, which is usually carried out by trial and error, and yet can be unreliable [9]. This task becomes even more complicated if individual users' preferences are to be taken into account. This applies especially to issues of a managerial nature, but more generally wherever a human being to some extent participates in the decision-making process. In practice, this applies to all issues except physical or technical ones, where only objective laws are in force [10]. The main intention of introducing the new measure is to make this stage of a research procedure more methodical. To the best of our knowledge, there are no clear guidelines for taking into account the parameters in the learning process and the selection of a classifier in conjunction with individual preferences. The next issue is the systematic classification into the following application areas, thus expanding the group using machine learning methods. Finally, some users need a tool to control the process of selecting a classifier for their own needs, which are more complex and related to many classes. For such users, a measure that allows them to simply, directly, and methodically include their preferences in selecting and training the classifier would be very beneficial.

To cope with the above drawbacks, and at the same time maintain the role of the decision maker in the process, we propose the idea of a new measure in which their preferences are vital to the importance of individual classes. We aim to propose a measure that balances a thorough analysis of the classifier's performance and the selection of the classifier that performs best under the conditions (preferences) specified by the user.

The main intention of introducing the new measure is to make this stage of the research procedure more methodical. Users need a tool with which they will be able to select a classifier for their own needs, which are more complex and related to many classes. A measure that allows them to directly and methodically include their preferences in selecting and training the classifier would be very beneficial for such users.

We undertook to propose such a measure because, to the best of our knowledge, there is no reasonable alternative where, for any number of decision classes, it is possible to aggregate the quality of classification depending on the weight for a particular decision class. First, our solution was discussed in detail on prepared examples for two decision classes. It was then tested on real-world data sets and re-examined for a more significant number of decision classes that occurred in these data sets.

This article is organized as follows. Section 1 introduces the subject of this article. Section 2 provides an overview of the work related to the classification, particularly the measures for assessing the quality of the classification. In Section 3, we describe the classification problem and the classification quality assessment measures based on the error matrix for binary and multi-class classification. In Section 4, we present a new measure for classification, in which it will be possible to control preferences. In Section 5, we present the analysis of our research on real data sets. Finally, in Sections 6 and 7, we discuss the results of the experiments and end with general remarks on this work and available directions for future research.

2. Related Works

Evaluating the classification performance is a difficult task, and the discussion on this topic arose from the beginning of work on automatic classification. The initial set of five measures (sensitivity, specificity, efficiency, positive and negative predictive value) was rapidly expanded [11]. The most often used measure of classification performance is accuracy. However, it is not the only measure of the quality of predictive models. Despite optimizing the classification error rate, high-accuracy models may fail to capture crucial information transfer in the classification task [12,13]. Despite the simplicity and intuitive interpretation, there are many reasons and situations in which accuracy should not be used [14]. Instead, the authors advocate for using Cohen's kappa as a better meter for measuring classifiers' own merits than accuracy. Moreover, [15] indicates that the most frequently used measures, which focus on correctly classified cases (precision, recall, or F-score), do not meet the needs of various decision-making situations, especially when more than one class is essential. The authors advocate for using three other measures—Youden's index, likelihood, and discriminant power—because they combine sensitivity and specificity and their complements.

While most studies concern binary classification, in [16], the authors focused on multi-class classification problems. The authors showed that the extension of measures to the classification of many classes is associated with averaging the results achieved for individual classes in most cases. Finally, in [17], the authors point out some shortcomings of the accuracy measure and list five conditions that the newly constructed discriminator metric should meet.

To solve the dilemma related to the choice of a measure for a given problem, a list of desired features of an ideal measure and analysis of the most known measures was proposed in [1]. More importantly, they proved that it is impossible to satisfy them simultaneously. They also proposed a new family of measures (Generalized Means) that meet all desirable properties except one, and a new measure called Symmetric Balanced Accuracy.

A comparative analysis and taxonomy of the quality of classification measures have been the subject of many studies. For instance [18], in their experimental comparison conducted on 30 data sets, proposed dividing the performance measures into three categories:

- Measures based on a threshold, such as accuracy, modified accuracy measure, F-measure, or Kappa statistic, which are used to minimize the number of errors in the model. They are based directly on a confusion matrix, and they are widely used in many classification tasks. One should note that the overall efficiency of these measures is strictly related to the quality of the data. However, some measures, such as accuracy, can be less effective in the case of unbalanced data sets.
- Measures based on a probabilistic approach to understanding error, i.e., measuring the deviation or entropy information between the actual and predicted probability, such as mean absolute error, mean square error (Brier score), LogLoss (cross-entropy). These measures are useful in measuring the probability of selecting the wrong class, which is essential in ensemble methods or for a committee of classifiers.
- Measures based on the model's ability to correctly rank cases include ROC, AUC, the Youden index, precision–recall curve, Kolmogorov–Smirnov, or lift chart. They are helpful when indicating the best n occurrences in a data set or when good class separation is needed. They are widely used in recommendation systems, design marketing campaigns, fraud detection, spam filtering, and more.

In a survey [19], the authors grouped the measures depending on the type of outcome on which a given measure is focused (correct or incorrect outcome). In turn, [20] presents an in-depth analysis of over twenty performance measures used in different classification tasks in the context of changes made to a confusion matrix and their relations with particular measures (measure invariance). In contrast, a comparative study of two or more classifiers based on statistical tests was presented in [21]. A comprehensive analysis of the methods and measures of classification assessment is also included in [22], where the relationships between all measures calculated based on the confusion matrix are shown. Finally, a different approach to the analysis of performance measures was presented in [23]. First, the authors grouped classification measures according to classification difficulty, which they defined in relation to a distance between the boundary lines and each correctly classified case. The authors later developed their idea and proposed an instance-based measure for calculating the performance of classification from the perspective of instances, called degree of credibility [24].

The set of measures is constantly growing. For instance, in [25] was proposed a measure that compares classifiers, which combines three measures from different groups: Matthews correlation coefficient as a measure, which is calculated from both true and false positives and negatives, and AUC (area under the curve), derived from ROC and accuracy. To overcome the shortcomings of the accuracy measure in evaluating multi-class classifiers and to improve the quality of classifiers, in [26], the authors proposed a metric based on the combined accuracy and dispersion values. They also showed experimentally that this two-dimensional metric is particularly suitable in complex, unbalanced data sets and with many classes.

The new measures are also proposed to supplement the already used measures that work better for specific tasks. For example, [27], as an alternative to measures used in medical diagnostics, which use only part of the values from the error matrix, define the measure AQM, which takes into account all values from the confusion matrix. Similarly, in [28], for image analysis, a new measure of classification performance called robust-and-balanced accuracy was introduced. It aims to connect balanced accuracy with measures of variations. In another proposition, to improve face recognition processes, a new classification measure, called the volume measure, based on the volume of the matrix, was proposed [29]. In turn, a measure dedicated to the analysis of imbalanced data sets based on the harmonic mean of recall and selectivity was proposed in [30].

Existing measures are also modified. For example, in [31], the F^* measure was proposed as a modification of the F-score, towards the more straightforward interpretation of this measure. The above short review shows that the issue of classification measures is constantly under the attention of researchers. Moreover, for specific applications, new measures are created that are better suited to the requirements of the domain or user preferences.

3. Classification and Quality Evaluation

Formally, the classification problem Q can be solved using empirical experience W, while the quality of the solution is estimated by the quality measure Y. The value of Y should be increasing, while the experience W rises as well [32].

In machine learning, classification refers to the prediction problem of determining the class to which samples from a data set will be assigned. A classifier algorithm (often shortened to "a classifier") must be provided with training data with labeled classes. Then, the classifier can predict classes for new test data based on the training data. This approach is called supervised machine learning, and classification is one example of such a method. The training set is selected as a subset of the whole data set. A typical approach is to divide the known examples into a train and test set, following some general principles about the ratio of the two. Eventually, the test set includes a far smaller number of samples than the training set (preferably, the test set and training set should be disjoint). Test set is used to evaluate classification quality. Every object (also called a sample or observation) from the training data is assigned some predefined label (decision class). The idea is to build such a classifier, which assigns the proper labels for the objects. In contrast, the evaluation is performed on the training set, where the difference between the assigned and the actual label is estimated.

Classification algorithms for a prediction problem are evaluated based on performance. Multiple measures for assessing classification quality can be used depending on the situation. One of the most commonly used measures is accuracy, which determines how many samples from the entire data set were correctly classified into the appropriate classes. Often, other measures are used in addition to accuracy that more accurately assess the classifier's performance. Such measures primarily include precision and recall. Unfortunately, there are many problems in which classical measures are insufficient, and it is necessary to look for new solutions.

Let us consider a set of all available samples (called the universe of objects) X, which will include n number of objects:

$$X = \{x_1, x_2, \ldots, x_i, \ldots, x_n\}, \tag{1}$$

A single observation x_i will be described by a finite set of attributes and the decision attribute:

$$a_1, a_2, \ldots, a_m, \tag{2}$$

where $a_j \in A_j, j = 1, \ldots, m$. In this context, A_j denotes the domain of the j-th attribute, while features $a_1, a_2, \ldots, a_m$ create a feature space $A_1 \times A_2 \times \ldots \times A_m$.

There are no general restrictions related to the values of attributes, which can be quantitative or categorical. However, one should note that preprocessing allows for discretizing selected quantitative attributes. The same procedure can be applied to the decision classes, where many labels can be merged into fewer during the discretization process (trade-off between the quality of classification and classification speed). Thus, the single sample can be described as follows:

$$x_i = (\vec{V}_i, c_k), v_i^j \in A_j, c_k \in \{1, \ldots, C\}, \tag{3}$$

where $\vec{V}_i = [v_i^1, \ldots, v_i^m]$ is a vector in an m-dimensional feature space, v_i^j is the value of attribute a_j for observation (sample) x_i, and c_i is the class label (also called the decision class) of this object. Thus, the universe X can be formally described as:

$$X : \{(\vec{V}_i, c_i)\}_{i=1}^n. \tag{4}$$

Hence, the classification problem can generally be understood as the assignment problem, where every element x_i from the universe X should have the decision class c_i assigned. Eventually, we end with the classifier capable of assigning the newly arrived objects from the test set into proper decision classes. However, even in the case of the binary classification problem, the above is not trivial. It can be challenging to achieve when, for example, we observe unbalanced data (i.e., the situation in which most of the objects from universe X are assigned to a single class). Therefore, many different estimation methods were proposed (both to binary and multiple-class classification problems) to cope with this. Next, this section discusses various classification measures capable of dealing with both mentioned classification problems.

3.1. Quality Assessment and Binary Confusion Matrix

A confusion matrix is among the most popular tools used in the process of the validation of the quality of performed classification. The confusion matrix for a binary decision class is defined as a table with different combinations of predicted and actual values related to the decision classes [33,34]. The rows contain existing classes, while the columns contain predicted classes. The diagonal of the confusion matrix includes the correctly classified samples for both classes, while the off-diagonal cells represent the errors (misclassified samples for both decision classes). The confusion matrix represents the errors that the classifier makes and shows the type of these errors. It represents a detailed breakdown of the answers considering the number of correct and incorrect classifications for both classes. It is imperative when the cost of misclassification is different for these classes and when the size of the classes is different [35].

For the binary classification (e.g., classification of emails and spam or sick and healthy patients), see Table 1, where the prediction of a positive class (labeled 1) and a negative class (labeled 2) can be uniquely determined. Such a situation indicates whether the classifier is more likely to incur an error by assigning a positive class as a negative class or vice versa. In Table 1, the confusion matrix for the binary classifier is shown, where:

- TP (true positive)–samples classified as predicted class 1, which are samples of actual class 1;
- FN (false negative)–samples classified as predicted class 2, which are samples of actual class 1;
- FP (false positive)–samples classified as predicted class 1, which are samples of actual class 2;
- TN (true negative)–samples classified as predicted class 2, which are samples of actual class 2.

Table 1. Confusion matrix for binary classification.

	Predicted Class 1	**Predicted Class 2**
Actual class 1	TP	FN
Actual class 2	FP	TN

The quality of classifiers is measured based on a confusion matrix (as shown in Table 1). Measures based on the confusion matrix include, but are not limited to, accuracy, precision, recall, and F-score (also called F1, which is the measure F_β for which $\beta = 1$). In addition, the Matthews correlation coefficient (MCC) measure and BalancedAccuracy are also often used for binary classification.

The accuracy measure indicates how often a classifier makes a correct prediction. It is the ratio of the number of accurate predictions to the total number of predictions and is calculated based on Formula (5).

$$accuracy\_binary = \frac{TP + TN}{TP + FP + FN + TN} \tag{5}$$

Precision determines how many samples, out of all those classified as positive, are samples of the positive class. Precision is expressed by Formula (6).

$$precision\_binary = \frac{TP}{TP + FP} \tag{6}$$

Recall is used to determine how many samples belonging to a positive class were classified as positive by the classifier. Recall is determined by Formula (7).

$$recall\_binary = \frac{TP}{TP + FN} \tag{7}$$

For binary classification, the measure F_β is defined as the harmonic mean of precision and recall, where additional precision or recall weights are used to obtain more accurate results. By setting the value of β, it is possible to control the effect of recall weight with respect to precision. F_β is specified by Formula (8), where β is the number of times the recall is as important as the precision [36]. The F_β value ranges from 0 to 1 (with 0 being the worst value and 1 being the optimal value). The most common value for β is 1, which simply means measure F1 (Equation (9)). Other frequently used values are 2 and 0.5. In the case of 2, recall weight is greater than precision; however, in the case of 0.5, recall weight is less than precision. The F_β measure is based on the Van Rijsbergen measure of effectiveness [37].

$$F_\beta = (1 + \beta^2) \times \frac{precision\_binary \times recall\_binary}{\beta^2 \times precision\_binary + recall\_binary} \tag{8}$$

The $F1$ measure is the instance of the measure F_β (Equation (8)), for which $\beta = 1$. The $F1$ measure is defined as the harmonic mean of precision and recall [38]. Therefore, $F1$ will only take on a high value if both of its components reach a high value. The $F1$ measure often replaces precision when class counts are unbalanced [39]. For example, if 97% of the data belong to class 1 and only 3% belong to class 2, then classifying all observations as class 1 would yield a misleading accuracy of 97%. The $F1$ measure is based on precision and recall, and is thus robust to such distortions. The measure is calculated from Formula (9).

$$F1\_binary = 2 \times \frac{precision\_binary \times recall\_binary}{precision\_binary + recall\_binary} = \frac{TP}{TP + \frac{1}{2}(FN + FP)} \tag{9}$$

The Matthews correlation coefficient (MCC) measure is often used for unbalanced data [40]. For the precision, recall, and $F1$ measures, the TN value is not used, which is very important if we are interested in both classes. Therefore, the Matthews correlation coefficient, calculated based on all terms from the confusion matrix, can be used. The MCC measure is calculated from Formula (10).

$$MCC\_binary = \frac{TP \times TN - FP \times FN}{\sqrt{(TP + FP)(TP + FN)(TN + FP)(TN + FN)}} \tag{10}$$

The MCC value ranges within $[-1 \ldots 1]$ (with -1 equal to the misclassification of all samples, while a value of 1 means that all samples are correctly classified). Therefore, the higher the correlation between actual and predicted values, the better the prediction. Furthermore, for the MCC measure, the two classes have the same importance weight, so when positive classes are swapped with negative classes, the MCC value will be the same, which means that the MCC measure is symmetric [40].

Another measure in binary classification is the BalancedAccuracy measure, which calculates balanced accuracy for anomaly or disease detection, where significant differences in the class size are observed [41]. Overestimated accuracy results can be avoided using the BalancedAccuracy measure for unbalanced classes. For binary classification, the balanced accuracy equals the arithmetic mean of the recall and specificity. The BalancedAccuracy measure is calculated by Formulas (11) or (12).

$$BalancedAccuracy\_binary = \frac{recall\_binary + specificity\_binary}{2} \tag{11}$$

$$BalancedAccuracy\_binary = \frac{1}{2}\left(\frac{TP}{TP + FN} + \frac{TN}{FP + TN}\right) \tag{12}$$

Specificity is used to determine the number of samples belonging to a negative class that have been classified as negative by the classifier [42]. Specificity is measured by Formula (13).

$$specificity\_binary = \frac{TN}{FP + TN} \tag{13}$$

3.2. Quality Assessment and Confusion Matrix for Multi-Class Classification

In addition to binary classification, it is common to classify samples into more than two classes. We are dealing with multi-class classification in such a situation—not to be confused with multi-label classification. The difference between multi-class and multi-label classification is that a sample can only be assigned to one class selected from multiple classes for multi-class classification. In contrast, a sample can be assigned to multiple classes for multi-label classification [43].

There is a need to construct an extended confusion matrix for a number of decision classes greater than 2. Thus, the number of rows and the number of columns in such an approach will equal the number of classes. For this purpose, we present the idea of a confusion matrix for C classes, which is shown in Table 2. Similarly, for the binary classification problem (see Table 1), we have listed four terms:

- TP (true positive) means that samples from the actual class have been classified into the same predicted class—denoted as $TP_1, TP_2, \ldots, TP_i, \ldots, TP_{C-1}, TP_C$;
- FN (false negative) indicates that samples in the actual class have been classified into other predicted classes. This is the sum of the values of the corresponding row of the real class except for the TP values for that real class—denoted as $FN_1, FN_2, \ldots, FN_i, \ldots, FN_{C-1}, FN_C$;
- FP (false positive) indicates that samples from other real classes have been classified into the selected predicted class. This is the sum of the values of the corresponding column of the predicted class except for the TP values for the actual class—denoted as $FP_1, FP_2, \ldots, FP_i, \ldots, FP_{C-1}, FP_C$;
- TN (true negative) indicates that, for the selected real class, samples from other actual classes were classified into predicted classes other than the predicted class corresponding to the chosen real class. For a given real class, it is the sum of the values of all columns and rows except for the row and column values of that real class for which we compute the values—denoted as $TN\backslash_{\{i\}}$ or $TN\backslash_{\{i,C\}}$.

To evaluate the quality of multi-class classification, as in binary classification, measures based on the confusion matrix shown in Table 2 were used. Therefore, referring to the measures described in Section 3.1, we wish to present them for the multi-class classification problem.

Accuracy is one of the most commonly used measures in a multi-class classification problem [16] and is calculated according to Formula (14) or (15); when we define the sum of all samples as s, this equation boils down to the form (16). To calculate accuracy, sum all correctly classified samples and then divide by the number of all classified samples. Correctly classified samples are shown in the confusion matrix (see Table 2) on the main diagonal (from upper left corner to lower right corner).

$$accuracy = \frac{\sum_{i=1}^{c} TP_i}{\sum_{i=1}^{c} TP_i + \sum_{i=1}^{c} FP_i} \tag{14}$$

$$accuracy = \frac{\sum_{i=1}^{c} TP_i}{\sum_{i=1}^{c} TP_i + \sum_{i=1}^{c} FN_i} \tag{15}$$

$$accuracy = \frac{\sum_{i=1}^{c} TP_i}{s} \tag{16}$$

Precision and recall are used for the multi-class approach as well. Two modifications can be distinguished in this case: *macro−* and *micro−* precision or recall [16]. For the *macro−* modification, to calculate the value of the measures for multiple classes, one must count precision and recall for each class separately and then calculate the arithmetic mean of these values. In this way, all classes during multi-class classification have the same validity, regardless of the class count. In multi-class classification, *macro_precision* is calculated by Formula (17) and *macro_recall* is calculated by Formula (18).

$$macro\_precision = \frac{1}{c} \sum_{i=1}^{c} \frac{TP_i}{TP_i + FP_i} \tag{17}$$

$$macro\_recall = \frac{1}{c} \sum_{i=1}^{c} \frac{TP_i}{TP_i + FN_i} \tag{18}$$

Table 2. Confusion matrix for multiple classes.

	Predicted Class 1	Predicted Class 2	Predicted Class 3	...	Predicted Class i	...	Predicted Class C-1	Predicted Class C
		FP_2	FP_3		FP_i		FP_{C-1}	FP_C
Actual class 1	TP_1 $TN\backslash_{\{1\}}$	$TN\backslash_{\{1,2\}}$	$TN\backslash_{\{1,3\}}$	...	$TN\backslash_{\{1,i\}}$	...	$TN\backslash_{\{1,C-1\}}$	$TN\backslash_{\{1,C\}}$
		FN_1	FN_1		FN_1		FN_1	FN_1
	FP_1		FP_3		FP_i		FP_{C-1}	FP_C
Actual class 2	$TN\backslash_{\{1,2\}}$	TP_2 $TN\backslash_{\{2\}}$	$TN\backslash_{\{2,3\}}$	...	$TN\backslash_{\{2,i\}}$	...	$TN\backslash_{\{2,C-1\}}$	$TN\backslash_{\{2,C\}}$
	FN_2		FN_2		FN_2		FN_2	FN_2
	FP_1	FP_2			FP_i		FP_{C-1}	FP_C
Actual class 3	$TN\backslash_{\{1,3\}}$	$TN\backslash_{\{2,3\}}$	TP_3 $TN\backslash_{\{3\}}$	...	$TN\backslash_{\{3,i\}}$	...	$TN\backslash_{\{3,C-1\}}$	$TN\backslash_{\{3,C\}}$
	FN_3	FN_3			FN_3		FN_3	FN_3
...	...	...	...	...	...	...	...	
	FP_1	FP_2	FP_3				FP_{C-1}	FP_C
Actual class i	$TN\backslash_{\{1,i\}}$	$TN\backslash_{\{2,i\}}$	$TN\backslash_{\{3,i\}}$	...	TP_i $TN\backslash_{\{i\}}$	...	$TN\backslash_{\{i,C-1\}}$	$TN\backslash_{\{i,C\}}$
	FN_i	FN_i	FN_i				FN_i	FN_i
...	...	...	...	...	...	...	...	...
	FP_1	FP_2	FP_3		FP_i			FP_C
Actual class C-1	$TN\backslash_{\{1,C-1\}}$	$TN\backslash_{\{2,C-1\}}$	$TN\backslash_{\{3,C-1\}}$	...	$TN\backslash_{\{i,C-1\}}$	...	TP_{C-1} $TN\backslash_{\{C-1\}}$	$TN\backslash_{\{C,C-1\}}$
	FN_{C-1}	FN_{C-1}	FN_{C-1}		FN_{C-1}			FN_{C-1}
	FP_1	FP_2	FP_3		FP_i		FP_{C-1}	
Actual class C	$TN\backslash_{\{1,C\}}$	$TN\backslash_{\{2,C\}}$	$TN\backslash_{\{3,C\}}$	...	$TN\backslash_{\{i,C\}}$	...	$TN\backslash_{\{C-1,C\}}$	TP_C $TN\backslash_{\{C\}}$
	FN_C	FN_C	FN_C		FN_C		FN_C	

In contrast, for *micro_* modification, to count the precision and recall values, one must look at all classes together. In this way, each correctly classified sample into a class is a component of all correctly classified samples. In other words, we calculate TP as the sum of all TP values for individual classes (the sum of the values from the main diagonal). The FP value will be the sum of all values off the main diagonal, equal to the FN value. Therefore, *micro_precision* and *micro_recall* are the same because they are the sum of TP values to all values in the confusion matrix [16]. In multi-class classification, *micro_precision* is calculated by Formula (19) and *micro_recall* is calculated by Formula (20).

$$micro\_precision = \frac{\sum_{i=1}^{c} TP_i}{\sum_{i=1}^{c} TP_i + \sum_{i=1}^{c} FP_i} \tag{19}$$

$$micro\_recall = \frac{\sum_{i=1}^{c} TP_i}{\sum_{i=1}^{c} TP_i + \sum_{i=1}^{c} FN_i} \tag{20}$$

The F_β measure is also used for multi-class classification problems; however, the results of this measure are obtained by macro-averaging or micro-averaging [20]. When we assume that all classes are of the same weight, we use macro-averaging, where two additional methods can be adopted. The first is the calculation of the F_β measure from the *macro_precision* (see Equation (17)) and *macro_recall* (see Equation (18)) measures, and the second is the arithmetic mean of the F-beta scores for each class separately, based on the F_β measure for binary classification (see Equation (8)). The *macro_F_β* measure is specified by Formula (21).

$$macro\_F_\beta = (1 + \beta^2) \times \frac{macro\_precision \times macro\_recall}{\beta^2 \times macro\_precision + macro\_recall} \tag{21}$$

In contrast, in the classification depending on the frequency of classes, the F_β results are calculated by micro-averaging based on *micro_precision* (see Equation (19)) and *micro_recall* (see Equation (20)). The *micro_F_β* measure is specified by Formula (22). In all of the discussed variants of the F_β measure, the beta parameter determines the weight of the reference in relation to the precision. When $\beta < 1$, precision has more weight, and when $\beta > 1$, recall has more weight [20].

$$micro\_F_\beta = (1 + \beta^2) \times \frac{micro\_precision \times micro\_recall}{\beta^2 \times micro\_precision + micro\_recall} \tag{22}$$

The F1 measure (the most common instance of F_β, where $\beta = 1$) is calculated based on recall and precision (present in the multi-class classification) [44], and thus can be used in the multi-class classification as well. For a problem where class imbalance does not matter and all classes are equally valid, we use the *macro_* modification and apply *macro_precision* (see Equation (17)) and *macro_recall* (see Equation (18)). Thus, the measure *macro_F1* can be calculated according to Formula (23).

$$macro\_F1 = 2 \times \frac{macro\_precision \times macro\_recall}{macro\_precision + macro\_recall} \tag{23}$$

In problems with unbalanced classes, where selected classes are more important than others, the *micro_* modification leads to the *micro_precision* (see Equation (19)) and *micro_recall* (see Equation (20)). Thus, the measure *micro_F1* can be calculated according to Formula (24).

$$micro\_F1 = 2 \times \frac{micro\_precision \times micro\_recall}{micro\_precision + micro\_recall} \tag{24}$$

From Formulas (19), (20) and (24), one can conclude that, for the multi-class classification problem, the values of the measures *micro_precision*, *micro_recall*, and *micro_F1* are equal to each other. At the same time, the values of these measures are equal to the *accuracy* of Formula (16).

The Matthews correlation coefficient (MCC) is only used in classifications of up to two classes (see Equation (10)). For classifications with more than two classes, it is often irrelevant to determine the division of multiple classes into two classes (positive and negative) [45]. Therefore, J. Gorodkin, in his work [46], proposed an extended correlation coefficient (called the R_K statistic, for K different classes) that can be used in multi-class classification. Based on this, we defined *MCC* for multiple classes denoted as Formula (25), where s is the sum of all samples, $TP_i + FP_i$ is the value of all samples in row i, and $TP_i + FN_i$ is the value of all samples in column i.

$$MCC\_multiclass = \frac{s \times \sum_{i=1}^{c} TP_i - \sum_{i=1}^{c}((TP_i + FP_i) \times (TP_i + FN_i))}{\sqrt{s^2 - \sum_{i=1}^{c}(TP_i + FP_i)^2} \times \sqrt{s^2 - \sum_{i=1}^{c}(TP_i + FN_i)^2}} \tag{25}$$

The last discussed measure is BalancedAccuracy, which could be used to calculate accuracy for unbalanced classes [41]. According to Formula (11), *BalancedAccuracy_binary* is the arithmetic mean of *recall_binary* and *specificity_binary*. For binary classification, the value of *specificity_binary* for the first class equals *recall_binary* for the second class. For this reason, in multi-class classification, to calculate *BalancedAccuracy*, one must count the recall for each class separately and then calculate the arithmetic mean of these values, according to Formula (26).

$$Balanced\,Accuracy = \frac{1}{c} \sum_{i=1}^{c} \frac{TP_i}{TP_i + FN_i} \tag{26}$$

4. Preference-Driven Quality Assessment for Multi-Class Classification

Considering the measures described above, the preferences for individual classes in the classifier quality assessment are insufficient because there is no place in their construction to indicate such preferences. Thus, we propose a new preference-driven classification quality evaluation measure to evaluate classification quality based on different weights for each decision class. The proposed measure works independently of the number of decision classes—its definition is the same in binary and multi-class classification. We also suggest default values for this measure that can be used, in the test case, without specifying exact preferences for each decision class.

4.1. Proposed Preference-Driven Classification Measure

According to the measures given in Section 3.2, the proposed measure was defined based on the confusion matrix given in Table 2. We aim to keep it as simple as possible while satisfying the assumption of adjustment to preferences (of each decision class). Therefore, the proposed measure is based on a confusion matrix and precision and recall measures with κ parameters determining their relative importance.

The preference-driven classification measure is denoted as preference-driven$_{\vec{\kappa}}$, where $\vec{\kappa}$ is the preference vector, whose length is equal to the number of decision classes (see Formula (27)). The κ weights for each of the subsequent measures are written on the subsequent positions of the vector. The higher the κ value for a given decision class, the greater the importance of precision (determined by Formula (28)) relative to recall (determined by Formula (29))—based on this class only. Therefore, changing the κ values of a given class makes it possible to control the relative importance of precision and recall. This is a multi-criteria process because the κ value can differ for each class.

Finally, the preference-driven$_{\vec{\kappa}}$ measure can be expressed by Formula (30). One should note that $\vec{\kappa}$ is a parameter related to the measure by which the relative importance between precision and recall can be established for each decision class separately. For example, $\vec{\kappa} = [0.2, 0.6, 0.3]$ means that, for the first class, 20% precision and 80% recall are used; for the second class, 60% precision and 40% recall are used, and for the third class, 30% precision and 70% recall are used. To keep the final value of the preference-driven$_{\vec{\kappa}}$ measure in the range [0.0, 1.0], the sum of all these values is divided by the number of classes.

$$preference\text{-}driven_{\vec{\kappa}} = \frac{1}{c} \sum_{i=1}^{c} \kappa_i \times \frac{TP_i}{TP_i + FP_i} + (1 - \kappa_i) \times \frac{TP_i}{TP_i + FN_i} \tag{27}$$

$$precision_i = \frac{TP_i}{TP_i + FP_i} \tag{28}$$

$$recall_i = \frac{TP_i}{TP_i + FN_i} \tag{29}$$

$$preference\text{-}driven_{\vec{\kappa}} = \frac{1}{c} \sum_{i=1}^{c} \kappa_i \times precision_i + (1 - \kappa_i) \times recall_i \tag{30}$$

4.2. Proposed Measure Analysis in the Test Case

The sample confusion matrix was prepared for a classification problem with two decision classes to demonstrate how the measure works depending on the preference and classification outcome. The two classes were chosen to visualize the measure's values.

Figure 1 presents nine confusion matrices for which preference-driven$_{\vec{\kappa}}$ values were determined for $\vec{\kappa}$, being a combination of all values from 0.0 to 1.0 with a step of 0.1, i.e.,

$$[0.0, 0.0], [0.0, 0.1], \ldots [0.5, 0.4], [0.5, 0.5], [0.5, 0.6], \ldots [1.0, 0.9], [1.0, 1.0].$$

It gives a total of 121 different combinations vectors of preferences. The results are visualized in the following figures.

To better capture the distribution of values for the preference-driven measure, an analysis based on the confusion matrix *cm*4 (see Figure 1) was presented. It was selected because the quality assessment in terms of classical measures always means the same values, i.e., precision (see Equation (17)) is 0.7500, and recall (see Equation (18)) is 0.8333. In Figure 2, one can see that the value of the preference-driven measure ranges from 0.5833 for $\vec{\kappa} = [1.0, 0.0]$ to 1.0000 for $\vec{\kappa} = [0.0, 1.0]$. It is possible to obtain exactly the same values for recall ($\vec{\kappa} = [0.0, 0.0]$) and precision ($\vec{\kappa} = [1.0, 1.0]$), but depending on the preference, the classifier will be evaluated differently. For example, preference-driven$_{[0.1,0.3]}$, i.e., for $\kappa_1 = 0.1$ and $\kappa_2 = 0.3$ is 0.8083 (green dot in Figure 2); similarly, preference-driven$_{[0.1,0.8]} = 0.9333$ (black dot in Figure 2), preference-driven$_{[0.4,0.4]} = 0.7833$ (red dot in Figure 2), preference-driven$_{[0.9,0.1]} = 0.6250$ (light blue dot in Figure 2), and preference-driven$_{[0.9,0.8]} = 0.8000$ (bright orange dot in Figure 2).

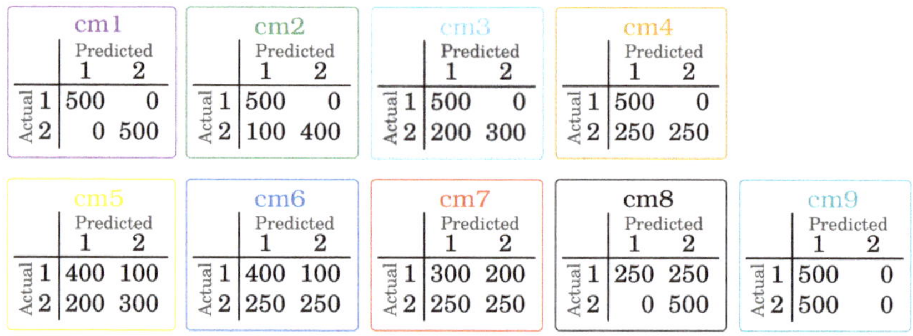

Figure 1. Confusion matrix used for measure analysis.

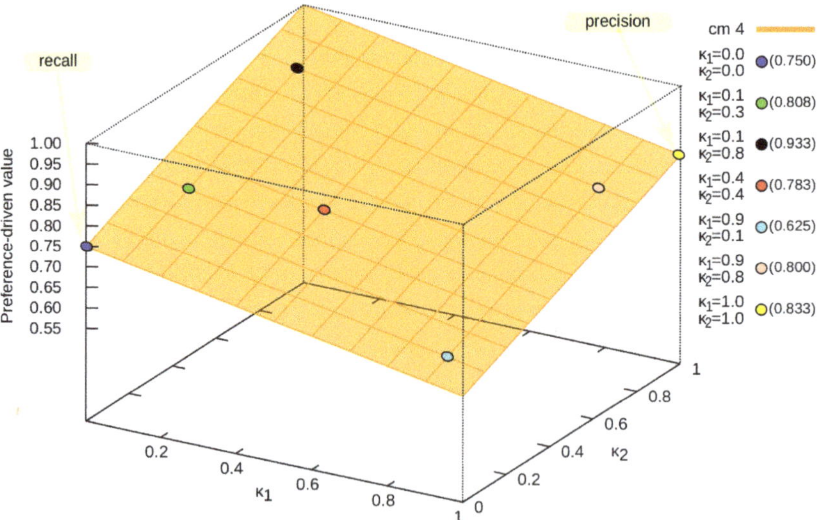

Figure 2. All possible values of the preference-driven measure for the confusion matrix *cm*4.

Next, we present different solution spaces obtained for successive confusion matrices. Figure 3 presents such solution spaces for the proposed measure for *cm*4 and *cm*8 confusion matrices (please refer to Figure 1, which can be described as an opposition to each other, and thus they were selected for the analysis. The value of κ_1 increases, and the value of κ_2 decreases (the value of the proposed measure for *cm*4 decreases, while that for *cm*8 increases—and vice versa).

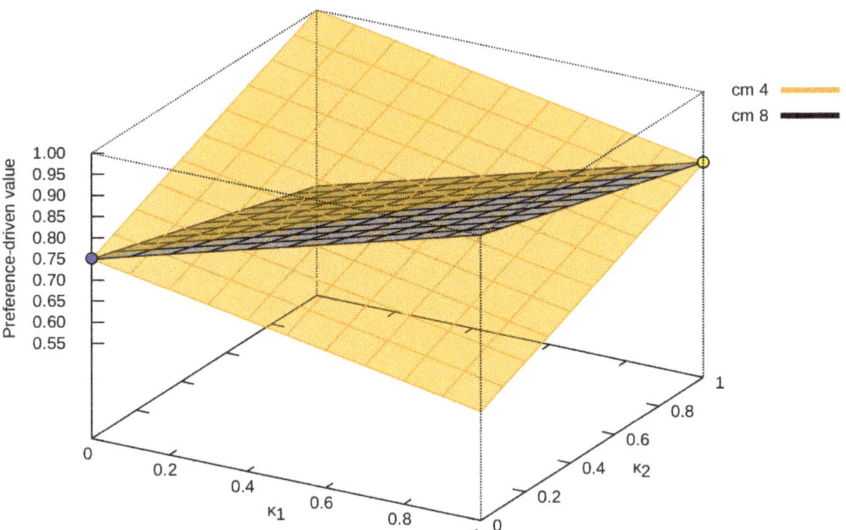

Figure 3. All possible values of the preference-driven measure for the confusion matrix *cm*4 and its opposite *cm*8.

Different solution spaces for the confusion matrices from Figure 1 are presented in Figures 4 and 5. In the first case (Figure 4), specific confusion matrices are selected:

- *cm*1, which represents the ideal classification—in Figure 4a, it can be seen that the value of the preference-driven$_{\overrightarrow{\kappa}}$ measure is the same for each $\overrightarrow{\kappa}$ vector;
- *cm*9, which represents a classification in which all samples are assigned to only one class (in this case, the first one)—in Figure 4b, it can be seen that κ_2 does not affect the value of the preference-driven$_{\overrightarrow{\kappa}}$ measure;
- *cm*4 and *cm*8, which were previously presented in Figure 3, but this time are shown separately in Figure 4c,d, respectively—they represent an example, not an extreme, case of classification.

Figure 5 presents the solution spaces for all (except *cm*9) confusion matrices from Figure 1. It allowed us to present the dynamics of the solution space based on the preference-driven$_{\overrightarrow{\kappa}}$, depending on the classification being evaluated and the value of the $\overrightarrow{\kappa}$ vector. Depending on the values of the $\overrightarrow{\kappa}$ vector, the proposed measure differently evaluates the classifier. It is also presented in Table 3, where the results for different values of the $\overrightarrow{\kappa}$ vector are presented. Examples $[0.3, 0.6]$, $[0.9, 0.4]$, and $[0.5, 0.5]$ have been chosen, as well as precision (which is equivalent to $\overrightarrow{\kappa} = [0.0, 0.0]$) and recall (equivalent to $\overrightarrow{\kappa} = [1.0, 1.0]$). As can be seen, depending on the given preferences, the evaluation of the classifier even in this case changes noticeably.

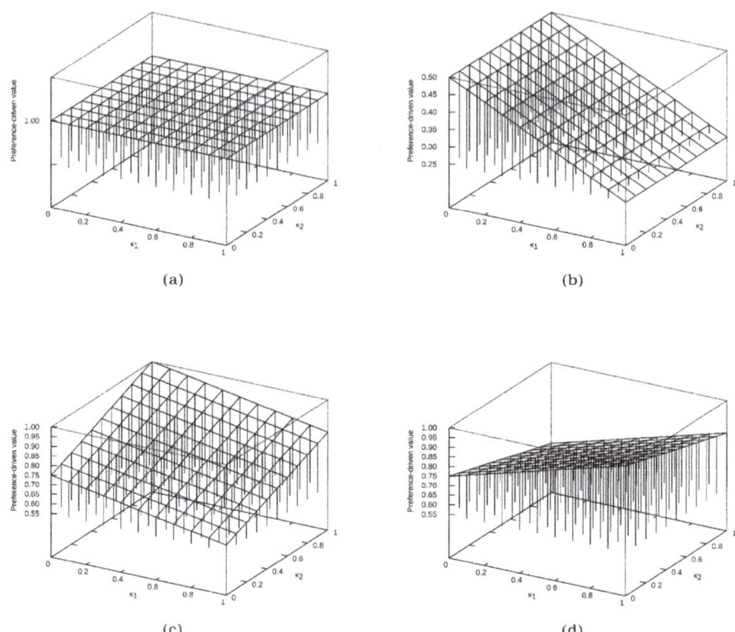

Figure 4. Example distributions of values of the preference-driven measure: (**a**) confusion matrix *cm*1; (**b**) confusion matrix *cm*9; (**c**) confusion matrix *cm*4; (**d**) confusion matrix *cm*8.

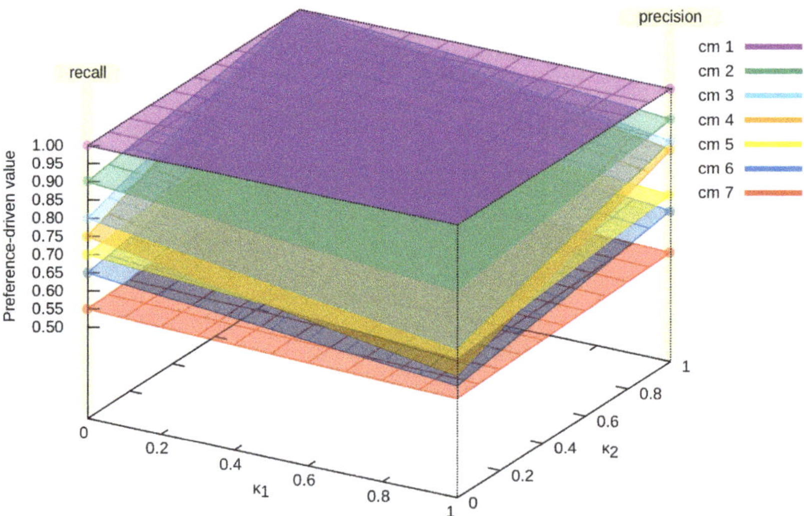

Figure 5. Solution spaces of proposed measure values for selected confusion matrices.

Analogously to Table 3, Figure 6 presents the classification quality assessment values for each of the prepared confusion matrices. Such a visualization allows us to observe the influence of the proposed measure on the classification evaluation (compare with the earlier discussed example for *cm*4 and *cm*8 for which precision, recall, and F1 have identical values), where diagonal lines allow a more straightforward analysis of changes between sample confusion matrices; in this case, we can see that the mentioned measures have identical values for these confusion matrices, but the proposed preference-driven measure obtains different scores for the same preference vectors. Please note that confusion matrices should not be interpreted as consecutive occurrences. Classification evaluation values should be compared regardless of the order in which they are presented.

To further compare the proposed preference-driven measure with the F1 measure, which is also based on recall and precision values, we analyzed the values of the preference vector ($\vec{\kappa}$) that produce the same classification score as the F1.

We performed careful analyses for all the confusion matrices described in this section, except for $cm1$, for which the value of all measures is always 1.0000 (this confusion matrix represents the ideal classification). Our observations indicate that there are (unlike the recall and precision measures) no classical values for the preference vector to find the equivalent value of the F1 measure. Therefore, for subsequent confusion matrices, the F1 counterparts are, respectively, preference-driven measures with preference vectors: for $cm2$, it is $\vec{\kappa} = [0.015, 0.095]$, then for $cm3$, it is $\vec{\kappa} = [0.010, 0.145]$; for $cm4$, it is $\vec{\kappa} = [0.189, 0.284]$; for $cm5$, it is $\vec{\kappa} = [0.045, 0.095]$; for $cm6$, it is $\vec{\kappa} = [0.06, 0.12]$; for $cm7$, it is $\vec{\kappa} = [0.5, 0.5]$; for $cm8$, it is $\vec{\kappa} = [0.284, 0.189]$, and for $cm9$, it is $\vec{\kappa} = [0.67, 0.00]$.

These observations indicate significant differences between the F1 and the proposed preference-driven measures. As we have already presented, F1 for a single confusion matrix (i.e., the classifier score) is always represented by a single value. In contrast, it is possible to control the quality score according to preferences in the preference-driven case.

Table 3. Example values of the proposed preference-driven measure in comparison with other measures for assessing the quality of classification (p-d is the abbreviation for the preference-driven measure). Results determined for all confusion matrices presented in Figure 1.

	p-d$_{[0.3,0.6]}$	p-d$_{[0.9,0.4]}$	p-d$_{[0.5,0.5]}$	Precision [1]	Recall [2]	F1 [3]
$cm1$ [4]	1.0000	1.0000	1.0000	1.0000	1.0000	1.0000
$cm2$ [4]	0.9350	0.8650	0.9083	0.9157	0.9000	0.9083
$cm3$ [4]	0.8771	0.7514	0.8286	0.8571	0.8000	0.8276
$cm4$ [4]	0.8500	0.7000	0.7917	0.8333	0.7500	0.7895
$cm5$ [4]	0.7250	0.6700	0.7042	0.7083	0.7000	0.7041
$cm6$ [4]	0.6866	0.6098	0.6574	0.6648	0.6500	0.6573
$cm7$ [4]	0.5585	0.5366	0.5503	0.5505	0.5500	0.5502
$cm8$ [4]	0.7250	0.9083	0.7917	0.8333	0.7500	0.7895
$cm9$ [4]	0.4250	0.2750	0.3750	0.2500	0.5000	0.3333

[1] *macro_precision* (Equation (17)) is equal to preference-driven$_{[1.0,1.0]}$. [2] *macro_recall* (Equation (18)) is equal to preference-driven$_{[0.0,0.0]}$. [3] *macro_F1* (Equation (23)). [4] See Figure 1.

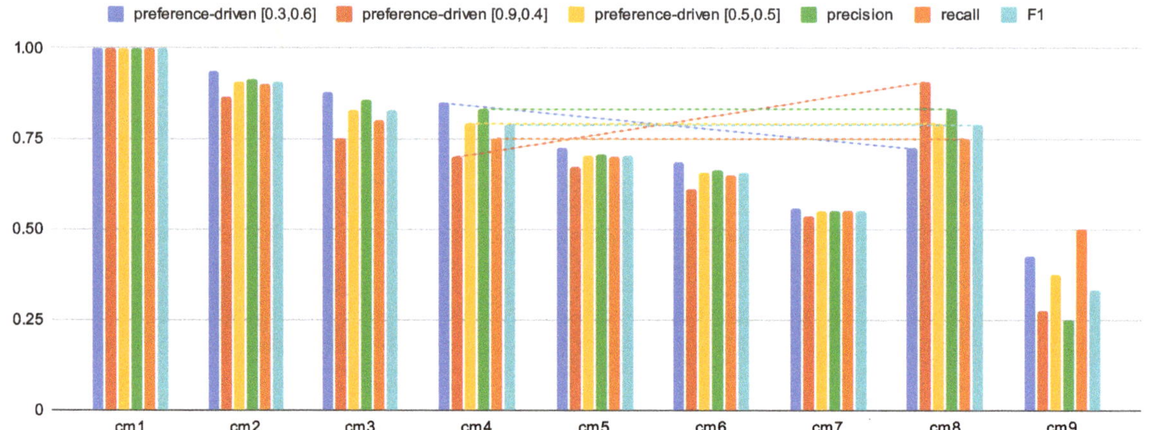

Figure 6. Example values of the proposed measure in comparison with other measures for assessing the quality of classification (the diagonal lines allow easier analysis of the changes between the sample confusion matrices).

4.3. Default Values of the Preference Vector

The proposed measure is used to evaluate the classifier's quality as closely as possible to the stated preferences (as long as the decision maker clearly describes these preferences). To make the measure more comparable with other measures, we propose, as an alternative, the default values of the preference vector for the preference-driven$_{\overrightarrow{\kappa}}$ measure.

The ratio of the number of objects in each class to the number of all samples is taken as the default value of the preference vector. This means that, for classes with many samples, higher weight is related to precision for this class, while, for classes with a relatively small number of samples, higher weight is related to recall of this class. Such a solution allows compensation for the situation in which the samples are more often classified into classes with many samples. When validating a classifier, there is always a learning set (whether for train-and-test or cross-validation), which each time allows the mentioned default values of the preference vector to be determined—the preference vector can correspond to the distribution of cases in the learning data.

In analogy with previously adopted designations, a notation for the default value of the $\overrightarrow{\kappa}$ vector in Formula (31) is proposed. The designations are the same as those contained in Table 2; additionally, s is the sum of all samples.

$$\overrightarrow{\kappa} = \left[\frac{TP_1 + FN_1}{s}, \frac{TP_2 + FN_2}{s}, \ldots, \frac{TP_i + FN_i}{s}, \ldots, \frac{TP_{C-1} + FN_{C-1}}{s}, \frac{TP_C + FN_C}{s}\right] \quad (31)$$

For example, in the situation described in Section 4.2, where the classes in the confusion matrix are at equilibrium, the implicit vector would be of the form $\overrightarrow{\kappa} = [0.5, 0.5]$. Its value is incidentally given in Table 3, in column preference-driven$_{[0.5,0.5]}$.

Extending the example, a confusion matrix is proposed, shown in Table 4. In this case, there are three classes, containing 50, 20, and 30 samples each, respectively. Therefore, the default values of the preference vector would be as follows:

$$\overrightarrow{\kappa} = \left[\frac{50}{100}, \frac{20}{100}, \frac{30}{100}\right],$$

so

$$\overrightarrow{\kappa} = [0.5, 0.2, 0.3].$$

Here, 50 out of 100 instances are in class 1, 20 out of 100 instances are in class 2, and 30 out of 100 instances are in class 3. The preference-driven measure value is 0.656, because

$$
\text{preference-driven}_{[0.5,0.2,0.3]} = \frac{1}{3} \times \left(0.5 \times \frac{TP_1}{TP_1 + FP_1} + (1 - 0.5) \times \frac{TP_1}{TP_1 + FN_1} \right) +
$$
$$
\frac{1}{3} \times \left(0.2 \times \frac{TP_2}{TP_2 + FP_2} + (1 - 0.2) \times \frac{TP_2}{TP_2 + FN_2} \right) +
$$
$$
\frac{1}{3} \times \left(0.3 \times \frac{TP_3}{TP_3 + FP_3} + (1 - 0.3) \times \frac{TP_3}{TP_3 + FN_3} \right)
$$

so

$$
\text{preference-driven}_{[0.5,0.2,0.3]} = \frac{1}{3} \times \left(0.5 \times \frac{40}{40 + 8 + 9} + 0.5 \times \frac{40}{40 + 7 + 3} \right) +
$$
$$
\frac{1}{3} \times \left(0.2 \times \frac{10}{10 + 7 + 1} + 0.8 \times \frac{10}{10 + 8 + 2} \right) +
$$
$$
\frac{1}{3} \times \left(0.3 \times \frac{20}{20 + 3 + 2} + 0.7 \times \frac{20}{20 + 9 + 1} \right)
$$

$$
\text{preference-driven}_{[0.5,0.2,0.3]} = \frac{1}{3} \times (0.5 \times 0.702 + 0.5 \times 0.8 + 0.2 \times 0.556 + 0.8 \times 0.5 + 0.3 \times 0.8 + 0.7 \times 0.667)
$$

$$
\text{preference-driven}_{[0.5,0.2,0.3]} = \frac{1}{3} \times (0.351 + 0.4 + 0.111 + 0.4 + 0.24 + 0.467)
$$

$$
\text{preference-driven}_{[0.5,0.2,0.3]} = 0.656
$$

The results for the default values of the preference vector for the preference-driven$_{\vec{\kappa}}$ measure are also analyzed in Section 5, when testing the proposed measure with different real data sets.

Table 4. Example confusion matrix used to demonstrate how to determine the proposed preference-driven measure.

	Predicted Class 1	Predicted Class 2	Predicted Class3
Actual class 1	40	7	3
Actual class 2	8	10	2
Actual class 3	9	1	20

5. Analysis on Real-World Data Sets

As the paper proposes a new classification quality assessment measure whose value depends on the stated preferences (called preference vector), we conducted experiments on real-world data sets. First, we checked the importance of the proposed measure depending on the given preference vector—while comparing the performance and ranks of the classifiers with classical measures of classification quality assessment.

5.1. Experiment Conditions

The proposed measure preference-driven$_{\vec{\kappa}}$ is compared with the classical measures described in this paper (see Section 3.2). As some of the measures, in the case of multi-class classification, reduce to the same measure, the following names are used in this section: accuracy (see Equation (16)), precision (see Equation (17)), recall (see Equation (18)), F1 (see Equation (23)), and MCC (see Equation (25)).

Four well-known data sets were selected for classification, whose structures are described in Table 5. As one can see, multi-class data sets were selected, in which, additionally, the distribution of samples in classes was uneven (see "percent of samples per class" in Table 5, where the ratio of samples of each class to the whole data set is given). On the other hand, as test classifiers, we selected classifiers available in the system Weka-3-6-11 [47]. More specifically, these were the following classifiers: Bagging [48], BayesNet [49], DecisionTable [50], C4.5 (J48) [51], and RandomForest [52]. In each case, 10-fold cross-validation was used so that each sample was subject to prediction.

Classification results from the Weka system (including confusion matrices) and the values of the proposed classification quality evaluation measure for fixed preference vectors are available on the website of the Department of Machine Learning of the University of Economics in Katowice (https://www.ue.katowice.pl/jednostki/katedry/wiik/katedra-uczenia-maszynowego/zrodla-danych/preference-driven.html (accessed on 6 February 2022)).

Table 5. Characteristics of the real data sets used to test the proposed preference-driven measure and comparison with classical measures.

Data Set	Number of Samples	Number of Attributes	Number of Classes	Percent of Samples per Class
car	1728	6	4	0.70 0.22 0.04 0.04
nursery	12960	8	5	0.00 0.33 0.33 0.03 0.31
dermatology	366	34	6	0.17 0.31 0.20 0.14 0.13 0.05
krkopt	28056	6	18	0.10 0.00 0.00 0.01 0.00 0.01 0.02 0.02 0.02 0.05 0.06 0.07 0.10 0.13 0.15 0.16 0.08 0.01

The number of decision classes in each data set is different, so the checked combinations of values in the preference vector determined different, possible values of subsequent elements of the vector. Thus, the number of combinations is close to 15,000 (except for the krkopt data set, which is too large and had to exceed this value). In this way, we obtained:

- car—[0, 0.1, 0.2, 0.3, 0.4, 0.5, 0.6, 0.7, 0.8, 0.9, 1], which gives 14,641 combinations (because: 4 decision classes and 11 values of κ, so $11^4 = 14{,}641$);
- nursery—[0, 0.166, 0.332, 0.498, 0.664, 0.83, 1], which gives $7^5 = 16{,}807$ combinations;
- dermatology—[0, 0.25, 0.5, 0.75, 1], which gives $5^6 = 15{,}625$; combinations;
- krkopt—[0.33, 0.66], which gives $2^{18} = 262{,}144$ combinations.

5.2. Experimental Results

Tables 6–9 present the results for all data sets and all classifiers. Ratings have been made for all measures, with preference-driven$_{\vec{\kappa}}$ determined each time for the default values and five different, selected preference vectors. Evaluation values were presented along with the ranking order of the classifier for each measure (in brackets). This approach allows us not only to evaluate the differences between the evaluation values but also to indicate whether, using the proposed measure, it could happen that another classifier would be better than those indicated by the classical classification quality evaluation measures. On the other hand, Figure 7 shows histograms of the preference-driven$_{\vec{\kappa}}$ measure values for all tested preference vectors.

Figure 7 presents 20 different histograms used to show the distribution of values (without any unnecessarily detailed information). In part (a), successive histograms are shown for the car data set, then (b) contains the nursery histograms, (c) is related to dermatology, and (d) is the krkopt results. This figure concerns the value of the preference-driven measure determined for different preference vectors ($\overrightarrow{\kappa}$). Therefore, the X-axis represents the value of the preference-driven measure, while the Y-axis represents the number of occurrences of this value.

This presentation in Figure 7 allows us to notice that in each case, using the proposed measure, it is possible to obtain a different score with the same confusion matrix. In addition, in most cases, the values of the measure are close to a normal distribution. A slightly more interesting case is (b), with the Bagging, BayesNet, and RandomForest classifiers. In this case, however, it should be noted that the classification each time gives a specific confusion matrix (detailed results are available on the UE Katowice website (https://www.ue.katowice.pl/jednostki/katedry/wiik/katedra-uczenia-maszynowego/zrodla-danych/preference-driven.html) (accessed on 6 February 2022)). Note that there are only two samples in the first class. The class with the most significant number of samples is always correctly identified. Additionally, in case (c), the particular histogram is obtained for the RandomForest classifier, where the classification was excellent, close to error-free. In contrast, the values were rounded to the second decimal for the histogram, hence such a high concentration of preference-driven measure values.

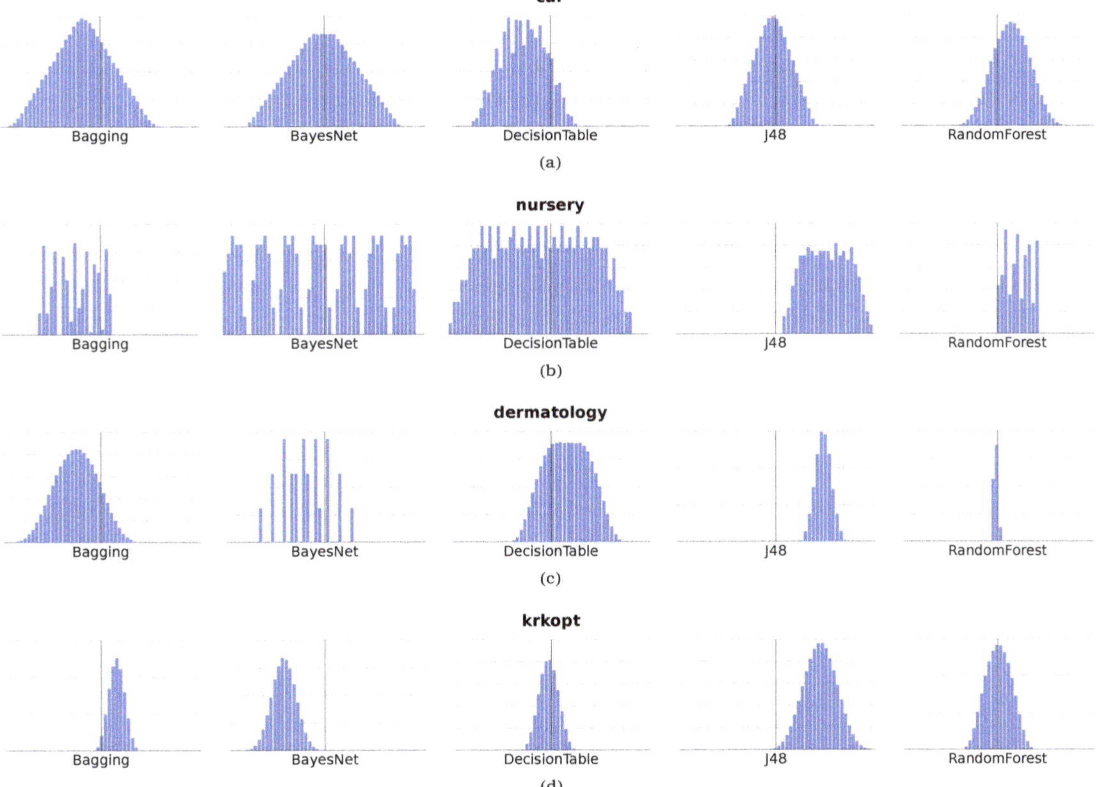

Figure 7. Histogram of preference-driven measure values in the (**a**) car data set; (**b**) nursery data set; (**c**) dermatology data set; (**d**) krkopt data set—the X-axis represents the value of the preference-driven measure, while the Y-axis represents the number of occurrences of this value.

Analyzing the experimental results, it is worth noting that by using different vectors, in the case of a preference-driven measure, it is possible to indicate a different classifier as more adapted to the problem. For example, in the case of the car data set (see Table 6), the classifier DecisionTable turns out to be the best for some preference vectors (e.g., $[0, 1, 1, 0.9]$); similarly, the classifier BayesNet turns out to be better than Bagging if the preference vector is, among others, $[0.3, 1, 1, 1]$. The situation is similar for the set of nursery data (see Table 7), where also DecisionTable turns out to be the best, while the preference vector is $[0.498, 1, 0.166, 1, 0]$. A similar difference, but on subsequent ranks (between 3 and 4), can be observed also in the case of the krkopt data set (see Table 9). Although, in the case of the dermatology data set (see Table 8), no change of ranks was observed, it could be observed that the disproportion between the assessment of specific classifiers changed a lot and, e.g., in the case of vector $[0, 0, 0, 1, 1, 0.75]$, the difference between the best BayesNet and the second RandomForest was less than 0.0083, where, with classical measures of classification quality assessment, this difference was around 0.02.

The aim of the experiment was to check whether the proposed measure will show different classifiers for the same problem depending on the value in the preference vector. It turned out that, indeed, the measure indicates different classifiers, even with a limited number of vectors tested.

Table 6. Results for the car data set—the value of the classification quality assessment (in brackets, we give the ranking of the classifier, according to the given measure). The ranking determines the order of the classifier depending on the classification quality rating measure used.

	Bagging	BayesNet	DecisionTable	C4.5 (J48)	RandomForest
accuracy	0.9167 (3)	0.8571 (5)	0.9149 (4)	0.9236 (2)	0.9462 (1)
precision	0.7586 (5)	0.7940 (4)	0.8557 (2)	0.8179 (3)	0.8565 (1)
recall	0.7665 (4)	0.6040 (5)	0.8177 (3)	0.8289 (2)	0.8581 (1)
F1	0.7625 (4)	0.6861 (5)	0.8363 (2)	0.8233 (3)	0.8573 (1)
MCC	0.8208 (3)	0.6737 (5)	0.8101 (4)	0.8345 (2)	0.8842 (1)
preference-driven [1]	0.7670 (4)	0.6031 (5)	0.8171 (3)	0.8291 (2)	0.8584 (1)
p-d$_{[0,0.8,0.6,1]}$	0.7757 (4)	0.7582 (5)	0.8600 (1)	0.8295 (3)	0.8600 (1)
p-d$_{[0,1,0,1]}$	0.8021 (4)	0.6853 (5)	0.8631 (2)	0.8475 (3)	0.8647 (1)
p-d$_{[0,1,1,0.9]}$	0.7494 (5)	0.7954 (4)	0.8601 (1)	0.8127 (3)	0.8499 (2)
p-d$_{[0.3,1,1,1]}$	0.7555 (5)	0.8012 (4)	0.8609 (1)	0.8160 (3)	0.8534 (2)
p-d$_{[0.6,1,0.8,1]}$	0.7664 (5)	0.7743 (4)	0.8587 (1)	0.8233 (3)	0.8572 (2)

[1] Default value of the preference vector (see Section 4.3).

Table 7. Results for the nursery data set—the value of the classification quality assessment (in brackets, we give the ranking of the classifier, according to the given measure). The ranking determines the order of the classifier depending on the classification quality rating measure used.

	Bagging	BayesNet	DecisionTable	C4.5 (J48)	RandomForest
accuracy	0.9737 (2)	0.9033 (5)	0.9470 (4)	0.9705 (3)	0.9909 (1)
precision	0.7518 (3)	0.7250 (5)	0.7661 (2)	0.7453 (4)	0.7849 (1)
recall	0.7226 (3)	0.5666 (5)	0.6722 (4)	0.7313 (2)	0.7765 (1)
F1	0.7369 (3)	0.6361 (5)	0.7160 (4)	0.7382 (2)	0.7806 (1)
MCC	0.9614 (2)	0.8579 (5)	0.9234 (4)	0.9568 (3)	0.9867 (1)
preference-driven [1]	0.7224 (3)	0.5671 (5)	0.6727 (4)	0.7312 (2)	0.7764 (1)
p-d$_{[0,0,0,1,0]}$	0.7550 (3)	0.7362 (5)	0.7733 (2)	0.7470 (4)	0.7857 (1)
p-d$_{[0,0,0,1,1]}$	0.7524 (2)	0.7412 (5)	0.7520 (3)	0.7428 (4)	0.7850 (1)
p-d$_{[0,0.83,0,1,0]}$	0.7545 (3)	0.7228 (5)	0.7850 (2)	0.7490 (4)	0.7856 (1)
p-d$_{[0.498,1,0.166,1,0]}$	0.7544 (3)	0.7200 (5)	0.7874 (1)	0.7495 (4)	0.7855 (2)
p-d$_{[0.83,0.332,0.166,0,1]}$	0.7198 (3)	0.5662 (5)	0.6555 (4)	0.7280 (2)	0.7757 (1)

[1] Default value of the preference vector (see Section 4.3).

Table 8. Results for the dermatology data set—the value of the classification quality assessment (in brackets, we give the ranking of the classifier, according to the given measure). The ranking determines the order of the classifier depending on the classification quality rating measure used.

	Bagging	BayesNet	DecisionTable	C4.5 (J48)	RandomForest
accuracy	0.3798 (5)	0.9727 (1)	0.8388 (4)	0.9454 (3)	0.9536 (2)
precision	0.3798 (5)	0.9692 (1)	0.8549 (4)	0.9372 (3)	0.9492 (2)
recall	0.3947 (5)	0.9707 (1)	0.8204 (4)	0.9368 (3)	0.9477 (2)
F1	0.3871 (5)	0.9699 (1)	0.8373 (4)	0.9370 (3)	0.9484 (2)
MCC	0.2170 (5)	0.9660 (1)	0.7982 (4)	0.9316 (3)	0.9418 (2)
preference-driven [1]	0.3923 (5)	0.9708 (1)	0.8217 (4)	0.9368 (3)	0.9477 (2)
p-d$_{[0,0,0,0,1,0]}$	0.4090 (5)	0.9589 (1)	0.8030 (4)	0.9400 (3)	0.9506 (2)
p-d$_{[0,0,0,1,1,0.75]}$	0.4014 (5)	0.9589 (1)	0.8430 (4)	0.9369 (3)	0.9506 (2)
p-d$_{[0,0,1,1,0,1]}$	0.3807 (5)	0.9707 (1)	0.8800 (4)	0.9315 (3)	0.9477 (2)
p-d$_{[0.5,0,0.25,0.5,0.5,1]}$	0.3823 (5)	0.9699 (1)	0.8575 (4)	0.9375 (3)	0.9492 (2)
p-d$_{[0.5,1,0.75,0.5,0.5,0]}$	0.3921 (5)	0.9699 (1)	0.8178 (4)	0.9364 (3)	0.9477 (2)

[1] Default value of the preference vector (see Section 4.3).

Table 9. Results for the krkopt data set—the value of the classification quality assessment (in brackets, we give the ranking of the classifier, according to the given measure). The ranking determines the order of the classifier depending on the classification quality rating measure used.

	Bagging	BayesNet	DecisionTable	C4.5 (J48)	RandomForest
accuracy	0.5872 (2)	0.3607 (5)	0.4908 (4)	0.5658 (3)	0.7025 (1)
precision	0.5735 (3)	0.3579 (5)	0.5784 (2)	0.5547 (4)	0.7377 (1)
recall	0.5406 (2)	0.2982 (5)	0.5187 (3)	0.5178 (4)	0.6628 (1)
F1	0.5566 (2)	0.3253 (5)	0.5469 (3)	0.5356 (4)	0.6982 (1)
MCC	0.5377 (2)	0.2784 (5)	0.4300 (4)	0.5135 (3)	0.6669 (1)
preference-driven [1]	0.5404 (2)	0.2967 (5)	0.5191 (3)	0.5177 (4)	0.6629 (1)
preference-driven$_{\vec{\kappa}}$ [2]	0.5776 (3)	0.3717 (5)	0.5806 (2)	0.5576 (3)	0.7426 (1)
preference-driven$_{\vec{\kappa}}$ [3]	0.5542 (2)	0.3176 (5)	0.5415 (3)	0.5288 (4)	0.6869 (1)
preference-driven$_{\vec{\kappa}}$ [4]	0.5554 (2)	0.3262 (5)	0.5462 (3)	0.5396 (4)	0.6925 (1)
preference-driven$_{\vec{\kappa}}$ [5]	0.5741 (2)	0.3467 (5)	0.5472 (4)	0.5485 (3)	0.7287 (1)
preference-driven$_{\vec{\kappa}}$ [6]	0.5638 (2)	0.3461 (5)	0.5587 (3)	0.5410 (4)	0.7033 (1)

[1] Default value of the preference vector (see Section 4.3). [2] $\vec{\kappa} = [0, 1, 1, 1, 1, 1, 0.9, 0.9, 1, 0.9, 0.9, 0.9, 0, 0, 0, 0, 0.9, 1]$. [3] $\vec{\kappa} = [0.33, 0.66, 0.33, 0.33, 0.66, 0.33, 0.66, 0.66, 0.33, 0.66, 0.66, 0.33, 0.66, 0.66, 0.66, 0.66, 0.33, 0.33]$. [4] $\vec{\kappa} = [0.66, 0.33, 0.33, 0.33, 0.33, 0.66, 0.66, 0.33, 0.66, 0.66, 0.66, 0.33, 0.33, 0.66, 0.33, 0.33, 0.33, 0.33]$. [5] $\vec{\kappa} = [1, 0.1, 1, 1, 1, 1, 0.1, 0.1, 0, 0.1, 0.1, 0.1, 0, 0, 0, 0, 0.9, 1]$. [6] $\vec{\kappa} = [1, 1, 0.9, 0.9, 0.8, 0.8, 0.7, 0.6, 0.5, 0.4, 0.3, 0.3, 0.2, 0.2, 0.1, 0.1, 0.0, 0.0]$.

6. Discussion

In supervised learning, one of the necessary steps in adequately designed research is teaching on training data different classifiers, sometimes with different parameters, and then evaluating their quality. In our work, we assume that the user has specific preferences. In addition, they relate to particular decision classes. Thus, we can immediately identify a suitable classifier.

Using a vector of preferences, it is possible to indicate whether precision or recall is more important. It is crucial for many problems to indicate that it is possible to select a classifier better adapted to the structure of the data and decision makers' expectations for each class separately. The best strategy is to examine the training of as many classifiers as possible (with different parameters). Such an approach allows the identification of the potentially best classifiers for a given problem.

The proposed preference-driven measure used in the experiments allowed for a better selection of the classifier most suitable for the task. With an approach allowing us to indicate whether precision or recall is more important—for each class separately—it is possible to select a classifier more adapted to the more important (from the evaluation point of view) decision class.

Despite the limited number of combinations, the conducted experiments indicated that the proposed measure, depending on the preferences conveyed in the form of a preference vector, points to different classifiers as the best choice for further prediction.

Since the set of values in the preference vector is infinite, the measurement values are also unlimited. Therefore, the calculation of a measure and comparison with other measures is possible only by calculating them at specific points. It should be noted that it is not necessary to test many combinations of the preference vector in a real application. Instead, these should be predetermined, indicating the relative importance of the recall–precision balance in each decision class.

This also distinguishes the preference-driven measure from the F_β measure, which also raised attempts to weigh between precision and recall. However, the proposed preference-driven measure is different. Note that, in the case of F_β, there is a weighting between precision and recall overall for the classifier. In contrast, there are weights for each decision class with the preference-driven measure. It allows us to change the emphasis on precision and recall depending on each decision class (differently). In the case of F_β, this possibility does not exist.

7. Conclusions

This article presents a new idea for a preference-driven classification measure. We tried to show that the measure works, i.e.,

- for different preference vectors, different classifiers are more advantageous then others;
- the obtained results of the comparison make it easier for the user to understand the effects of classification and make the right decision as to which classifier to use.

In the "objective" approach (without preferences), the result is unambiguous and comparable, but the best classifier will not necessarily be adjusted to the subjective needs of the user. In the "subjective" approach (preference-driven), comparability is difficult, but in return, users can acquire a classifier better suited to their requirements.

The concept of this measure results from the shortcomings of measures related to the multi-class classification. Nowadays, we observe a large number of classification methods. There is no single versatile classification measure capable of catching up on concepts related to both: overall good classification quality and a particular focus on the selected decision classes. The whole idea of different classification measures is mostly extended for the binary decision classes, which often fail to achieve good results for real-world data. At the same time, multi-class classification measures are based on averaging the results, which can be fair for general cases but fails to include decision makers' preferences related to the particular classes.

Similarly, for unbalanced cases, where there is a need to focus on particular classes, the proposed preference-driven measure fits well for this gap. To be more precise, our proposed preference-driven measure can be aligned with the decision makers' preferences regarding the relative importance of precision and recall. We also present the idea of setting the default values for the vector of preferences based on the overall number of samples assigned to every decision class. The most important advantage of the proposed idea is the good fit between the well-known measures such as precision and recall. Moreover, the κ preference vector allows us to direct the focus to a particular decision class, or even to express the importance of selected decision classes in terms of the precision measure (and others in terms of recall).

At the same time, we show that even for potentially trivial cases, such preference-driven measures could lead to entirely different results based on the κ selection. It opens a discussion for multi-class classification and leads to an interesting situation. The solution for the classification problem should not be considered a single scalar value.

In the future, a preference-driven measure can be used in line with the proposed approach. Alternatively, the factors of which the measure is composed could be scrutinized, and other measures could be used instead of the class's relative precision and recall values.

Author Contributions: Conceptualization, J.K.; methodology, J.K., B.P., K.K. and P.J.; software, J.K.; validation, J.K. and B.P.; formal analysis, J.K., B.P., K.K. and P.J.; investigation, J.K. and B.P.; resources, J.K.; data curation, J.K.; writing—original draft preparation, J.K., B.P., K.K. and P.J.; writing—review and editing, J.K., B.P., K.K. and P.J.; visualization, J.K.; supervision, J.K.; project administration, J.K., B.P. and P.J. All authors have read and agreed to the published version of the manuscript.

Funding: This research received no external funding.

Data Availability Statement: The classification results (including confusion matrices) obtained in the Weka program, as well as the values of the preference-driven measure for the analyzed preference vectors, can be found on the website of the Department of Machine Learning of the University of Economics in Katowice: https://www.ue.katowice.pl/jednostki/katedry/wiik/katedra-uczenia-maszynowego/zrodla-danych/preference-driven.html (accessed on 6 February 2022). The implementation of the proposed preference-driven measure was prepared in Python language. The source code of this implementation is publicly available via GitHub: https://github.com/jankozak/preference-driven (accessed on 6 February 2022).

Conflicts of Interest: The authors declare no conflict of interest.

References

1. Gösgens, M.; Zhiyanov, A.; Tikhonov, A.; Prokhorenkova, L. Good Classification Measures and How to Find Them. In Proceedings of the 35th Conference on Neural Information Processing Systems (NeurIPS 2021), Online, 6–14 December 2021.
2. Seijo-Pardo, B.; Porto-Díaz, I.; Bolón-Canedo, V.; Alonso-Betanzos, A. Ensemble feature selection: Homogeneous and heterogeneous approaches. *Knowl.-Based Syst.* **2017**, *118*, 124–139. [CrossRef]
3. Lewis, D.D.; Catlett, J. Heterogeneous uncertainty sampling for supervised learning. In *Machine Learning Proceedings 1994*; Elsevier: Amsterdam, The Netherlands, 1994; pp. 148–156.
4. Campagner, A.; Sconfienza, L.; Cabitza, F. H-accuracy, an alternative metric to assess classification models in medicine. In *Digital Personalized Health and Medicine*; IOS Press: Amsterdam, The Netherlands, 2020; pp. 242–246.
5. Gilli, M.; Schumann, E. Accuracy and precision in finance. *Available SSRN 2698114* **2015**. [CrossRef]
6. Canbek, G.; Taskaya Temizel, T.; Sagiroglu, S. BenchMetrics: A systematic benchmarking method for binary classification performance metrics. *Neural Comput. Appl.* **2021**, *33*, 14623–14650. [CrossRef]
7. Amershi, S.; Cakmak, M.; Knox, W.B.; Kulesza, T. Power to the people: The role of humans in interactive machine learning. *Ai Mag.* **2014**, *35*, 105–120. [CrossRef]
8. Wu, X.; Xiao, L.; Sun, Y.; Zhang, J.; Ma, T.; He, L. A Survey of Human-in-the-loop for Machine Learning. *arXiv* **2021**, arXiv:2108.00941.
9. Talbot, J.; Lee, B.; Kapoor, A.; Tan, D.S. EnsembleMatrix: Interactive visualization to support machine learning with multiple classifiers. In Proceedings of the SIGCHI Conference on Human Factors in Computing Systems, Boston, MA, USA, 4–9 April 2009; pp. 1283–1292.
10. Green, B.; Chen, Y. The principles and limits of algorithm-in-the-loop decision making. *Proc. ACM Hum. -Comput. Interact.* **2019**, *3*, 1–24. [CrossRef]
11. Kononenko, I.; Bratko, I. Information-Based Evaluation Criterion for Classifier's Performance. *Mach. Learn.* **1991**, *6*, 67–80. [CrossRef]
12. Valverde-Albacete, F.J.; Peláez-Moreno, C. 100% classification accuracy considered harmful: The normalized information transfer factor explains the accuracy paradox. *PLoS ONE* **2014**, *9*, e84217. [CrossRef]
13. Saito, T.; Rehmsmeier, M. The precision-recall plot is more informative than the ROC plot when evaluating binary classifiers on imbalanced datasets. *PLoS ONE* **2015**, *10*, e0118432. [CrossRef]
14. Ben-David, A. A lot of randomness is hiding in accuracy. *Eng. Appl. Artif. Intell.* **2007**, *20*, 875–885. [CrossRef]
15. Sokolova, M.; Japkowicz, N.; Szpakowicz, S. Beyond accuracy, F-score and ROC: A family of discriminant measures for performance evaluation. In *Australasian Joint Conference on Artificial Intelligence*; AAAI Workshop-Technical Report; Springer: Berlin/Heidelberg, Germany, 2006; Volume 4304, pp. 24–29. [CrossRef]
16. Grandini, M.; Bagli, E.; Visani, G. Metrics for multi-class classification: An overview. *arXiv* **2020**, arXiv:2008.05756.
17. Hossin, M.; Sulaiman, M. A Review on Evaluation Metrics for Data Classification Evaluations. *Int. J. Data Min. I Knowl. Manag. Process* **2015**, *5*, 1–11. [CrossRef]
18. Ferri, C.; Hernández-Orallo, J.; Modroiu, R. An experimental comparison of performance measures for classification. *Pattern Recognit. Lett.* **2009**, *30*, 27–38. [CrossRef]
19. Emmert-Streib, F.; Moutari, S.; Dehmer, M. A comprehensive survey of error measures for evaluating binary decision making in data science. *Wiley Interdiscip. Rev. Data Min. Knowl. Discov.* **2019**, *9*, 1–15. [CrossRef]
20. Sokolova, M.; Lapalme, G. A systematic analysis of performance measures for classification tasks. *Inf. Process. Manag.* **2009**, *45*, 427–437. [CrossRef]
21. Demšar, J. Statistical comparisons of classifiers over multiple data sets. *J. Mach. Learn. Res.* **2006**, *7*, 1–30.

22. Tharwat, A. Classification assessment methods. *Appl. Comput. Inform.* **2018**, *17*, 168–192. [CrossRef]
23. Zhang, X.; Li, X.; Feng, Y. A classification performance measure considering the degree of classification difficulty. *Neurocomputing* **2016**, *193*, 81–91. [CrossRef]
24. Yu, S.; Li, X.; Feng, Y.; Zhang, X.; Chen, S. An instance-oriented performance measure for classification. *Inf. Sci.* **2021**, *580*, 598–619. [CrossRef]
25. Gong, M. A Novel Performance Measure for Machine Learning Classification. *Int. J. Manag. Inf. Technol.* **2021**, *13*, 11–19. [CrossRef]
26. Carbonero-Ruz, M.; Martínez-Estudillo, F.J.; Fernández-Navarro, F.; Becerra-Alonso, D.; Martínez-Estudillo, A.C. A two dimensional accuracy-based measure for classification performance. *Inf. Sci.* **2017**, *382–383*, 60–80. [CrossRef]
27. Kasperczuk, A.; Dardzinska, A. Automatic system for IBD diagnosis. *Procedia Comput. Sci.* **2021**, *192*, 2863–2870. [CrossRef]
28. Bac, C.W.; Hemming, J.; Van Henten, E.J. Robust pixel-based classification of obstacles for robotic harvesting of sweet-pepper. *Comput. Electron. Agric.* **2013**, *96*, 148–162. [CrossRef]
29. Meng, J.; Zhang, W. Volume measure in 2DPCA-based face recognition. *Pattern Recognit. Lett.* **2007**, *28*, 1203–1208. [CrossRef]
30. Burduk, R. Classification Performance Metric for Imbalance Data Based on Recall and Selectivity Normalized in Class Labels. *arXiv* **2020**, arXiv:2006.13319.
31. Hand, D.J.; Christen, P.; Kirielle, N. F*: An interpretable transformation of the F-measure. *Mach. Learn.* **2021**, *110*, 451–456. [CrossRef] [PubMed]
32. Mitchell, T.M. *Machine Learning, International Edition*; McGraw-Hill Series in Computer Science; McGraw-Hill Education: New York, NY, USA. 1997.
33. Townsend, J.T. Theoretical analysis of an alphabetic confusion matrix. *Percept. Psychophys.* **1971**, *9*, 40–50. [CrossRef]
34. Provost, F.; Kohavi, R. Glossary of terms. *J. Mach. Learn.* **1998**, *30*, 271–274. [CrossRef]
35. Room, C. Confusion Matrix. *Mach. Learn.* **2019**, *6*, 27.
36. Lee, N.; Yang, H.; Yoo, H. A surrogate loss function for optimization of F_β score in binary classification with imbalanced data. *arXiv* **2021**, arXiv:2104.01459.
37. Van Rijsbergen, C.J. *Information Retrieval*; Butterworth-Heinemann: Newton, MA, USA, 1979.
38. Buckland, M.; Gey, F. The relationship between recall and precision. *J. Am. Soc. Inf. Sci.* **1994**, *45*, 12–19. [CrossRef]
39. Chicco, D.; Jurman, G. The advantages of the Matthews correlation coefficient (MCC) over F1 score and accuracy in binary classification evaluation. *BMC Genom.* **2020**, *21*, 1–13. [CrossRef]
40. Matthews, B.W. Comparison of the predicted and observed secondary structure of T4 phage lysozyme. *Biochim. Biophys. Acta (BBA)-Protein Struct.* **1975**, *405*, 442–451. [CrossRef]
41. Brodersen, K.H.; Ong, C.S.; Stephan, K.E.; Buhmann, J.M. The Balanced Accuracy and Its Posterior Distribution. In Proceedings of the 2010 20th International Conference on Pattern Recognition, Istanbul, Turkey, 23–26 August 2010; pp. 3121–3124. [CrossRef]
42. Parikh, R.; Mathai, A.; Parikh, S.; Sekhar, G.C.; Thomas, R. Understanding and using sensitivity, specificity and predictive values. *Indian J. Ophthalmol.* **2008**, *56*, 45–50. [CrossRef] [PubMed]
43. Tsoumakas, G.; Katakis, I. Multi-label classification: An overview. *Int. J. Data Warehous. Min. (IJDWM)* **2007**, *3*, 1–13. [CrossRef]
44. Takahashi, K.; Yamamoto, K.; Kuchiba, A.; Koyama, T. Confidence interval for micro-averaged F1 and macro-averaged F1 scores. *Appl. Intell.* **2022**, *28*, 4961–4972. [CrossRef] [PubMed]
45. Jurman, G.; Riccadonna, S.; Furlanello, C. A comparison of MCC and CEN error measures in multi-class prediction. *PLoS ONE* **2012**, *7*, e41882. [CrossRef] [PubMed]
46. Gorodkin, J. Comparing two K-category assignments by a K-category correlation coefficient. *Comput. Biol. Chem.* **2004**, *28*, 367–374. [CrossRef]
47. Hall, M.; Frank, E.; Holmes, G.; Pfahringer, B.; Reutemann, P.; Witten, I.H. The WEKA data mining software: An update. *ACM SIGKDD Explor. Newsl.* **2009**, *11*, 10–18. [CrossRef]
48. Breiman, L. Bagging predictors. *Mach. Learn.* **1996**, *24*, 123–140. [CrossRef]
49. Bouckaert, R.R. *Bayesian Network Classifiers in Weka*; Working Paper No. 14/2004; University of Waikato: Hamilton, New Zealand, 2004.
50. Kohavi, R. The Power of Decision Tables. In Proceedings of the 8th European Conference on Machine Learning, Crete, Greece, 25–27 April 1995; Springer: Berlin/Heidelberg, Germany, 1995; pp. 174–189.
51. Quinlan, R. *C4.5: Programs for Machine Learning*; Morgan Kaufmann Publishers: San Mateo, CA, USA, 1993.
52. Breiman, L. Random Forests. *Mach. Learn.* **2001**, *45*, 5–32. [CrossRef]

Article

A Novel Method for Fault Diagnosis of Rotating Machinery

Meng Tang, Yaxuan Liao, Fan Luo * and Xiangshun Li

School of Automation, Wuhan University of Technology, Wuhan 430070, China; tangmeng@whut.edu.cn (M.T.);
lyx19971020@whut.edu.cn (Y.L.); lixiangshun@whut.edu.cn (X.L.)
* Correspondence: dr_luofan@whut.edu.cn

Abstract: When rotating machinery fails, the consequent vibration signal contains rich fault feature information. However, the vibration signal bears the characteristics of nonlinearity and nonstationarity, and is easily disturbed by noise, thus it may be difficult to accurately extract hidden fault features. To extract effective fault features from the collected vibration signals and improve the diagnostic accuracy of weak faults, a novel method for fault diagnosis of rotating machinery is proposed. The new method is based on Fast Iterative Filtering (FIF) and Parameter Adaptive Refined Composite Multiscale Fluctuation-based Dispersion Entropy (PARCMFDE). Firstly, the collected original vibration signal is decomposed by FIF to obtain a series of intrinsic mode functions (IMFs), and the IMFs with a large correlation coefficient are selected for reconstruction. Then, a PARCMFDE is proposed for fault feature extraction, where its embedding dimension and class number are determined by Genetic Algorithm (GA). Finally, the extracted fault features are input into Fuzzy C-Means (FCM) to classify different states of rotating machinery. The experimental results show that the proposed method can accurately extract weak fault features and realize reliable fault diagnosis of rotating machinery.

Keywords: fast iterative filtering; parameter adaptive refined composite multiscale fluctuation-based dispersion entropy; rotating machinery; fault diagnosis

Citation: Tang, M.; Liao, Y.; Luo, F.; Li, X. A Novel Method for Fault Diagnosis of Rotating Machinery. *Entropy* **2022**, *24*, 681. https://doi.org/10.3390/e24050681

Academic Editors: Jan Kozak and Przemysław Juszczuk

Received: 6 April 2022
Accepted: 8 May 2022
Published: 12 May 2022

Publisher's Note: MDPI stays neutral with regard to jurisdictional claims in published maps and institutional affiliations.

Copyright: © 2022 by the authors. Licensee MDPI, Basel, Switzerland. This article is an open access article distributed under the terms and conditions of the Creative Commons Attribution (CC BY) license (https://creativecommons.org/licenses/by/4.0/).

1. Introduction

Rotating machinery, such as electric motors, centrifugal pumps, and turbine engines, represent the most widely used mechanical equipment in industrial processes [1]. The mechanical equipment usually operate under unstable loads and harsh working conditions, thus various failures of their critical components, such as bearing damage and impeller damage, are inevitable. The operating states of rotating machinery directly affect the productivity and safety of the industrial sector. Therefore, accurate and reliable fault diagnosis of rotating machinery is of great practical significance [2].

The key to fault diagnosis of rotating machinery is to extract fault features from vibration signals. Vibration signals are nonlinear and nonstationary [3], and are easily interfered by noise, thus it is difficult to extract hidden features. Therefore, it is necessary to combine the appropriate time–frequency analysis method with the entropy measurement method to extract the hidden tiny fault features. The first step is to choose the appropriate signal processing method. Studies have shown that when the fault signal is disturbed by noise, traditional time–frequency analysis techniques, such as Fourier transform (FFT) and Wavelet Transform (WT) cannot accurately extract fault features [4,5]. The more commonly used method is the Empirical Mode Decomposition (EMD) method proposed by Huang et al. in 1998 [6]. The EMD can adaptively decompose the signal into the sum of finite intrinsic mode functions (*IMF*), each *IMF* component represents a set of characteristic scale signals, and the feature extraction of each component can better reveal the fault information intrinsic characteristics. However, EMD suffers from modal aliasing, end-point effects, and a lack of rigorous mathematical framework for using envelopes in an iterative manner [7]. Although the Ensemble Empirical Mode Decomposition (EEMD) [8] optimized on the basis of EMD can effectively improve the problem of mode aliasing, and the Fast

Ensemble Empirical Mode Decomposition (FEEMD) [9] further improves the calculation speed, neither escape the drawbacks of using envelopes in an iterative fashion without a rigorous mathematical framework. Subsequently, Dragomiretskiy K et al. proposed a new adaptive Variational Mode Decomposition (VMD) method. The method is a non-recursive variational decomposition model, and the optimal solution of the variational model is iteratively searched by the alternating direction multiplier method, thereby determining the center frequency and bandwidth of each mode. It avoids mode mixing in EMD, and has better robustness to noise [10]. However, VMD suffers from relatively slow computational efficiency, and its performance depends heavily on its two input parameters, namely the penalty factor and the number of decomposition modes [11]. The Iterative Filtering (IF) method proposed by Lin et al. and its derivatives [12], such as the Adaptive Local Iterative Filtering (ALIF) method [13], the Fast Iterative Filtering (FIF) method [14] can produce results similar to EMD-based algorithms, with the important advantage that their convergence and stability are guaranteed. Moreover, the FIF method uses a fixed low-pass filter function to replace the envelope mean curve in the EMD method, which solves the problem of EMD lacking a strict mathematical framework. Meanwhile, the FIF method is unaffected by mode aliasing, and mode splitting can be easily avoided by adjusting the value of the stopping criterion parameter [4]. Furthermore, FIF greatly improves the calculation speed on the basis of ensuring decomposition accuracy, with small decomposition error, good noise robustness, and can achieve efficient and accurate signal decomposition [15]. Therefore, this paper adopts the FIF method to decompose the vibration signal of rotating machinery.

The components of the vibration signal following decomposition by FIF contain rich fault information. Moreover, the components of vibration signals in different states of rotating machinery show different complexity, so the entropy parameter can be used to extract the fault information [16]. Approximate Entropy (ApEn), Sample Entropy (SampEn), Fuzzy Entropy (FE), and Permutation Entropy (PE) are widely used in the field of rotating machinery fault diagnosis to measure the complexity of vibration signals [17–19]. However, ApEn and Multiscale Approximate Entropy include the comparison of their own data segments in the calculation process, and their calculation depends on the data length. If the data length is short, the obtained value is usually smaller than the actual value. The SampEn is an improvement on the approximate entropy. It does not include the comparison of its own data segments, and has higher calculation accuracy and better consistency. However, SampEn and its improvements also have clear shortcomings: Firstly, SampEn and its improvements use Heaviside functions to measure the complexity of time series, resulting in inaccurate estimates in practical applications [20]. Secondly, SampEn and its improvements are computationally inefficient, especially for long time series. FE and its improvements replace the Heaviside function with a fuzzy membership function that is insensitive to background noise and highly sensitive to dynamic changes, but it is computationally inefficient [16]. PE is a method to measure the complexity of chaotic time series. PE has high computational efficiency, can be used to calculate huge datasets, and exhibits good anti-noise performance. However, the main disadvantage of PE is that it is prone to generating undefined entropy values for short-term time series and cannot classify well-defined patterns for a specific design [21]. In order to overcome the above problems, Hamed Azami et al. proposed a nonlinear time complexity evaluation method of Dispersion Entropy (DE). DE can generate reliable entropy values, is insensitive to noise interference, can accurately capture signal characteristics, and calculate with high efficiency [22]. Subsequently, in order to improve the extraction ability of hidden fault features, Hamed Azam et al. continued to propose the Refined Composite Multiscale Fluctuation-based Dispersion Entropy (RCMFDE), which can more accurately analyze the complexity of nonlinear time series under various scale factors, with more stable entropy values [23].

However, in the RCMFDE method, there are two key parameters (i.e., embedding dimension and class number) that need to be manually selected in advance. Furthermore, the parameter setting of the RCMFDE algorithm will affect the final processing result. If the parameter settings are unreasonable, the hidden tiny fault features may not be accurately extracted, resulting in misclassification. Aiming at the determination of the embedding dimension m and the class number c in the RCMFDE algorithm, this paper proposes a Parameter Adaptive Refined Composite Multiscale Fluctuation-based Dispersion Entropy (PARCMFDE). The method takes skewness as the objective function, and uses a Genetic Algorithm (GA) to optimize parameters of RCMFDE. PARCMFDE can automatically and effectively determine the important parameters of RCMFDE, so as to describe the complexity and uncertainty of time series more accurately, and achieve the purpose of extracting the features of hidden faults. In view of the shortcomings of existing methods, relevant research is carried out, and the main contributions are as follows:

(1)   PARCMFDE based on GA is proposed, which overcomes the insufficiency of experience-based parameter selection. PARCMFDE can more accurately extract tiny fault features hidden in vibration signals of rotating machinery.
(2)   A fault diagnosis method for rotating machinery based on FIF, PARCMFDE and Fuzzy C-Means (FCM) is proposed, which can classify rotating machinery faults accurately and automatically without depending on the length of data samples.
(3)   The effectiveness of the method is verified by the bearing data of Case Western Reserve University and the experimental data of centrifugal pumps obtained by building a water circulation experimental system. Compared with other methods, it shows that feature extraction of PARCMFDE is more accurate and stable, and the rotating machinery fault diagnosis method based on FIF, PARCMFDE and FCM exhibits better classification effect.

This paper is mainly divided into the following sections: Section 2 briefly introduces the basic principles and characteristics of the FIF algorithm. In Section 3, the principle of PARCMFDE is introduced and compared with RCMFDE and Multiscale Sample Entropy (MSE) and Multiscale Fluctuation-based Dispersion Entropy (MFDE). Section 4 briefly introduces the principle and evaluation index of FCM. Section 5 presents the method of fault diagnosis of rotating machinery. Section 6 verifies the effectiveness of the method and compares it with other vibration signal fault diagnosis methods through the bearing data of Case Western Reserve University and experimental data from centrifugal pumps obtained by building a water circulation experimental system. Section 7 provides the conclusion.

2. Fast Iterative Filtering

The key idea of Fast Iterative Filtering is to iteratively subtract the simple oscillatory components contained in the signal from the signal itself, the so-called IMFs, by approximating the moving average of the signal, thereby separating the simple oscillatory components in the signal [14]. The approximate moving average is computed by convolution with the window/filter function w. Consider a raw vibration signal $s(x)$, define a window/filter function w is a non-negative even function in the range of $C^0([-L, L]), L > 0$. The Fokker–Plank filter is used here, and $\int_R w(z)z = \int_{-L}^{L} w(z)z = 1$, $\hat{s}$ denotes the Fourier transform of s, DFT denotes the discrete Fourier transform, and $IDFT$ denotes the inverse discrete Fourier transform. The specific implementation process of FIF is as follows:

(1)   Calculate the length L of the corresponding filter w of the signal $s(x)$:

$$L := 2 \left\lfloor \xi \frac{N}{k} \right\rfloor \tag{1}$$

where N is the total number of sampling points of the signal $s(x)$, k is the number of its extreme points, and ξ is a tuning parameter, which is usually fixed around 1.6 for the Fokker–Plank filter.
(2)   Calculate the discrete Fourier transform of the signal $s(x)$ and the corresponding filter w, denoted as $DFT(s)$ and $DFT(w)$, respectively.

(3)   Calculate $\hat{s}_{m+1}$:

$$\hat{s}_{m+1} = (I - diag(DFT(w)))^m DFT(s) \tag{2}$$

(4)   Calculate $N_0 \in N$ and IMF:

$$\frac{N_0{}^{N_0}}{(N_0 + 1)^{N_0+1}} < \frac{\delta}{||s_m||_2} \tag{3}$$

$$IMF = \sum_{k=0}^{n-1} u_k (1 - \lambda_k)^{N_0} \sigma_k = IDFT((I - D)^{N_0} DFT(s)) \tag{4}$$

where $\delta > 0$, represents the required precision; N_0 represents the number of iterations required to achieve the required precision δ when calculating a specific IMF; σ_k represents the kth element of the Fourier transform of the signal s; λ_k represents the kth eigenvalue; u_k is the kth eigenvector; I is the identity matrix; D is the diagonal matrix, whose diagonal is the eigenvalue.

(5)   Judgment of inner loop stop condition: if the stop standard SD is met, then stop the inner loop, otherwise let $m = m + 1$ repeat steps (3)–(5), the stop standard SD is calculated by the following formula:

$$SD := \frac{||s_{m+1} - s_m||_2}{||s_m||_2} < \delta, \forall m \geq N_0. \tag{5}$$

(6)   Calculate the IMF component and the new s:

$$IMF = IMF \cup \{IDFT(\hat{s}_m)\} \tag{6}$$

$$s = s - IDFT(\hat{s}_m). \tag{7}$$

(7)   Judgment of outer loop stop condition: Calculate the extreme point of s, if there is only one extreme point of s or less, the outer loop stops, otherwise repeat steps (1)–(7).

(8)   Extract the final IMF component

$$IMF = IMF \cup \{s\}. \tag{8}$$

In short, the FIF method includes two processes: inner loop and outer loop. The purpose of the inner loop is to filter out the IMF components of each order. The purpose of the outer loop is to end the process of extracting the IMF component of the inner loop. When the residual obtained by removing all IMF components from the original signal $s(x)$ contains only one or less extreme points, the outer loop stops.

3. Parameter Adaptive Refined Composite Multiscale Fluctation Based Dispersion Entropy

3.1. Refined Composite Multiscale Fluctuation-Based Dispersion Entropy

Refined Composite Multiscale Fluctuation-based Dispersion Entropy (RCMFDE) accounts for the shortcomings of Multiscale Fluctuation-based Dispersion Entropy (MFDE) in the process of coarse-graining, which has low computational efficiency and a high probability of invalid entropy values. The entropy value is more stable, the operation speed is faster, and the probability of invalid entropy occurrence is greatly reduced. The specific process of RCMFDE is as follows:

(1)   For a given univariate signal $L : v = \{v_1, v_2, \ldots, v_L\}$. Dividing v into non-overlapping segments of length τ is called the scale factor. Construct a composite coarse-grained time series:

$$x_k^{(\tau)}(i) = \frac{1}{\tau} \sum_{c=(i-1)\tau+k}^{i\tau+k-1} v_c, 1 \leq i \leq \left\lfloor \frac{L}{\tau} \right\rfloor = n, k = 1, 2, \ldots, \tau \tag{9}$$

where k represents the coarse-grained sliding number of the scale factor under τ.

(2)   Map $X = \{x_1, x_2, \ldots, x_n\}$ to $Y = \{y_1, y_2, \ldots, y_n\}$ through the normal cumulative distribution function (NCDF) as follows:

$$y_k(i) = \frac{1}{\sigma\sqrt{2\pi}} \int_{-\infty}^{x_k(i)} e^{\frac{-(t-\mu)^2}{2\sigma^2}} dt \tag{10}$$

where σ is the standard deviation of the time series X and μ is the mean.

(3)   Linearly assign $y_k(i)$ to an integer $z_k(i)$ from 1 to c as follows:

$$z_k^c(i) = round(c \times y_k(i) + 0.5) \tag{11}$$

(4)   Time series $Z_k^{m,c}(j) = \{z_k^c(j), z_k^c(j+d), \ldots, z_k^c(j+(m-1)d)\}, j = 1, 2, \ldots, n-(m-1)d$, $(m-1)$ is the embedding dimension and d is the time delay.

(5)   Each time series $Z_k^{m,c}(j)$ maps to a fluctuation-based dispersion pattern $\pi_{u_0 u_1 \ldots u_{m-1}}$, where $z_k^c(j) = u_0, z_k^c(j+d) = u_1, z_k^c(j+(m-1)d) = u_{m-1}$. The number of fluctuation-based dispersion modes assignable to each time series $Z_k^{m,c}(j)$ is equal to $(2c-1)^{m-1}$.

(6)   For each fluctuation-based dispersion pattern $\pi_{u_0 u_1 \ldots u_{m-1}}$, the relative frequency is obtained by Equation (12).

$$W(\pi_{v_0 \ldots v_{m-1}}) = \frac{\#\{j | j \leq n-(m-1)d, Z_k^{m,c}(j) \, has \, type \, \pi_{v_0 \ldots v_{m-1}}\}}{n-(m-1)d} \tag{12}$$

where # means cardinality.

(7)   The Refined Composite Multiscale Fluctuation-based Dispersion Entropy (RCMFDE) is obtained by the following formula:

$$RCMFDE = -\sum_{\pi=1}^{(2c-1)^{m-1}} \sum_{k=1}^{\tau} W(\pi_{v_0 \ldots v_{m-1}}) \times \ln(\sum_{k=1}^{\tau} W(\pi_{v_0 \ldots v_{m-1}})). \tag{13}$$

The RCMFDE algorithm has four parameters, which are the embedding dimension m, the class number c, the delay time d, and the maximum scale factor $\tau_{\max}$. A study [23] pointed out that the results of the RCMFDE do not change significantly with the time delay d, and a different embedding dimension m and class number c have influence on RCMFDE. The higher the number of dispersion modes based on potential fluctuations $(ln((2c-1)m-1))$, the higher the RCMFDE value [23]. When m and c are too small, the ability of RCMFDE to detect signal mutations is lower, but the larger m and c are, the longer the algorithm runs. For samples of the same category, the feature vectors should be as similar as possible; for samples of different categories, the feature vectors should be significantly different. If the parameter selection is not suitable, it may cause instability of entropy value, incomplete extraction of hidden feature information or excessively long operation time, rendering it difficult to classify correctly. Therefore, it is necessary to select appropriate values of the class number c and the embedding dimension m.

3.2. Genetic Algorithm

Genetic Algorithm (GA) is a computational model that simulates the biological evolution process of natural selection and genetic mechanism of Darwin's theory of biological evolution. It is a method to search for optimal solutions by simulating the natural evolution process [24]. When solving more complex combinatorial optimization problems, this algorithm can usually obtain better optimization results faster than some conventional optimization algorithms. The specific process of GA is as follows:

(1)   Set the evolutionary generation counter $t = 0$, set the maximum evolutionary generation T, and randomly generate M individuals as the initial population $P(0)$.

(2)   Determine the fitness function and calculate the fitness of each individual in the population $P(t)$.

(3)   Apply the selection operator, the crossover operator, and the mutation operator to the population $P(t)$, and then obtain the next generation population $P(t+1)$.

(4) If $t = T$, or the change of the fitness function value reaches the given threshold, the optimal fitness individual is used as the optimal solution output. If $t < T$, and the change of the fitness function value is greater than the given threshold, define $t = t + 1$, and repeat steps (2)–(4).

3.3. Parameter Adaptive Refined Composite Multiscale Fluctuation-Based Dispersion Entropy

The settings of the embedding dimension m and the class number c of RCMFDE affect its final entropy value, entropy value stability and operation time. If the parameter settings are unreasonable, the best processing effect will not be achieved. Therefore, a suitable method is needed to adaptively select the embedding dimension m and the class number c in RCMFDE. For the above problems, this paper proposes a Parameter Adaptive Refined Composite Multiscale Fluctuation-based Dispersion Entropy (PARCMFDE). The method performs parameter optimization through Genetic Algorithm (GA) to determine the optimal parameter combination of the embedding dimension m and the class number c in RCMFDE. Figure 1 shows the flowchart of using GA to optimize the parameters of RCMFDE. The steps of parameter optimization in PARCMFDE are described as follows:

(1) Determine the approximate range and encoding length of the embedding dimension m and the class number c, and perform real encoding. The constraint function of the parameters is $(2c - 1)^{m-1} < \left\lfloor \frac{L}{\tau_{\max}} \right\rfloor$, where L represents the data length, $\tau_{\max}$ is the maximum scale factor, and $\lfloor . \rfloor$ represents rounding.

(2) Initialization: Set the evolutionary generation counter $t = 0$, set the maximum evolutionary generation T to 200, and randomly generate M individuals as the initial population $P(0)$.

(3) Calculate the fitness of each individual in the population $P(t)$. Skewness can characterize the overall profile of a set of data. The larger the absolute value of skewness, the more problematic the performance of the mean, and the smaller the absolute value of skewness, the more reliable the mean [25]. Therefore, this paper selects the square function of RCMFDE skewness (ske) as the fitness function and finds its minimum value. The RCMFDE at all scales of the time series $S = \{s_1, s_2, \ldots, s_n\}$ are composed of the series $H_P(x) = \{H_p(1), \ldots, H_p(m)\}$, and the skewness ($ske$) is calculated by the following formula:

$$ske = E[H_P(X) - H_p^m(X)]^3 / [H_p^d(X)]^3 \tag{14}$$

where $H_p^m(X)$ is the mean of $H_p(X)$, $H_p^d(X)$ is the standard deviation of $H_p(X)$, and $E[.]$ represents the mathematic expectation. The fitness function is taken as $f = ske^2$.

(4) Apply selection operator, crossover operator and mutation operator to the population. After the population $P(t)$ is selected, crossed and mutated, the next generation population $P(t + 1)$ is obtained.

(5) Judgment of termination condition: If $t \geq T$, or the change of fitness function value is less than 10^{-6}, then the individual with the smallest fitness obtained in the evolution process is used as the optimal solution, and the optimal parameter combination m, c is obtained. If $t < T$, and the change of the fitness function value is greater than 10^{-6}, define $t = t + 1$, and repeat steps (3)–(5).

(6) Use the parameter-optimized RCMFDE to extract the features of the reconstructed rotating machinery vibration signal.

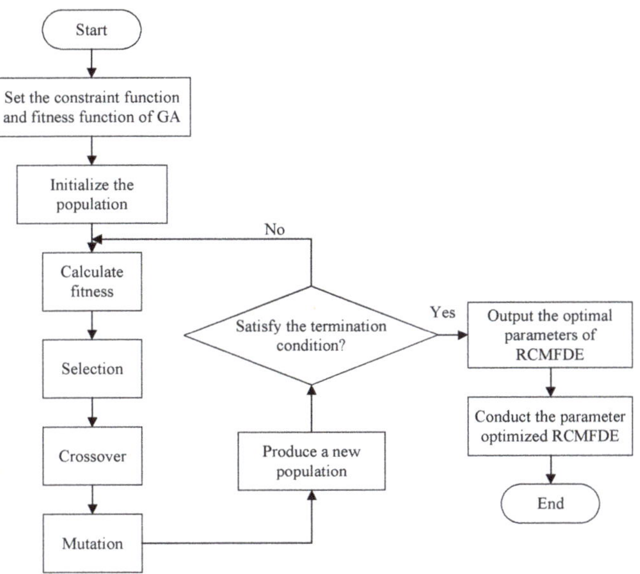

Figure 1. The flowchart of using GA to optimize the parameters of RCMFDE.

3.4. Error Analysis and Comparison Results

To demonstrate the effectiveness of PARCMFDE in assessing the complexity and irregularity of time series, PARCMEDE of white Gaussian and pink noise signals are calculated and compared with RCMFDE, Multiscale Fluctuation-based Dispersion Entropy(MFDE), and Multiscale Sample Entropy (MSE). Furthermore, to compare the accuracy of the complexity measures with different entropies, 10 groups of white Gaussian noise and 10 groups of pink noise were randomly generated. For unified comparison, the maximum scale factor τ_{max} of the entropy is set to 10, and the time delay d is set to 1. Among them, the embedding dimension m and the class number c of PARCMFDE take 2 and 3, respectively, in white Gaussian noise, and take $m = 2$, $c = 21$ in pink noise. Based on experience, the embedding dimension m and the class number c of RCMFDE and MFDE take the default values of $m = 3$ and $c = 6$, respectively. The MSE takes the default value of $m = 2$, $r = 0.15 \times \sigma$, and σ represents the standard deviation of the signal. Figure 2a,b show the time-domain waveforms of white Gaussian noise and pink noise. Figure 3a,b plot the error bars of different entropy algorithms for white Gaussian noise and pink noise, respectively. The entropy value of pink noise time series should remain almost constant, while the entropy value of white Gaussian noise data should decrease monotonically [26]. It can be seen from Figure 3a that with the increase of the scale factor τ, the average curves of the four entropies of white Gaussian noise (i.e., RCMFDE, PARCMFDE, MFDE and MSE) bear a downward trend which indicates that the four algorithms have good sensitivity in detection complexity. Furthermore, the standard deviation of PARCMFDE for white Gaussian noise at each scale is smaller than that of RCMFDE, MFDE and MSE, indicating that PARCMFDE has higher accuracy than the other three algorithms on the complexity measure of white Gaussian noise. It can be seen from Figure 3b that the MSE of pink noise exhibits a slight downward trend with large fluctuations, but the PARCMFDE remains almost unchanged, indicating that the RCMFDE is better than the MSE. Furthermore, the standard deviation of PARCMFDE for pink noise at each scale is smaller than that of RCMFDE, MFDE and MSE, indicating that PARCMFDE can provide a more accurate complexity estimate for pink noise [27]. That is, PARCMFDE is effective in complexity measurement and feature extraction of nonstationary signals. The standard deviation of MSE is much larger than the other three methods, indicating that MSE is insufficiently accurate regarding the complexity measurement and feature extraction of nonstationary signals.

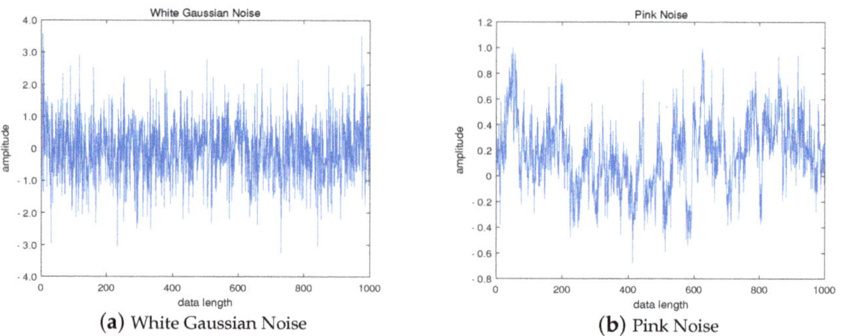

(a) White Gaussian Noise

(b) Pink Noise

Figure 2. Time Domain Waveform of (**a**) White Gaussian Noise and (**b**) Pink Noise.

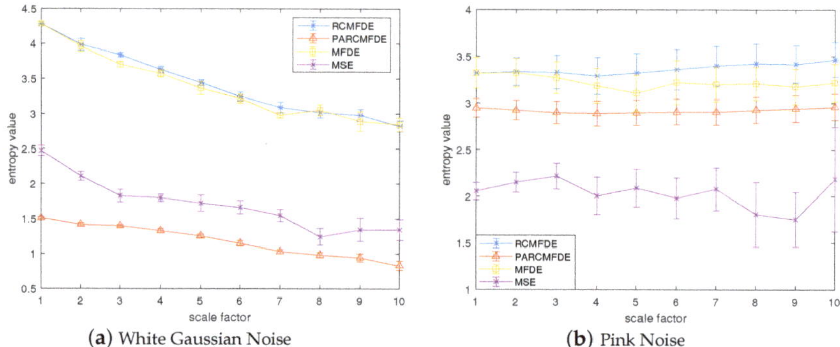

(a) White Gaussian Noise

(b) Pink Noise

Figure 3. Entropy of different algorithms of (**a**) White Gaussian Noise and (**b**) Pink Noise.

4. Fuzzy C-Means Clustering

Fuzzy C-means (FCM) clustering algorithm is the most widely used fuzzy clustering algorithm based on objective function. It obtains the membership degree of each sample point to all class centers by optimizing the objective function, so as to determine the class of the sample point to achieve the purpose of automatically classifying the sample data [28].

Let $R = \{r_1, r_2, \ldots, r_n\}$ be the set of data samples, and n is the number of samples. $C = \{c_1, c_2, \ldots, c_t\}$ is the cluster center vector, and t is the total number of clusters. The FCM clustering algorithm minimizes the objective function shown in Equation (15) through continuous iteration of the least squares method, and its constraints are shown in Equation (16).

$$J_m = \sum_{i=1}^{t} \sum_{k=1}^{n} [\mu_{ik}(r_k)]^m ||r_k - c_i|| \tag{15}$$

$$\sum_{i=1}^{t} \mu_{ik}(r_k) = 1 \tag{16}$$

where r_k is the kth sample point to be clustered, μ_{ik} is the degree of membership of r_k to the ith cluster center c_i, m is the weight index of the degree of membership, generally $m = 2$.

The cluster center c_i and the membership matrix μ_{ik} are randomly selected initially. Then iteratively calculate through Equations (17) and (18), and stop until the change of the objective function is less than the threshold.

$$C_i = \frac{\sum\limits_{k=1}^{n} r_k \mu_{ik}^2}{\sum\limits_{k=1}^{n} \mu_{ik}^2}, i = 1, 2, \ldots, t \tag{17}$$

$$\mu_{ik} = \frac{1}{\sum\limits_{j=1}^{t} \frac{||r_k - c_i||_2^{2/(m-1)}}{||r_k - c_j||_2^{2/(m-1)}}} \tag{18}$$

The average fuzzy entropy E, classification coefficient S and classification accuracy Acc are used to analyze and evaluate the clustering effect of the fuzzy C-means, which are, respectively, defined as:

$$E = -\frac{1}{n} \sum_{i=1}^{t} \sum_{k=1}^{n} \mu_{ik} \ln \mu_{ik} \tag{19}$$

$$S = \frac{1}{n} \sum_{i=1}^{t} \sum_{k=1}^{n} \mu_{ik}^2 \tag{20}$$

$$Acc = \frac{1}{n} \left(\sum_{i=1}^{n} 1\{R_v == \hat{R}_V\} \right) \tag{21}$$

where R_v and $\hat{R}_V$ denote the actual class and the class assigned by FCM on the test dataset, n is the number of samples in the test dataset.

The ambiguity of clustering is represented by the average fuzzy entropy E, which reflects the distribution characteristics of the clustering dataset, so it can be used as an index to judge the clustering effect and correctness. The smaller the ambiguity, the higher the order of the system. The classification coefficient S measures the overlap between clusters, and the closer it is to 1, the more effective the clustering result [29]. Therefore, the closer E is to 0, the closer S is to 1, and the closer Acc is to 100%, the better the sample clustering effect is.

5. Proposed Fault Diagnosis Method

In order to quickly and reliably extract the characteristic information of the vibration signal and realize the automatic classification of the working state of the rotating machinery, a new fault diagnosis method of the rotating machinery based on FIF-PARCMFDE and Fuzzy C-means (FCM) is proposed. The specific process is as follows:

(1) Use the accelerometer to collect the original vibration signal $y(x)$ of the rotating machinery in different states.
(2) The FIF algorithm decomposes the collected vibration signal $y(x)$ to obtain a series of IMFs.
(3) Calculate the correlation coefficient of each order *IMF*, and select components with a correlation coefficient greater than 0.4 for reconstruction.
(4) The PARCMFDE of the reconstructed signal $S(x)$ is calculated, and the corresponding entropy value is used as the characteristic information reflecting the working state of the rotating machinery.
(5) Input the training set into FCM to obtain the cluster centers.
(6) Input the testing set and cluster centers into FCM to automatically classify the working state of rotating machinery.

The block diagram of the proposed fault diagnosis method is shown in Figure 4.

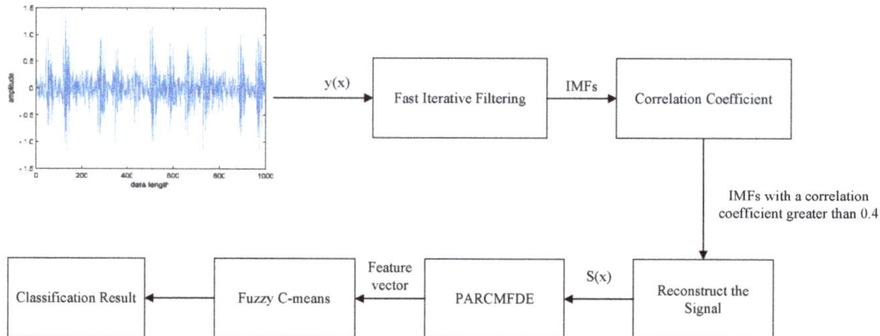

Figure 4. Block diagram of the proposed fault diagnosis method.

6. Experimental Verification

In this section, we apply the proposed fault diagnosis method to the bearing vibration signal of Case Western Reserve University and centrifugal pump vibration signals obtained by building a water circulation experimental system. It is compared with some similar commonly used methods to evaluate the effectiveness and superiority of our method.

6.1. Experiment 1: Bearing Data From Cwru

6.1.1. Experimental Setup

The experimental data adopts the vibration signal from the Bearing Data Center of Case Western Reserve University [30]. Vibration data are collected using accelerometers, which are mounted on the drive end bearing housing. The outer race, rolling element and inner race of rolling bearings are machined using electric sparks to simulate single point crack failure of the outer race, rolling element and inner race. The selected test bearing model is 6205-2RS, the rotational speed is 1750 r/min, and the sampling frequency is the vibration data of 12 kHz. Analyze the vibration data of normal, inner ring failure, outer ring failure, and rolling element failure. Twenty samples for each of the four bearing conditions were obtained through a non-overlapping sliding window of length 5500 points, that is, each sample contained 5500 points. The first 10 samples for each of the four bearing conditions are selected as the training set, and the remaining 10 samples for each of the four bearing conditions are selected as the testing set.

6.1.2. Comparison And Analysis

To verify the effectiveness of the FIF-PARCMFDE-FCM method for bearing fault diagnosis, under the same test conditions, the vibration data of four operating states of normal bearing, inner race fault, rolling element fault and outer race fault were classified and identified.

The first 1000 points of the original vibration signal in the four states of the bearing are selected, as shown in Figure 5. It can be seen from Figure 5 that the vibration signals in the four states of the bearing are quite different and bear distinct characteristics, but they are not enough to be directly classified according to the waveform.

FIF decomposes the vibration signals in the four states of the bearing, and selects the first five-order *IMF* components for comparison. The *IMF* components with larger correlation coefficients can well retain the fault characteristic information of the signals [31]. The correlation coefficient between the first five-order *IMF* components and the original signal is shown in Table 1, and the component with the correlation coefficient greater than 0.4 is selected to reconstruct the signal. Therefore, the outer race fault selects IMF1 as the reconstruction signal, the inner race fault and rolling element fault select IMF1 and IMF2 for reconstruction, and the normal signal selects IMF1, IMF2 and IMF4 for reconstruction.

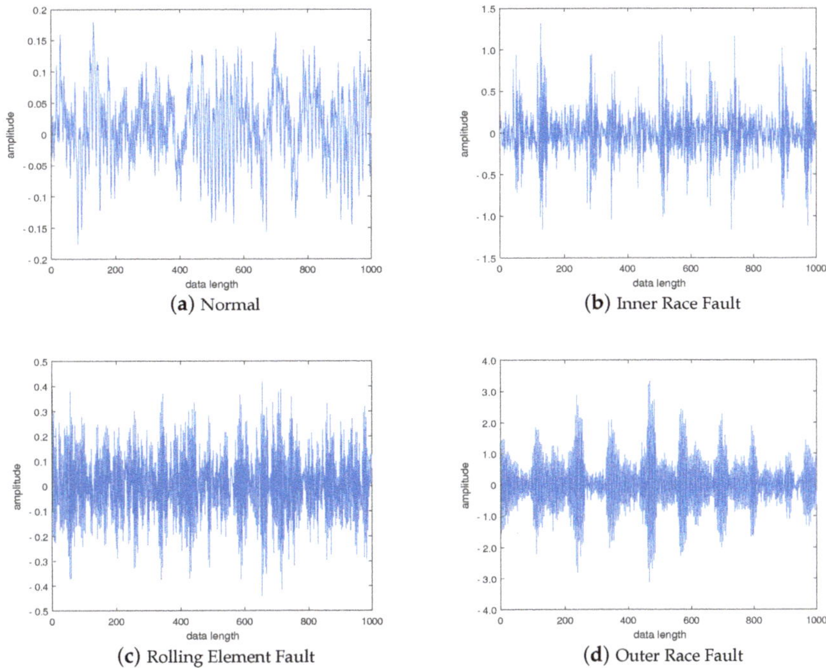

Figure 5. Time domain waveform of vibration signal under (**a**) Normal, (**b**) Inner Race Fault, (**c**) Rolling Element Fault and (**d**) Outer Race Fault of bearing.

Table 1. Correlation coefficients of bearings in different states.

Bearing States	Correlation Coefficients				
	IMF 1	**IMF 2**	**IMF 3**	**IMF 4**	**IMF 5**
Normal	0.62	0.653	0.3923	0.5282	0.33
Outer Race Fault	0.9992	0.2421	0.0525	0.0203	0.0117
Inner Race Fault	0.9057	0.5787	0.2222	0.0509	0.0062
Rolling Element Fault	0.9708	0.4376	0.2246	0.1306	0.0661

The reconstruction signals s of different states of the bearing are selected. The skewness is used as the objective function in GA, and set the maximum evolutionary generation T to 200, the threshold for the fitness function to change is 10^{-6}. The parameters of RCMFDE are optimized by GA. Calculate the PARCMFDE, RCMFDE and MSE of the reconstructed signal s, where the scale factor is 10, and the embedding dimension m and class number c of PARCMFDE under different conditions are shown in Table 2. Based on experience, RCMFDE takes default values $m = 3, c = 6$, MSE takes default value $m = 2$, $r = 0.15 \times \sigma$, σ represents the standard deviation of the signals s. It can be seen from Figures 6–8 that PARCMFDE, RCMFDE and MSE can all distinguish the four states of the bearing, indicating that the three methods can effectively extract the hidden features of different states of the bearing. However, compared with RCMFDE and MSE, PARCMFDE can distinguish the four states of the bearing more clearly, and is more suitable for further classification of bearing faults as a feature vector.

Table 2. PARCMFDE parameters of bearings in different states.

Bearing Status	Embedding Dimension m	Class Number c
Outer Race Fault	2	232
Inner Race Fault	3	7
Rolling Element Fault	3	5
Normal	4	4

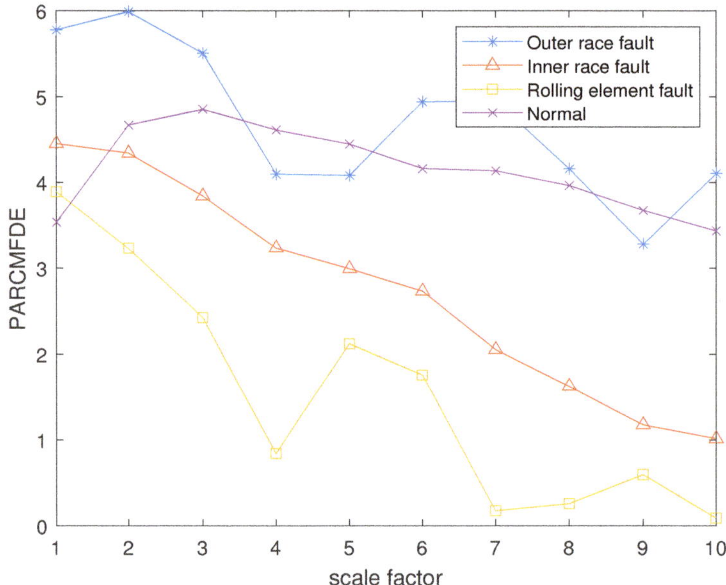

Figure 6. PARCMFDE in different states of bearing.

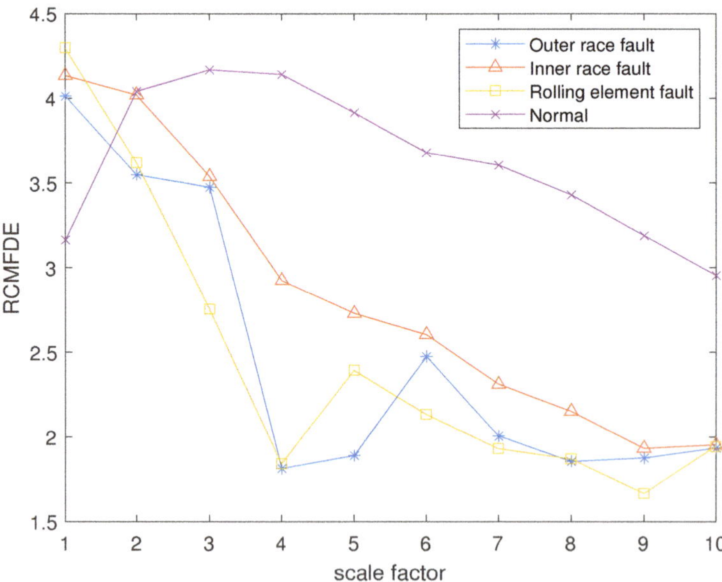

Figure 7. RCMFDE in different states of bearing.

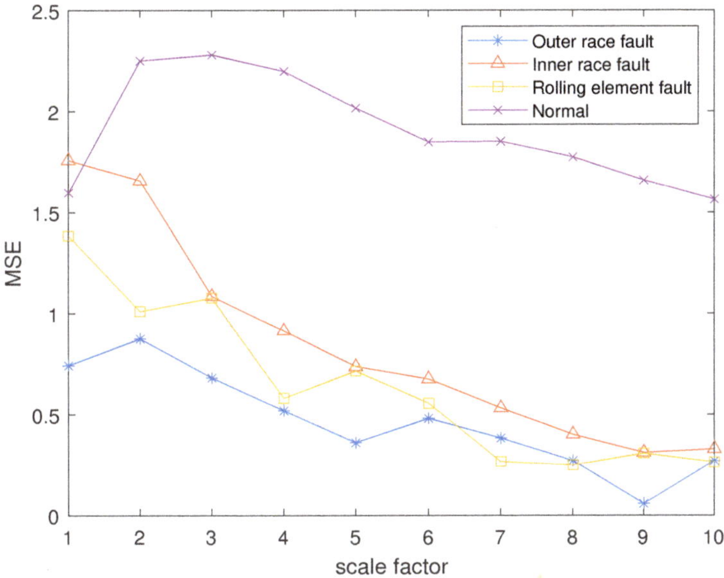

Figure 8. MSE in different states of bearing.

Take PARCMFDE, RCMFDE and MSE as the eigenvector matrix. Perform FCM cluster analysis on the eigenvector matrix of the training samples, and four cluster centers can be obtained. Then the obtained cluster centers and testing sample eigenvector matrix are input into FCM clustering. The clustering results are shown in Figures 9–11.

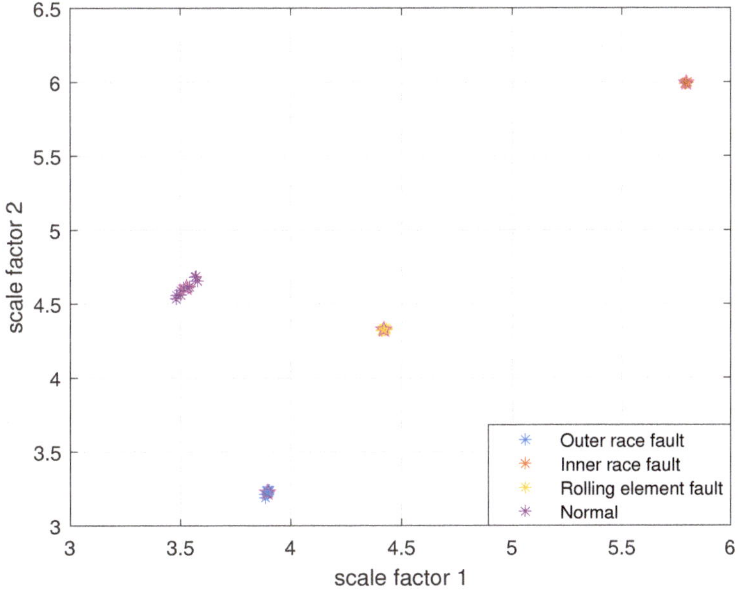

Figure 9. FIF-PARCMFDE-FCM clustering results of different bearing states.

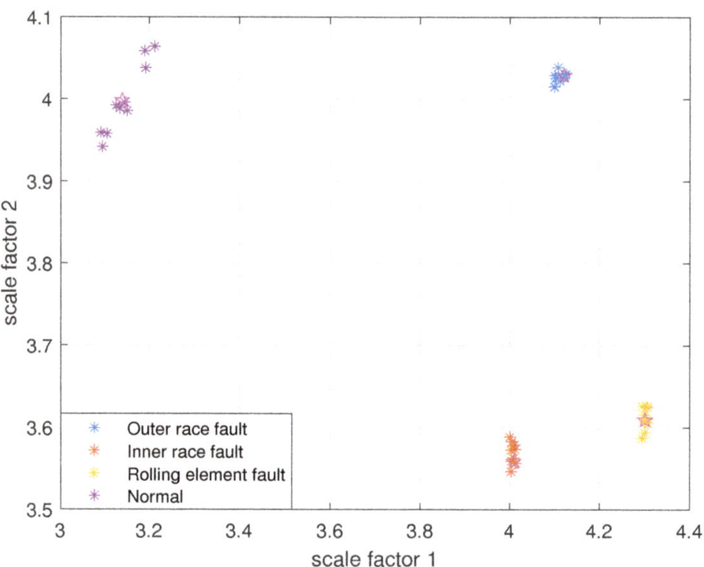

Figure 10. FIF-RCMFDE-FCM clustering results of different bearing state data.

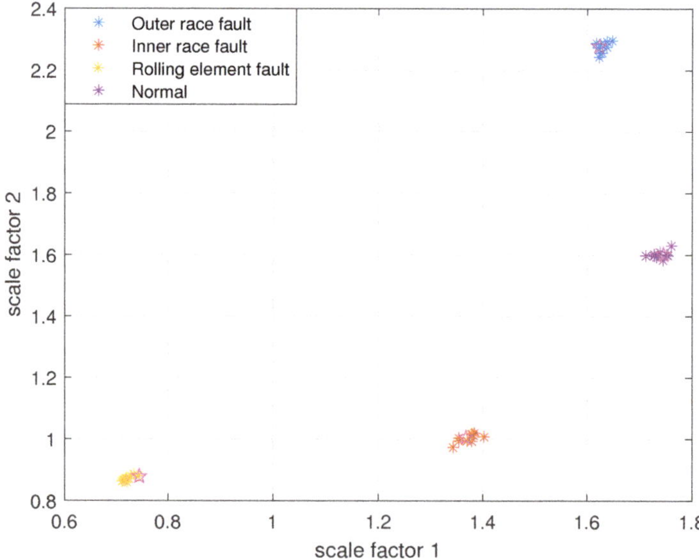

Figure 11. FIF-MSE-FCM clustering results of different bearing state data.

It can be seen from Figures 9–11 that the data points of the same state are concentrated around their respective cluster centers, and the data points of different states are separated from each other. In addition, the positions of the cluster centers obtained by different methods are different, and the degree of closeness of the data points distributed around the cluster centers is also different. In comparison, FIF-PARCMFDE-FCM have the best clustering effect, that is, the categories are most distinct, the clustering centers of various signals are far apart, and the data points of various types are compactly clustered around the clustering centers. Compared with FIF-RCMFDE-FCM and FIF-MSE-FCM, the class center distance of FIF-PARCMFDE-FCM method is larger, and the different signals are more clearly distinguished, indicating that the method has a better classification effect on various fault signals of rolling bearings.

The classification coefficient S, the average fuzzy entropy E and the classification accuracy Acc of each clustering result were calculated, respectively, and the clustering effect of the above three algorithms and the fault recognition rate were quantitatively compared. The clustering results of the above three algorithms are shown in Table 3. It can be seen from Table 3 that the classification accuracy Acc of the above four methods are all 100%. The classification coefficient S of the FIF-MSE-FCM method is 0.9913, the average fuzzy entropy E is 0.0306, and the clustering effect is poor. The classification coefficient based on FIF-PARCMFDE-FCM method is 0.9967, the average fuzzy entropy is 0.0123, and the clustering effect is the best, indicating that this method can achieve a more accurate and reliable fault diagnosis.

Table 3. FCM clustering effect of different bearing entropy algorithms.

Algorithms	Classification Coefficient S	Average Fuzzy Entropy E	Classification Accuracy Acc
FIF-PARCMFDE-FCM	0.9967	0.0123	100%
FIF-RCMFDE-FCM	0.9935	0.0239	100%
FIF-MSE-FCM	0.9913	0.0306	100%

6.2. Experiment 2: Experimental Data of Centrifugal Pump

6.2.1. Experimental Setup

To verify the effectiveness of the method, a water circulation system was built in Wuhan University of Technology, and the vibration signals of centrifugal pumps in different states were collected. In this experiment, a model CDL1-11FSWPG light-duty vertical multistage centrifugal pump was selected, with a rated speed of 2900 r/min, a lift of 61 m, and a rated flow of 1 m^3/h. According to GBT-29531-2013 pump vibration measurement and evaluation method, the vibration sensor measurement points of centrifugal pump are arranged, and vibration data in three directions of x, y, and z are collected at the same time. The measuring point in the x-axis direction is arranged on the pump casing, the measuring point in the z-axis direction is arranged on the base, and the measuring point in the y-axis direction is arranged at the outlet flange. The structure of the pump body and the arrangement of measuring points are shown in Figure 12.

Figure 12. Layout of measuring points of centrifugal pump.

According to the actual situation during operation of the centrifugal pump, there are rotor unbalance and air binding faults. The impeller is the main component of the rotor in the centrifugal pump. During actual operation, the impeller is in contact with the working medium, thus it is the rotor part most prone to failure. The laboratory conditions will simulate a centrifugal pump rotor unbalance fault with impeller damage, as shown in Figure 13. The centrifugal pump is not filled with the liquid to be conveyed before starting, or air will infiltrate the pump during operation, because the density of the gas is less than the density of the liquid, the centrifugal force generated is small, and the air cannot be expelled. The negative pressure generated by the fluid in the pump casing during centrifugal motion with the motor is not enough to suck the liquid into the pump casing, which is called the air binding phenomenon of the centrifugal pump. In this experiment, by tightening the exhaust screw of the centrifugal pump, and then removing the centrifugal pump, the air can enter the inner chamber of the centrifugal pump. After installing the centrifugal pump, the residual air cannot be discharged from the centrifugal pump through the exhaust screw, so as to set the air binding fault of the centrifugal pump, as shown in Figure 14.

Figure 13. Rotor unbalance fault setup: (**a**) Impeller in normal condition, (**b**) Impeller in damaged condition.

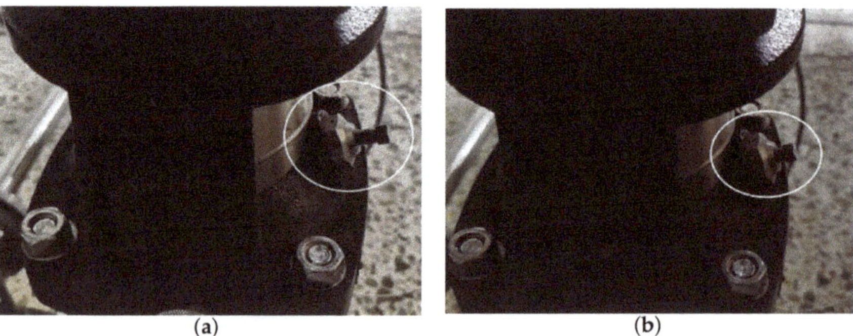

Figure 14. Air binding fault setting: (**a**) Exhaust screw loose, (**b**) Exhaust screw tightened.

After building the experimental platform, the sampling frequency was set to 1 kHz, the motor speed to 1015 r/min, and the acquisition system developed by Labview was employed to collect the vibration signals of the centrifugal pump in normal state. Then, the vibration signals of rotor unbalance and air bind state of the centrifugal pump were collected. A total of 25 samples for each of the three conditions of the centrifugal pump were obtained through a non-overlapping sliding window of length 4000 points. This means there are 4000 points per sample. The first 10 samples for each of the three conditions of the centrifugal pump are selected as the training set, and the remaining 15 samples for each of the three conditions are selected as the testing set.

6.2.2. Comparison and Analysis

To verify the superiority of FIF-PARCMFDE-FCM clustering for fault diagnosis of centrifugal pump, under the same test conditions, the vibration data of three operating states of centrifugal pump normal, rotor unbalance fault and air binding fault were classified and identified. Furthermore, these data were compared with the cluster analysis results of FIF-RCMFDE-FCM and FIF-MSE-FCM.

The first 1000 points of the original vibration signal in the three states of the centrifugal pump are selected, as shown in Figure 15. It can be seen from Figure 15 that the characteristics of the vibration signals in the three states of the centrifugal pump are not clear and cannot be directly classified according to the waveform. The vibration signals of the centrifugal pump in three states were decomposed by FIF, and the first five-order *IMF* components were selected for comparison. The *IMF* components with larger correlation coefficients can well preserve the fault characteristic information of the signal. The correlation coefficients between the first five-order *IMF* components and the original signal are shown in Table 4. The components with a correlation coefficient greater than 0.4 are selected to reconstruct the signal. Therefore, IMF1 and IMF2 are selected for reconstruction for normal, rotor unbalance fault and air bind fault.

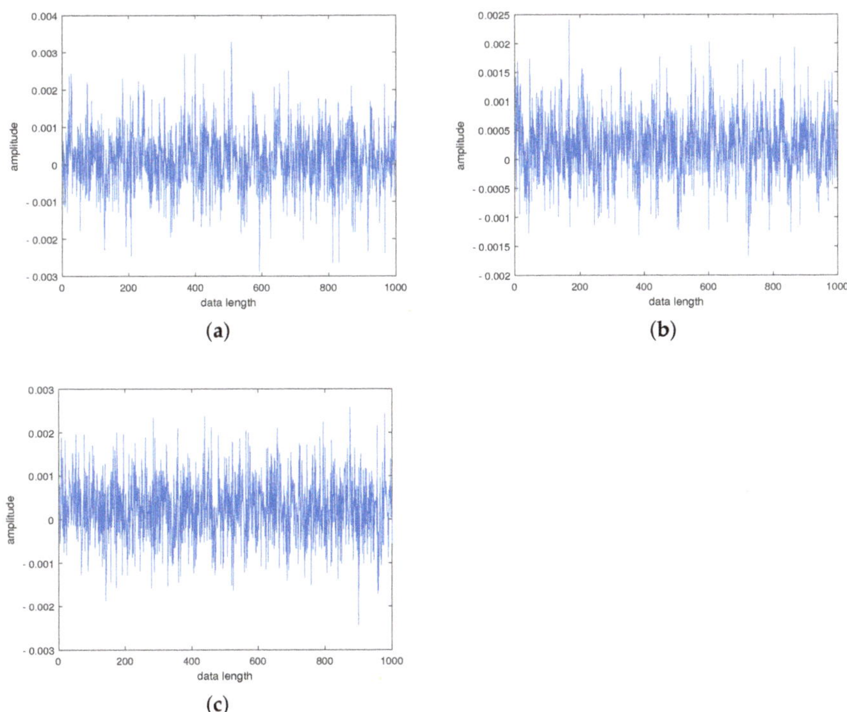

Figure 15. Time domain waveform of vibration signal of centrifugal pump in (**a**) Normal, (**b**) Air Bind Fault and (**c**) Rotor Unbalance Fault.

Table 4. Correlation coefficients of centrifugal pumps in different states.

States	Correlation Coefficients				
	IMF 1	**IMF 2**	**IMF 3**	**IMF 4**	**IMF 5**
Normal	0.8241	0.5885	0.3266	0.3127	0.2019
Air Bind Fault	0.7979	0.5945	0.3538	0.3561	0.2545
Rotor Unbalance Fault	0.8230	0.5870	0.3244	0.2722	0.1289

The reconstructed signals of different states of the centrifugal pump are selected. The skewness is used as the objective function in GA and the maximum evolutionary generation T is set to 200, the threshold for the fitness function to change is 10^{-6}. The parameters of RCMFDE are optimized by GA. The PARCMFDE, RCMFDE and MSE of the reconstructed signal s, are calculated where the scale factor is 10. The embedding dimension m and the class number c of PARCMFDE in different situations are shown in Table 5. Based on experience, RCMFDE takes default values $m = 3$, $c = 6$, MSE takes default values $m = 2$, $r = 0.15 \times \sigma$, σ represents the standard deviation of signal s. As shown in Figure 16, PARCMFDE can clearly separate the three states of the centrifugal pump, indicating that PARCMFDE can effectively extract the hidden features of the three states of the centrifugal pump, which is suitable as a feature vector to further classify the states of the centrifugal pump. As shown in Figures 17 and 18, RCMFDE and MSE are almost inseparable from the three states of the centrifugal pump, indicating that RCMFDE and MSE may not be able to effectively extract the small fault features hidden in rotating machinery, and are not suitable for further classification of centrifugal pump states as feature vectors.

Table 5. PARCMFDE parameters of centrifugal pump in different states.

Status	Embedding Dimension m	Class Number c
Normal	2	7
Air Bind Fault	2	192
Rotor Unbalance Fault	2	3

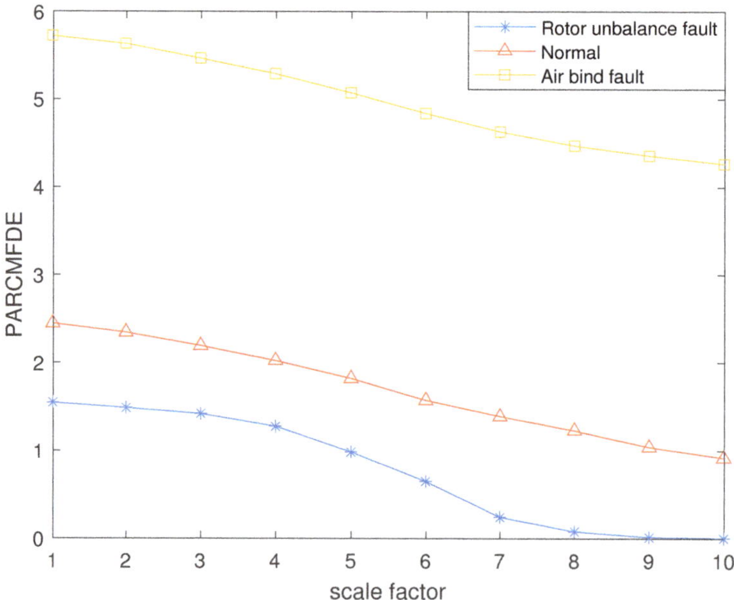

Figure 16. PARCMFDE in different states of centrifugal pump.

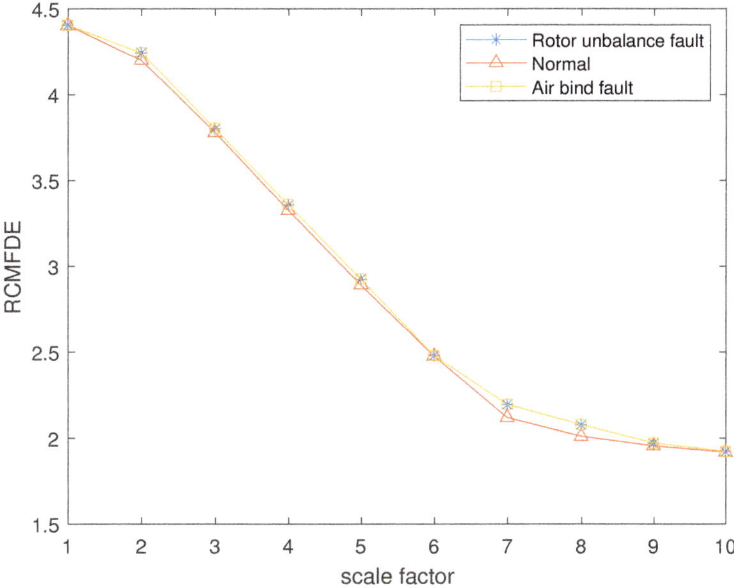

Figure 17. RCMFDE in different states of centrifugal pump.

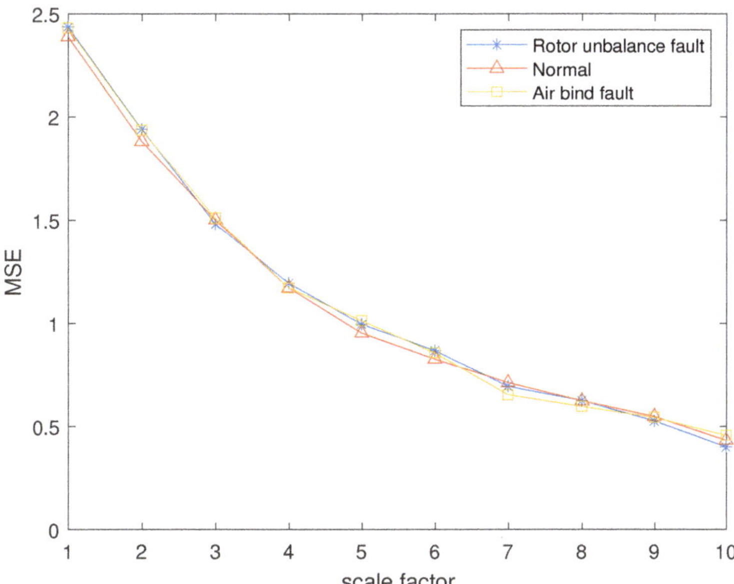

Figure 18. MSE in different states of centrifugal pump.

Taking PARCMFDE as the eigenvector matrix. Perform FCM cluster analysis on the eigenvector matrix of the training samples, and three cluster centers can then be obtained. Next, the obtained cluster centers and testing sample eigenvector matrix are input into FCM clustering. The clustering results are shown in Figure 19. Similarly, RCMFDE and MSE are separately input into FCM clustering as eigenvector matrices. The clustering results are shown in Figures 20 and 21.

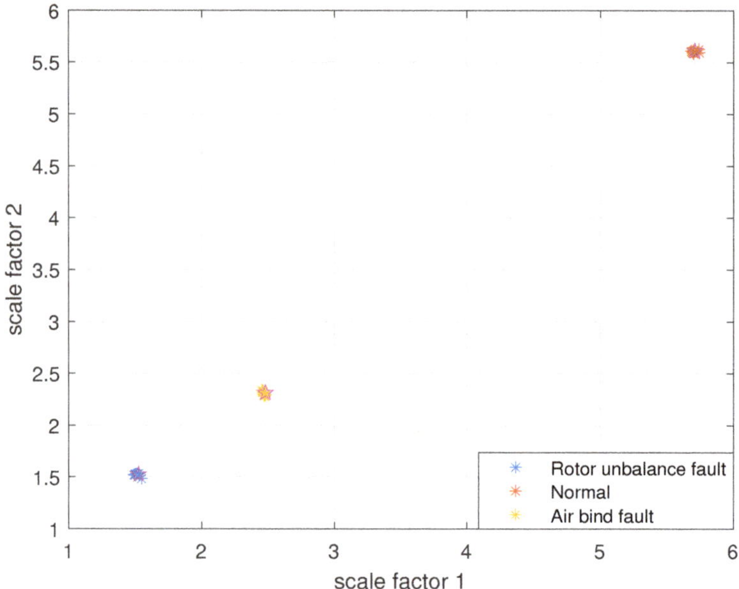

Figure 19. FIF-PARCMFDE-FCM clustering results of centrifugal pump data in different states.

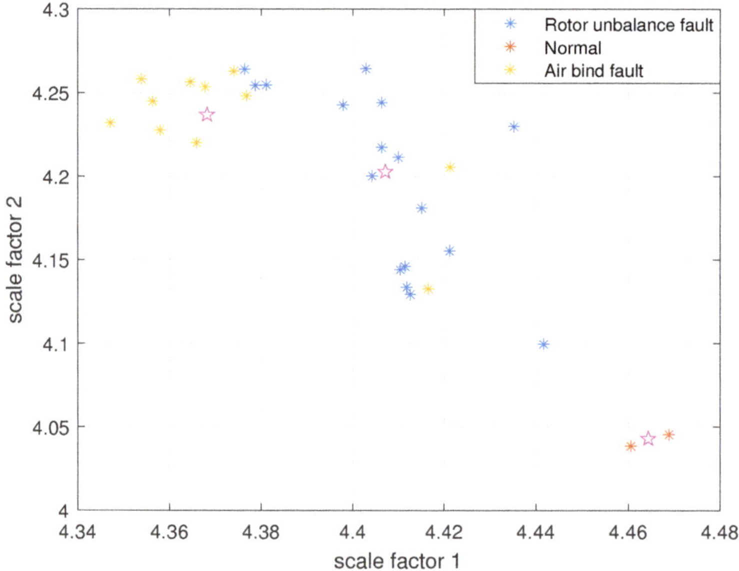

Figure 20. FIF-RCMFDE-FCM clustering results of centrifugal pump data in different states.

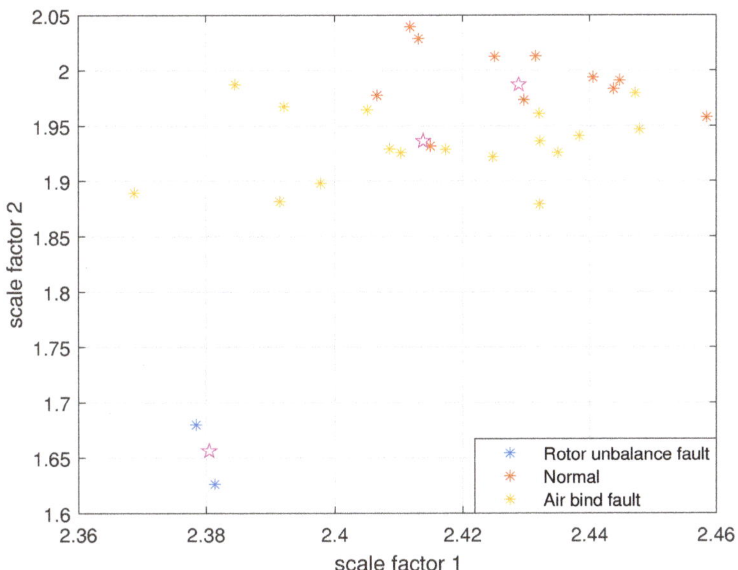

Figure 21. FIF-MSE-FCM clustering results of centrifugal pump data in different states.

The clustering effect and classification accuracy of the above three algorithms are quantitatively compared, and the classification coefficient S, average fuzzy entropy E and classification accuracy Acc of each method are calculated, respectively, as shown in Table 6. From Figure 19 and quantitative analysis, it can be seen that FIF-PARCMFDE-FCM clearly distinguished the fault categories, the cluster centers of various signals are far apart, and the data points of various types are compactly clustered around the cluster centers. Moreover, the classification accuracy Acc is 100%. This shows that the method has a good classification effect on the fault signals of the centrifugal pump in different states. However, the methods based on FIF-RCMFDE-FCM and based on FIF-MSE-FCM exhibit poor clustering effect, serious misclassification, and low classification accuracy Acc.

It can be seen from the above two experiments that the accuracy of the three methods in experiment 1 is 100%, but the accuracy of FIF-MSE-FCM and FIF-RCMFDE-FCM is greatly reduced in Experiment 2. There are two reasons: 1. The vibration signal fault feature of Experiment 1 are more evident and easy to distinguish. However, the fault features of the vibration signal in Experiment 2 are relatively weak and difficult to extract accurately. 2. By using GA to optimize the parameters of RCMFDE, the problem regarding the selection of m and c depends on experience is solved, and the performance of feature extraction is improved. This reflects that the FIF-PARCMFDE-FCM can adaptively select parameter combinations according to different application scenarios, which has better adaptability for signals that are more difficult to classify and with less obvious fault features.

Table 6. FCM clustering effects of different entropy algorithms for centrifugal pumps.

Algorithms	Classification Coefficient S	Average Fuzzy Entropy E	Classification Accuracy Acc
FIF-PARCMFDE-FCM	0.9933	0.0215	100%
FIF-RCMFDE-FCM	0.7819	0.3849	57%
FIF-MSE-FCM	0.6313	0.5962	63%

7. Conclusions

To overcome the shortcomings of traditional feature extraction methods that bear difficulty in extracting tiny fault features hidden in vibration signals, and the shortcomings of RCMFDE to select parameters based on experience, a PARCMFDE is proposed. PARCMFDE takes the skewness of RCMFDE as the objective function, and uses genetic algorithm to optimize parameters. PARCMFDE can more accurately extract tiny fault features hidden in vibration signals of rotating machinery. At the same time, a new fault diagnosis method for rotating machinery based on FIF-PARCMFDE-FCM is proposed, which can classify rotating machinery faults accurately and automatically without depending on the length of data samples. FIF quickly decomposes the original vibration signal, and selects components with large correlation coefficients for reconstruction. The reconstructed signal features are extracted by PARCMFDE, and the feature vector is formed into FCM for automatic label-free classification. The bearing experiments with clear fault characteristics prove that the classification performance of this method is superior to other methods. Experiments on centrifugal pumps with weak fault features demonstrate that this method can extract hidden weak fault features from vibration signals and perform accurate and reliable automatic classification. Therefore, the proposed diagnostic method can achieve reliable diagnosis performance for rotating machinery.

However, the proposed method only identifies single faults of rotating machinery, and does not consider the identification of compound faults. Furthermore, PARCMFDE is slower than RCMFDE. Therefore, the identification of compound faults in rotating machinery and the improvement of the computing speed of PARCMFDE will be regarded as the focus of our future work.

Author Contributions: Methodology, software, validation, writing—original draft preparation, M.T.; data curation, Y.L.; experimental system construction, F.L.; writing—review and editing, X.L. All authors have read and agreed to the published version of the manuscript.

Funding: This research received no external funding.

Institutional Review Board Statement: Not applicable.

Informed Consent Statement: Not applicable.

Data Availability Statement: The data used in this study are all owned by the research group and will not be transmitted.

Conflicts of Interest: The authors declare no conflict of interest.

References

1. Liu, R.; Yang, B.; Zio, E.; Chen, X.F. Artificial intelligence for fault diagnosis of rotating machinery: A review. *Mech. Syst. Signal Process.* **2018**, *108*, 33–47. [CrossRef]
2. Peng, B.; Wan, S.; Bi, Y.; Xue, B.; Zhang, M.J. Automatic feature extraction and construction using genetic programming for rotating machinery fault diagnosis. *IEEE Trans. Cybern.* **2021**, *51*, 4909–4923. [CrossRef]
3. Rodriguez, N.; Cabrera, G.; Lagos, C.; Cabrera, E. Stationary Wavelet Singular Entropy and Kernel Extreme Learning for Bearing Multi-Fault Diagnosis. *Entropy* **2017**, *19*, 541. [CrossRef]
4. Cicone, A. Nonstationary signal decomposition for dummies. In *Advances in Mathematical Methods and High Performance Computing*; Singh, V., Gao, D., Fischer, A., Eds.; Springer: Cham, Switzerland , 2019; pp. 69–82.
5. Rubini, R.; Meneghetti, U. Application of the envelope and wavelet transform analyses for the diagnosis of incipient faults in ball bearings. *Mech. Syst. Signal Process.* **2001**, *15*, 287–302. [CrossRef]
6. Huang, N.E.; Shen, Z.; Long, S.R.; Wu, M.C.; Shih, H.H.; Zheng, Q.; Yen, N.C.; Tung, C.C.; Liu, H.H. The empirical mode decomposition and the Hilbert spectrum for nonlinear and non-stationary time series analysis. *Proc. R. Soc. Lond. Ser. Math. Phys. Eng. Sci.* **1998**, *454*, 903–995. [CrossRef]
7. Feng, Z.; Zhang, D.; Zuo, M.J. Adaptive mode decomposition methods and their applications in signal analysis for machinery fault diagnosis: A review with examples. *IEEE Access* **2017**, *5*, 24301–24331. [CrossRef]
8. Wu, Z.; Huang, N.E. Ensemble empirical mode decomposition: A noise-assisted data analysis method. *Adv. Adapt. Data Anal.* **2009**, *1*, 1–41. [CrossRef]
9. Wang, Y.H.; Yeh, C.H.; Young, H.V.; Hu, K.; Lo, M.T. On the computational complexity of the empirical mode decomposition algorithm. *Phys. A Stat. Mech. Appl.* **2014**, *400*, 159–167. [CrossRef]

10. Dragomiretskiy, K.; Zosso, D. Variational mode decomposition. *IEEE Trans. Signal Process.* **2014**, *62*, 531–544. [CrossRef]
11. Yi, C.; Lv, Y.; Dang, Z. A fault diagnosis scheme for rolling bearing based on particle swarm optimization in variational mode decomposition. *Shock Vib.* **2016**, *2016*, 9372691. [CrossRef]
12. Lin, L.; Wang, Y.; Zhou, H. Iterative filtering as an alternative algotithm for empirical mode decomposition. *Adv. Adapt. Data Anal.* **2009**, *1*, 543–560. [CrossRef]
13. Cicone, A.; Liu, J.; Zhou, H. Adaptive local iterative filtering for signal decomposition and instantaneous frequency analysis. *Appl. Comput. Harmon. Anal.* **2016**, *41*, 384–411. [CrossRef]
14. Cicone, A.; Zhou, H. Numerical analysis for iterative filtering with new efficient implementations based on FFT. *Numerische Mathematik* **2021**, *147*, 1–28. [CrossRef]
15. Xu, Y.; Fan, F.; Jiang, X. A fast iterative filtering decomposition and symmetric difference analytic energy operator for bearing fault extraction. *ISA Trans.* **2021**, *108*, 317–332. [CrossRef]
16. Zhu, K.H.; Chen, L.; Hu, X. Rolling element bearing fault diagnosis by combining adaptive local iterative filtering, Modified Fuzzy Entropy and Support Vector Machine. *Entropy* **2019**, *20*, 926. [CrossRef]
17. Deng, W.; Zhang, S.J.; Zhao, H.M.; Yang, X. A novel fault diagnosis method based on integrating empirical wavelet transform and fuzzy entropy for motor bearing. *IEEE Access* **2018**, *6*, 35042–35056. [CrossRef]
18. Ding, J.; Xiao, D.; Huang, L.; Li, X. Gear fault diagnosis based on VMD sample entropy and discrete hopfield neural network. *Math. Probl. Eng.* **2020**, *2020*, 8882653. [CrossRef]
19. Lv, Y.; Zhang, Y.; Yi, C. Optimized adaptive local iterative filtering algorithm based on permutation entropy for rolling bearing fault diagnosis. *Entropy* **2018**, *20*, 920. [CrossRef]
20. Li, Y.; Xu, M.; Wang, R.; Huang, W. A fault diagnosis scheme for rolling bearing based on local mean decomposition and improved multiscale fuzzy entropy. *J. Sound Vib.* **2016**, *360*, 277–299. [CrossRef]
21. Li, Y.; Wang, X.; Liu, Z.; Liang, X.; Si, S. The entropy algorithm and its variants in the fault diagnosis of rotating machinery: A review. *IEEE Access* **2018**, *6*, 66723–66741. [CrossRef]
22. Rostaghi, M.; Azami, H. Dispersion entropy: A measure for time-series analysis. *IEEE Signal Process. Lett.* **2016**, *23*, 610–614. [CrossRef]
23. Azami, H.; Arnold, S.E.; Sanei, S.; Chang, Z.; Sapiro, G.; Escudero, J.; Gupta, A.S. Multiscale fluctuation-based dispersion entropy and its applications to neurological diseases. *IEEE Access* **2019**, *7*, 68718–68733. [CrossRef]
24. Kenneth, D.J. Learning with genetic algorithms: An overview. *Mach. Learn.* **1988**, *3*, 121–138.
25. Mardia, K.V. Measures of multivariate skewness and kurtosis with applications. *Biometrika* **1970**, *57*, 519–530. [CrossRef]
26. Gan, X.; Lu, H.; Yang, G.Y. Fault diagnosis method for rolling bearings based on composite multiscale fluctuation dispersion entropy. *Entropy* **2019**, *21*, 290. [CrossRef]
27. Yan, X.A.; Xu, Y.D.; She, D.M.; Zhang, W. A bearing fault diagnosis method based on PAVME and MEDE. *Entropy* **2021**, *23*, 1402. [CrossRef]
28. Cannon, R.L.; Dave, J,V.; Bezdek, J.C. Efficient implementation of the Fuzzy c-Means Clustering algorithms. *IEEE Trans. Pattern Anal. Mach. Intell.* **1986**, *8*, 248–255. [CrossRef]
29. Bezdek, J.C. Cluster validity with fuzzy sets. *J. Cybern.* **1973**, *3*, 58–73. [CrossRef]
30. Bearing Data Center. Case Western Reserve University. Available online: http://csegroups.case.edu/bearingdatacenter/pages/download-data-file (accessed on 19 October 2021).
31. An, X.; Zeng, H. Demodulation analysis based on adaptive local iterative filtering for bearing fault diagnosis. *Measurement* **2016**, *94*, 554–560. [CrossRef]

Article

A Short-Term Hybrid TCN-GRU Prediction Model of Bike-Sharing Demand Based on Travel Characteristics Mining

Shenghan Zhou, Chaofei Song, Tianhuai Wang, Xing Pan, Wenbing Chang and Linchao Yang *

School of Reliability and Systems Engineering, Beihang University, Beijing 100191, China
* Correspondence: yanglinchao@buaa.edu.cn

Abstract: This paper proposes an accurate short-term prediction model of bike-sharing demand with the hybrid TCN-GRU method. The emergence of shared bicycles has provided people with a low-carbon, green and healthy way of transportation. However, the explosive growth and free-form development of bike-sharing has also brought about a series of problems in the area of urban governance, creating a new opportunity and challenge in the use of a large amount of historical data for regional bike-sharing traffic flow predictions. In this study, we built an accurate short-term prediction model of bike-sharing demand with the bike-sharing dataset from 2015 to 2017 in London. First, we conducted a multidimensional bike-sharing travel characteristics analysis based on explanatory variables such as weather, temperature, and humidity. This will help us to understand the travel characteristics of local people, will facilitate traffic management and, to a certain extent, improve traffic congestion. Then, the explanatory variables that help predict the demand for bike-sharing were obtained using the Granger causality with the entropy theory-based MIC method to verify each other. The Multivariate Temporal Convolutional Network (TCN) and Gated Recurrent Unit (GRU) model were integrated to build the prediction model, and this is abbreviated as the TCN-GRU model. The fitted coefficient of determination R2 and explainable variance score (EVar) of the dataset reached 98.42% and 98.49%, respectively. Meanwhile, the mean absolute error (MAE) and root mean square error (RMSE) were at least 1.98% and 2.4% lower than those in other models. The results show that the TCN-GRU model has strong efficiency and robustness. The model can be used to make short-term accurate predictions of bike-sharing demand in the region, so as to provide decision support for intelligent dispatching and urban traffic safety improvement, which will help to promote the development of green and low-carbon mobility in the future.

Keywords: short-term demand prediction; bike-sharing; travel characteristics analysis; hybrid TCN-GRU model

Citation: Zhou, S.; Song, C.; Wang, T.; Pan, X.; Chang, W.; Yang, L. A Short-Term Hybrid TCN-GRU Prediction Model of Bike-Sharing Demand Based on Travel Characteristics Mining. *Entropy* **2022**, *24*, 1193. https://doi.org/10.3390/e24091193

Academic Editors: Jan Kozak and Przemysław Juszczuk

Received: 5 July 2022
Accepted: 24 August 2022
Published: 26 August 2022

Publisher's Note: MDPI stays neutral with regard to jurisdictional claims in published maps and institutional affiliations.

Copyright: © 2022 by the authors. Licensee MDPI, Basel, Switzerland. This article is an open access article distributed under the terms and conditions of the Creative Commons Attribution (CC BY) license (https://creativecommons.org/licenses/by/4.0/).

1. Introduction

With the gradual improvement of people's living standards and the enhancement of environmental awareness, the series of negative social impacts brought about by rapid economic growth, such as traffic congestion, environmental degradation and noise pollution caused by overloaded motor vehicle usage, have undoubtedly led to an increasing demand for green and low-carbon means of travel. Bike-sharing has not only made a contribution to low-carbon environmental protection, but also alleviated the problem of "human transportation" in the area of public transportation to a certain extent. However, the explosive growth and "free-range" development of bike-sharing has also brought about a series of problems: first, given the lack of supervision, the excessive proliferation of bike-sharing has caused a waste of resources and urban "bicycle pollution"; second, the lack of overall layout planning for bike-sharing parking has led to the occupation of crowded public land; third, the free-moving bikes are unevenly distributed in time and space, and their operation and maintenance is not timely.

Building a prediction model based on the historical data of bike-sharing demand can effectively explain the time series characteristics of this phenomenon, but the influence of other elements in the bike-sharing system is not considered; thus, there is a certain one-sidedness, and a limit to the ability to explain and predict the fluctuation mechanism of bicycle travel demand [1]. Related studies have found that factors affecting the demand for bike-sharing rides include external factors such as weather, air quality, spatial location, user price sensitivity, and chance events, in addition to historical data on travel demand [2]. Through a survey of bike-sharing programs in Beijing, Campbell et al. [3] pointed out that the main factors affecting the demand for bike-sharing are distance, temperature, precipitation, and air quality, and that the users' own demographic characteristics (including income, gender, and occupation) have no significant effect on the demand for bicycles. Matton et al. [4] pointed out that climatic conditions such as temperature, wind, and precipitation are the main factors affecting the demand for bike-sharing, and Faghih et al. [5] suggested that point-in-time factors are also important variables affecting the demand for bike-sharing, including the time of day, whether it is a weekend, and peak hours. In addition, weather factors and point-in-time factors [6–8], population density [9,10], the availability of bicycle lane facilities [10–12], and distance to the urban CBD and universities [5,10,13] are also related to the demand for bike-sharing.

Therefore, some studies have started to incorporate external factors such as weather, time factors and holiday factors into the independent variables of bike-sharing demand prediction. Li et al. [14] established an LSTM linear regression model considering the distance variable of users' rides, and the results of the study show that the prediction accuracy was improved compared with the existing time series prediction models. Li et al. [15] proposed a prediction method based on a clustering algorithm with an augmented regression tree model based on weather conditions, temperature, and wind speed, so as to predict the number of rentals and returns of bicycles at stations separately. Chen et al. [16] argued that the demand for bike-sharing is affected not only by general factors such as time and weather, but also by contingent factors such as traffic events, and proposed a dynamic cluster-based forecasting framework.

From the perspective of forecasting model development, statistical methods such as the Autoregressive Integrated Moving Average model (ARIMA) were first applied to solve the bike-sharing cycling demand forecasting problem. Statistical inferential forecasting methods based on statistics include traditional models such as ARIMA models, regression analysis and Markov chains [17]. Andreas et al. [18] developed a prediction model based on a differential sliding average autoregressive model, using operational data from bicycle companies and data from bike-sharing in the Barcelona community, to forecast the number of available bicycles at each bicycle station. To investigate the characteristics and patterns of peak bicycle demand hours, Lin et al. built an ARIMA model [19]. Yan et al. [20] considered both the temporal and spatial dependence of bicycle borrowing and returning demand. For the time series, the cyclicality and trend of bicycle travel demand were obtained by building an ARIMA model considering seasonal patterns; for the spatio-temporal dependence, the inter-cluster transfer characteristics were portrayed by building a Bayesian transfer network model. Zhou et al. proposed a prediction method based on the Markov chain model. The study evaluated the model using data from the public bicycle system in Zhongshan City. The results of the case analysis verify the high prediction accuracy and generalization ability of the Markov chain model [21].

The traditional statistical methods are more sensitive to data, and the presence of data noise can greatly reduce the reliability of model parameter estimation. At the same time, there is a certain degree of spatial and temporal dependence between the demand for bike-sharing trips and external influences such as weather, and the prediction models based on statistical methods have weak explanatory power for the complex nonlinear relationships between bicycle demand and the influencing factors. In the era of big data, nonparametric methods can handle massive traffic trip data and discover the dynamic characteristics of the bicycle system.

Nonparametric methods include machine learning methods and deep learning methods. Using machine learning methods such as random forests [22], Bayesian networks [23],

GBDT [24] and artificial neural networks (ANN) [25], nonlinear prediction models can be built based using a large amount of bike-share historical travel data to predict future bike-share demand at any time interval. In addition, deep learning methods are gradually being used to predict short-term bikeshare demand. Wang et al. [26] used a long short-term memory (LSTM) neural network and gated recursive units (GRU) to predict short-term bicycle availability. Chen et al. [27] proposed a recurrent neural network (RNN) using time, weather, and seasonal data to predict the rental and return demand for each station in the system. Zhang et al. [28] proposed a deep learning model for the short-term prediction of bike-sharing demand, considering the correlation between bike-sharing users and public transportation riders. He et al. [29] proposed a bike-share demand prediction (BDP) model that incorporates a temporal convolutional network (TCN) and a self-attention mechanism. The BDP model extracts feature information with multiple inputs of multiple sources of data, and uses the parallelism of the self-attention mechanism to improve the training speed. A better prediction accuracy is obtained in comparison with other models. Ma et al. [30] proposed a Spatio-Temporal Graphical Attention Long-Term Memory (STGA-LSTM) neural network framework for predicting demand for bike-sharing at the station level using a multi-source dataset. This short-term prediction model can be used to help bike-sharing users make better route choices, and help operators implement dynamic redistribution strategies. Mehdizadeh et al. [31] proposed a hybrid CNN-LSTM model for the short-term prediction of mountain biking demand, which had considerable prediction accuracy during the COVID-19 pandemic after adding additional variables such as weather conditions and time of day.

The research for this thesis includes two main aspects: (1) mining the travel pattern of bike-sharing users, analyzing the travel characteristics of residents, and providing references for bicycle demand prediction; (2) making accurate predictions of bike-sharing demand, improving the bicycle turnover rate, and providing a decision basis for the intelligent scheduling of regional bike-sharing.

The study is divided into the following sections: Section 1 focuses on the study background, study content and literature review. Section 2 mainly concerns data description and pre-processing, including a preliminary correlation analysis. Section 3 mines the bike-sharing trip characteristics through multiple dimensions, such as time, temperature, humidity, and weather. Section 4 introduces the TCN model, MIC variable selection method, GRU model, hybrid time series model and evaluation indicators. This is followed by multiple rounds of comparison experiments for validation. Sections 5 and 6 are the discussion and conclusions sections, respectively.

2. Data Overview and Preprocessing

2.1. Data Overview

This paper used the London bike-sharing public dataset as the subject of the study. The dataset recorded a total of 17,414 data points (one data point generated every hour, i.e., 24 data points per day) for the London area from 4 January 2015 to 3 January 2017. The dataset recorded the influencing factors, such as weather and travel time, related to the demand of bike-sharing; we performed a data background gain by adding data nouns such as "Hour" and "Month" with timestamp information. The descriptions of the data terms and examples are shown in Table 1.

2.2. Data Preprocessing

Since the dimensionality and magnitude of each variable are not uniform, to eliminate the influence of magnitude and to speed up model training, the normalization method was used to normalize the data. This involves a linear transformation of the original data that maps the data values to the [0, 1] interval. The transformation function is shown in Equation (1):

$$x^* = \frac{x - min}{\max - min} \tag{1}$$

where *max* is the maximum value of the data, and *min* is the minimum value.

Table 1. Data set fields description.

Field Name	Description	Example
timestamp	Timestamp for grouping data together	4 January 2015, 12:00
demand	Counting of new bike share	182
t1	Actual temperature (°C)	3.0
t2	Subjective perception of temperature (°C)	2.0
hum	Humidity percentage (%)	93.0
wind_speed	Wind speed value (km/h)	6.0
weather_code	Sunny: 1, Less Cloudy: 2, Cloudy: 3, Overcast:4, Rainy: 7, Storms: 10, Snowy: 26	3
is_holiday	Holiday: 1, Non-holiday: 0	0
is_weekend	Weekend: 1, Non-weekend: 0	1
season	Spring: 0; Summer: 1; Autumn: 2; Winter: 3	3
hour	24 h per day	12
day_of_month	Natural days per month	1
day_of_week	Monday: 0, . . . , Sunday: 6	1
month	January: 1, . . . , December: 12	6

2.3. Correlation Analysis

There is correlation between different features in the data, resulting in feature redundancy. In addition, not all influencing factors are related to the demand for bike-sharing. The correlation analysis aimed to investigate the correlation between bike-sharing variables, i.e., a preliminary analysis of other variables that are correlated with the demand for bike-sharing. After the normality test, the data of most of the variables used in this study did not conform to a normal distribution. Therefore, we used Spearman's rank correlation coefficient for measuring the linear correlation between the variables [32].

The rank is the average descending position of a number in the overall data. If X and Y are two observed variables with sample size n, and for each sample (X_i, Y_i), the corresponding rank is (x_i, y_i), then the Spearman's rank correlation coefficient ρ between these two variables is determined via Equation (2).

$$\rho = 1 - \frac{6 \sum (x_i - y_i)^2}{n(n^2 - 1)} \tag{2}$$

The Spearman's rank correlation coefficient ranges within $[-1, 1]$. When the absolute value is close to 1, this indicates that the two variables are more strongly correlated. When the value is positive, if one of the two characteristics shows an increasing trend, the other also tends to increase, and when the value is 1, it indicates a perfect positive correlation; when the value is negative, if one of the two characteristics tends to increase, the other tends to decrease, and when the value is -1, it indicates a perfect negative correlation; when the value is 0, this indicates a perfect non-correlation (the tendency of one to change does not change with that of the other). In general, the absolute value of the correlation coefficient in the range of (0.8, 1.0) is considered as very strong correlation, while the range (0.6, 0.8) is considered strong correlation, (0.4,0.6) moderate correlation, (0.2, 0.4) weak correlation, and (0, 0.2) very weak or no correlation.

The results of the correlation analysis between demand and each variable are shown in Figure 1, which shows that the actual temperature t1 is highly correlated with the subjectively perceived temperature t2, and there is a problem of feature redundancy. In addition, temperature demand shows a weak positive correlation with temperature, while demand shows a moderate negative correlation with humidity, a very weak positive correlation with temperature, and a very weak negative correlation with weather and season. The correlation analysis can roughly determine the linear relationship between demand and its influencing factors. In order to obtain the trend of demand under its different influencing factors, data mining methods can be used to analyze the travel characteristics of bike-sharing.

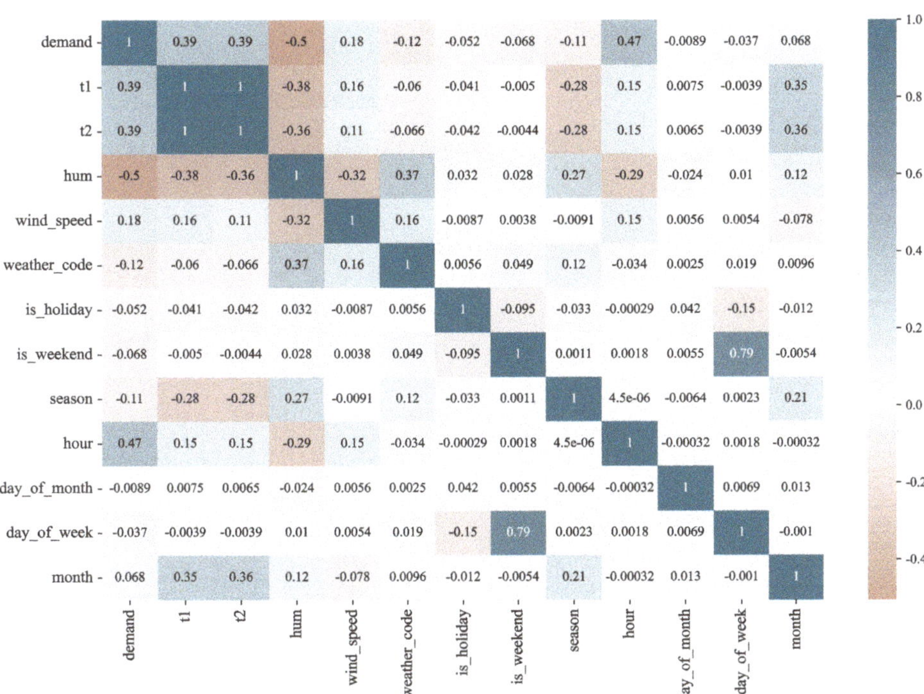

Figure 1. Bike-sharing demand correlation analysis heat map.

3. Bike-Sharing Travel Characteristics Analysis

As an important means of transportation for urban residents, bike-sharing often presents different characteristics due to a variety of factors, which must be explored for the purpose of traffic management. Therefore, based on the considered dataset, we explored bike-sharing travel characteristics via several dimensions such as time, temperature, humidity, and weather [33].

3.1. Bike-Sharing Travel: Time Characteristics Analysis

3.1.1. Demand Varies with the Hours and Months

First, we assessed the distribution of the demand for bike-sharing in different months, and the results are shown in Figure 2. The demand shows an obvious single hump shape that develops with the month, i.e., the demand for bike-sharing in the area gradually increases from January until it peaks in July, and it then starts to decrease month by month.

Next, we determined the distribution of bike-sharing demand at different times of the day, and the results are shown in Figure 3. The demand shows an obvious double-hump shape that develops with the time of the day, that is, the demand for bike-sharing in the area is high at 7 and 8 a.m. and 5 and 6 p.m. This result coincides perfectly with people's commuting time to and from work on weekdays, and also reflects that bike-sharing is in the highest demand when people commute to and from work, suggesting that bike-sharing can provide convenience for people's work travel.

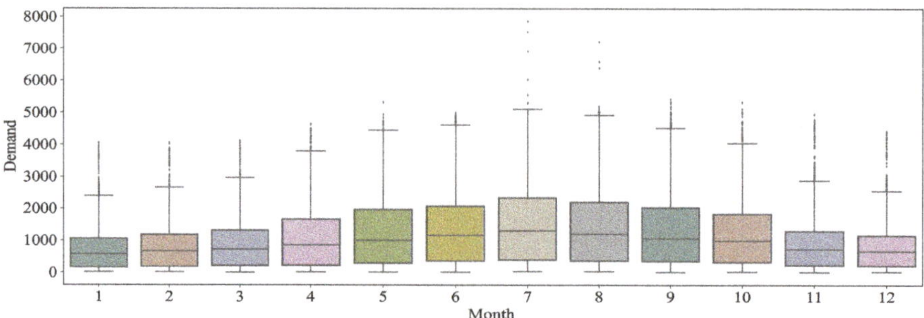

Figure 2. Box plot of demand for bike-sharing in different months.

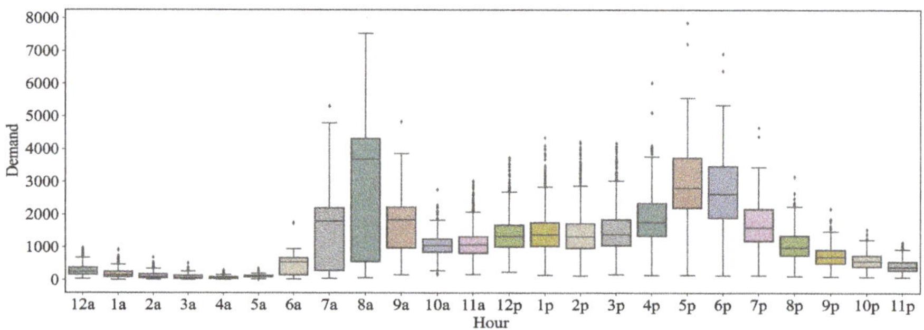

Figure 3. Box plot of demand for bike-sharing at different hours.

We analyzed the distribution of bike-sharing demand by month at different moments of the day with bubble chart statistics, where in larger bubbles indicate higher demand. The statistical results are shown in Figure 4. As can be seen, the vast majority of months show a double-hump distribution of demand. However, in December, demand for shared bikes increases when people are at work, while demand is roughly the same throughout the afternoon from 12:00 to 6:00, with no clear trend. This may have more to do with the local climate as well as holidays.

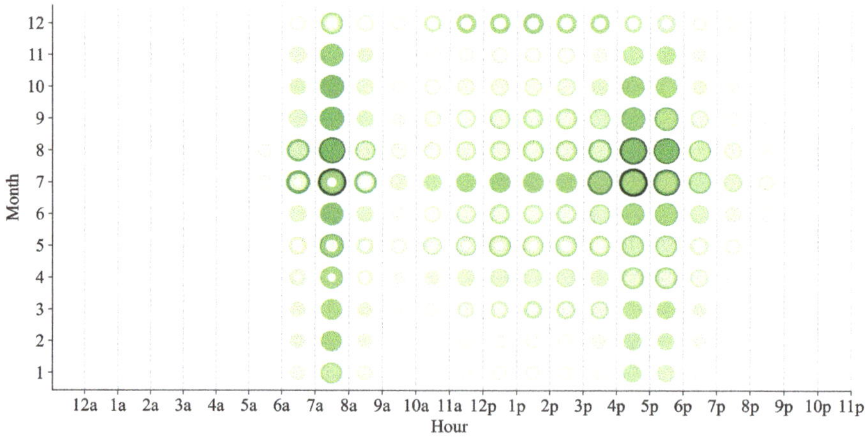

Figure 4. Time bubble map of bike-sharing demand.

It is known from the previous analysis that July is the month with the highest demand for bike-sharing; so, we took July 2016 as the research object and analyzed the daily demand changes in this month using heat maps, and the statistical results are shown in Figure 5. It can be seen that there is an obvious cycle pattern in the demand distribution, with every seven days being a cycle, and the demand distribution on five of the days corresponds to the weekday travel characteristics, i.e., the obvious double-hump feature of on and off work. This also reflects the obvious difference in the distribution of demand on weekdays and non-working days. There is a high demand for bike-sharing on July 30 and 31, which may be related to the local Prudential Ride London event, a popular ride that locals say turned London into a bicycle-centric environment.

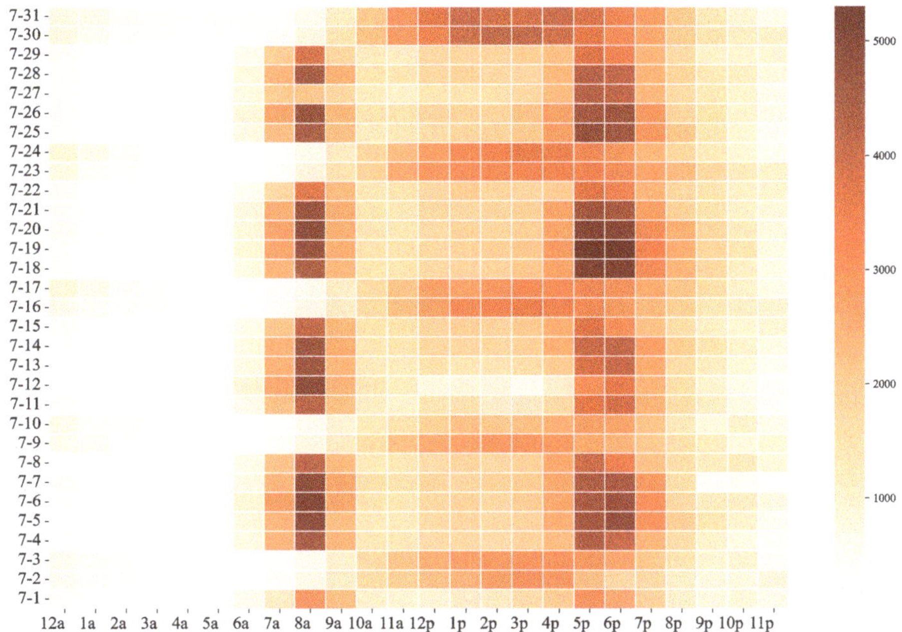

Figure 5. Bike-sharing demand time heat map.

3.1.2. Demand Varies with Working and Nonworking Days

We found in the previous analysis that there is a significant difference in the distribution of bike-sharing demand between weekdays and non-weekdays. Therefore, we took weekends and holidays as the research object and used weekday data for comparative analysis, and the analysis results are shown in Figure 6. It can be seen that the distribution of people's travel characteristics on holidays and weekends is roughly the same. On weekdays, 8:00 and 17:00 and 18:00 are the peak times for car use, which coincides with the time points for going to and leaving work. In the case of nonworking days 14:00–15:00 is the real peak period of car usage. This reflects people's preference for using shared bikes to travel in the afternoon during nonworking days.

3.1.3. Demand Varies with the Season

In addition, we analyzed the distribution of bike-sharing demand by season at different moments of the day through line graph statistics. The results are shown in Figure 7. It can be seen that the trend of bike-sharing demand is more or less the same in different seasons, with higher demand in summer and autumn, and the lowest in winter, which is obviously related to the seasonal climate.

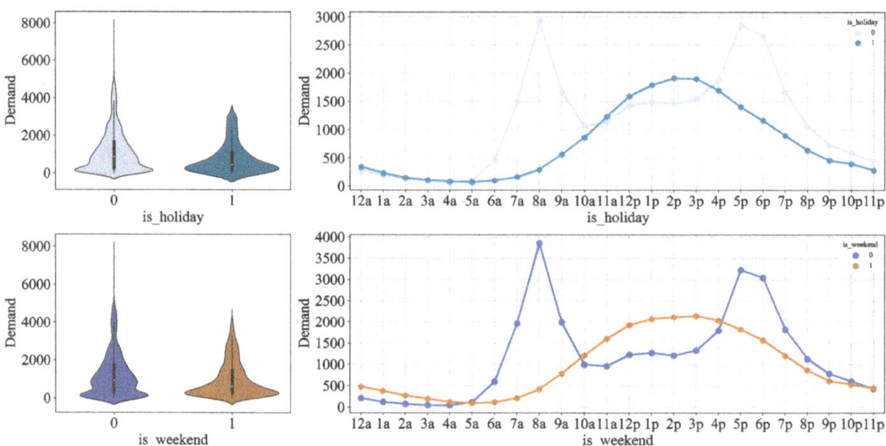

Figure 6. Bike-sharing demand on working and nonworking days.

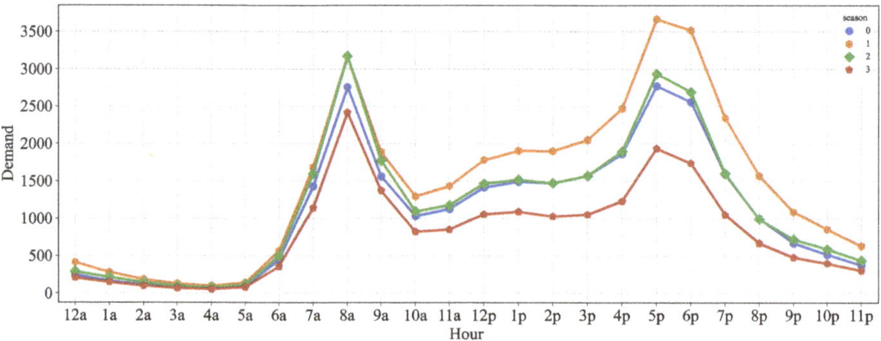

Figure 7. Bike-sharing demand in different seasons.

3.2. Bike-Sharing Travel: Meteorology Characteristics Analysis

Many studies have shown that, as external environmental factors, weather type [34], temperature [35], and air quality [36], also have direct and indirect effects on travel characteristics. To investigate the influence of weather characteristics on bike-sharing trips, we obtained the trends of bike-sharing demand with wind speed, humidity, weather type, and temperature using line plots as well as box plots. The results are shown in Figure 8. It can be seen that there is a local peak at a wind speed of 25 km/h, and the demand decreases at higher and lower wind speeds. There is a negative correlation between air humidity and demand; that is, with greater air humidity, the overall demand shows a decreasing trend. In weather codes 2 and 3, that is, when the weather type is either less cloudy or cloudy, the demand is larger; when the weather is more severe, the demand gradually decreases, and when the weather code is 26 (snow), the demand is almost 0. The demand shows a trend of increasing first and then decreasing with the rise in temperature; that is, below the temperature is 25 °C, the demand shows a relatively strong positive correlation with temperature, and after the temperature exceeds 25 °C, the demand shows a relatively weak negative correlation with temperature.

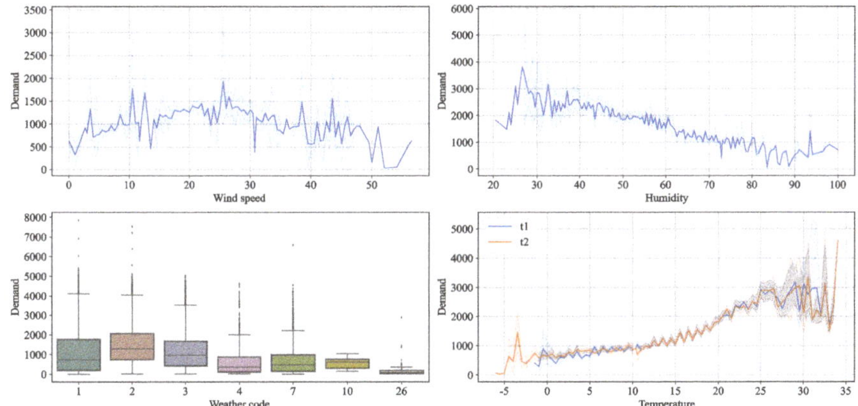

Figure 8. Bike-sharing demand under different meteorological conditions.

3.3. Bike-Sharing Travel: Characteristics Analysis Based on Granger Causality Test

The correlation analysis lacks an explanation for the causal mechanism of the fluctuation in bike-sharing demand, and we next explore the impact of weather and other characteristics on the demand for bike-sharing travel from the causality perspective. Weather data indicators include t1, hum, weather_code and wind_speed. In order to further screen the indicators that help predict the demand for bike-sharing travel, this paper uses the Granger causality test method for weather and other features' screening. The basic idea of the method is that [37], if a series X helps to explain the future trend of series Y—that is, in the regression model of series Y regarding its own historical information, adding the historical information of X will significantly improve the explanatory power of the regression model—then series X is the Granger cause of series Y.

Before Granger causality tests were performed on the weather indicator grid, the unit root method was used to perform a smoothness test. For non-stationary series, differencing was performed until it passed the stationarity test. The results of the causality test for each variable at the significance level $\alpha = 0.05$ are presented in Table 2.

Table 2. Results of causality tests for each variable.

Variable	Original Hypothesis	F-Statistic	Probability (p)
t1	t1 is not a bike-sharing demand Granger reason	230.8794	8.275×10^{-8}
hum	hum is not a bike-sharing demand Granger reason	257.9023	1.296×10^{-9}
weather_code	windspeed is not a bike-sharing demand Granger reason	20.1423	0.0728
wind_speed	weather code is not a bike-sharing demand Granger reason	5.1211	0.2036

When $p < 0.05$ rejects the original hypothesis, this indicates that there is a Granger causality with statistical significance between weather indicators t1, hum and the demand for bike-sharing, i.e., adding weather indicators t1 and hum to the model helps predict the demand.

The analysis of bike-sharing travel characteristics in London reveals that both point-in-time factors [5] and weather conditions [4] affect the variation in bike-sharing demand to varying degrees. There is consistency and interoperability between our analysis and the results of other literature analyses. In addition, we found that the factors influencing bike-sharing demand were roughly the same across regions, i.e., differences in regional attributes, culture, climate, and ethnicity do not affect travel characteristics. A survey of the

Beijing [3] bike-sharing program also found that users' own demographic characteristics do not have a significant effect on bicycle demand.

4. Bike-Sharing Short-Term Demand Prediction

The bike-sharing demand data are susceptible to the influence of time, climate and traffic management policies, showing strong volatility and nonlinearity. The bike-sharing demand data used in this paper are hourly, and the sample size is relatively small. The deep neural network has a strong fitting ability for nonlinear data but is prone to the risk of overfitting in the case of small samples. Based on the above analysis, this paper has tried to combine the typical models of deep learning temporal prediction, GRU and TCN, with the principle of the least-squared error sum. In so doing we aimed to reduce the possibility of overfitting and to take advantage of the fitting of deep learning models on nonlinear and non-stationary data, in order to improve the prediction ability of the models.

4.1. Temporal Convolutional Network (TCN)

TCN is a novel architecture based on a Convolutional Neural Network (CNN). Unlike general CNNs, TCNs use structures such as expanded causal convolution and residual blocks [38–40]. This gives them the ability to extract features and achieve prediction from large sample time series, and TCNs can effectively address the performance degradation of deep networks during network training. TCN consists of dilated, causal 1D fully convolutional layers with the same input and output lengths. The convolution in the TCN model is causal convolution, wherein the layers are causally related to each other, thus ensuring that no historical information or future data will be missed. In addition, TCN can map sequences of arbitrary length to output sequences of the same length, using residual modules and dilation convolution to better control the memory length of the model and improve the predictive power.

4.1.1. TCN Modeling

Supposing that the input sequence is given as $\{x_1, x_2, \cdots, x_t\}$, and the expected predicted output is $\{\hat{y}_1, \hat{y}_2, \cdots, \hat{y}_t\}$, the equation of the predicted output versus the input sequence can be presented by Equation (3):

$$(\hat{y}_1, \hat{y}_2, \cdots, \hat{y}_t) = f(x_1, x_2, \cdots, x_t) \tag{3}$$

where $\hat{y}_t$ is only related to the input sequence at time t and in the past, and is independent of any future input. The purpose of TCN modeling is to establish a mapping relationship f between the input and output sequences, and its objective function is to minimize the error loss between the actual output $\{y_1, y_2, \cdots, y_t\}$ and the predicted values $\{\hat{y}_1, \hat{y}_2, \cdots, \hat{y}_t\}$.

4.1.2. Extended Causal Convolution

The causal convolutions were originally proposed in the WaveNets [41] networks for learning the input audio data before moment t to predict the output at moment $t + 1$. Compared to RNNs, no circular connections are used in models using causal convolutions, so time series data can be input in parallel, which allows for faster network training, especially for large-sample time series [42]. However, standard causal convolution requires increasing the receptive field of neurons in the neural network by stacking many network layers or using very large convolutional kernels when dealing with large sample time series. For this reason, TCN uses the Dilated Causal Convolution (DCC) technique to achieve an increase in the perceptual field without a significant increase in computational cost. DCC is a convolution operation that performs a step-skipping operation on the input sequence, and its expression is given by Equation (4):

$$F(i) = \sum_{j=1}^{k} h(j)x(i - dj) \tag{4}$$

where $F(i)$ is the convolution result for the ith element in the sequence $\{x_1, x_2, \cdots, x_t\}$; $h(j)$ is the convolution kernel, and for a one-dimensional sequence its convolution kernel size $K = 1 \times k$; d is the expansion factor (when $d = 1$, that is the standard causal convolution).

The structure of DCC is shown in Figure 9 ($K = 1 \times 2$ and $d = 2^l - 1$, l is the number of hidden layers). Compared with standard causal convolution, DCC allows the output to be associated with as many inputs as possible with the same number of network layers. Multilayer stacking combined with extended causal convolution also allows deep learning networks to achieve very large sensory fields with fewer network layers [43]. Moreover, the sliding operation of the convolution kernel on the input data allows the TCN to handle inputs of variable length. Thus, in conjunction with the updating of the model's input data (i.e., the predicted values from the previous moment are added to the input as information), new predictions can be continuously computed and output.

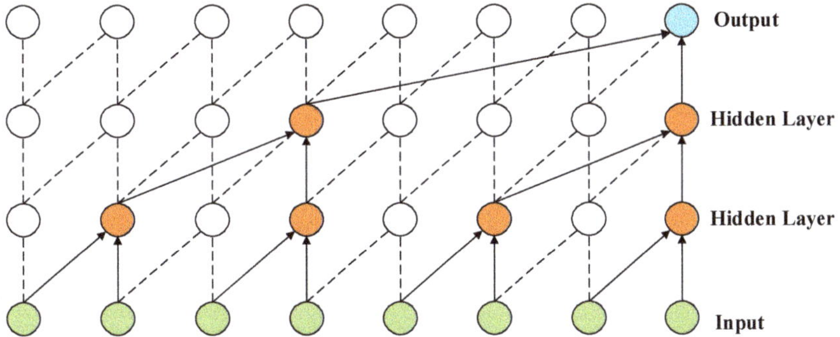

Figure 9. Schematic diagram of extended causal convolution.

4.1.3. Residual Block

Residual Block (RB) is proposed to solve the degradation problem of deep learning networks, and its core idea is to introduce a "jump connection" operation that skips one or more layers [44]. Assuming that x is the input of the residual block, the output o of the residual block is shown in Equation (5), which is the result of linear variation and mapping through the activation function. Since the residual $\kappa(x)$ will not be zero in practice, the stacked layers in the deep learning network can always learn new features, so the learning performance of the deep network will not degrade [45].

In TCN modeling, using a network structure combining RB and DCC can effectively improve the feature learning capability and robustness of TCN models.

$$o = Activation(x + \kappa(x)) \tag{5}$$

4.2. Gated Recurrent Unit (GRU)

LSTM [46] and GRU [47] show strong potential applicability in the data prediction problem studied in this paper, with GRU performing slightly better. Compared with the LSTM method, GRU requires fewer training parameters, is easier to converge and can reduce the risk of model overfitting in the case of limited time series data. GRU optimizes the three gate functions of LSTM, turning the set of forgetting gates and input gates into a single update gate, and mixing the neuron states with the hidden states. This can effectively alleviate the problem of "gradient disappearance" in RNN networks and reduce the number of parameters of LSTM network units, shortening the training time of the model. The basic structure is shown in Figure 10, and the mathematical description is shown in Equations (6)–(10):

$$r_t = \sigma(W_r \cdot [h_{t-1}, x_t]) \tag{6}$$

$$u_t = \sigma(W_u \cdot [h_{t-1}, x_t]) \tag{7}$$

$$\widetilde{h}_t = \tanh(W_{\widetilde{h}} \cdot [r_t * h_{t-1}, x_t]) \tag{8}$$

$$h_t = (1 - u_t) * h_{t-1} + u_t * \widetilde{h}_t \tag{9}$$

$$y_t = \sigma(W_o \cdot h_t) \tag{10}$$

where x_t, h_{t-1}, h_t, r_t, u_t, $\widetilde{h}_t$ and y_t are the input vector, the state memory variable of the previous moment, the state memory variable of the current moment, the state of the update gate, the state of the reset gate, the state of the current candidate set, and the output vector of the current moment, respectively. W_r, W_u, $W_{\widetilde{h}}$ and W_o are the weight parameters used for multiplying the update gate, reset gate, candidate set, and output vector with the connection matrix composed of x_t and h_{t-1}, respectively; I denotes unit matrix; $\cdot$ denotes the matrix dot product; $*$ denotes the matrix product; and σ denotes the *sigmoid* activation function.

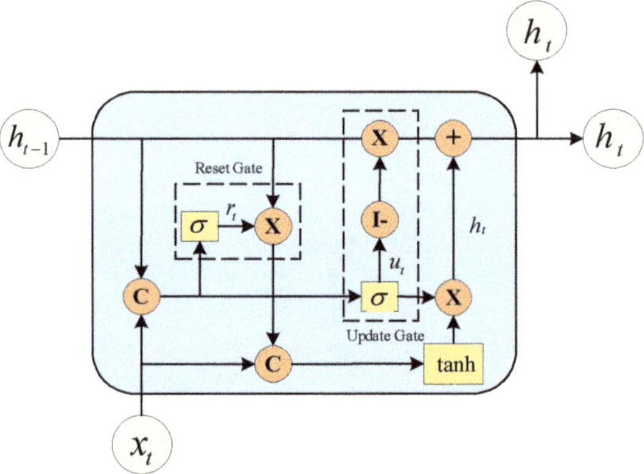

Figure 10. GRU model' internal structure.

GRU uses update and reset gates as core modules. The splicing matrix of the input variable x_t and the state memory variable h_{t-1} of the previous moment, are input into the update gate after *sigmoid* nonlinear transformation, which determines the extent to which the state variable of the previous moment is brought into the current state. The reset gate controls the amount of information that was written to the candidate set at the previous moment, stores the information at the previous moment by $I - u_t$ times h_{t-1}, records the information at the current moment by u_t times $\widetilde{h}_t$, and sums the two as the output of the current moment.

4.3. Hybrid Multivariate Bike-Sharing Demand Prediction Model

Hybrid model forecasting is used to try to combine different forecasting models and the information they provide to derive a hybrid forecasting model in the form of an appropriate weighted average. The key to hybrid model forecasting is how to find out the weighting coefficients, which makes the hybrid forecasting model more effective in improving the forecasting accuracy.

Different forecasting models have their own strengths, and a better linear hybrid forecasting model can be obtained by the linear combination of different forecasting models. The linear hybrid forecasting model's form is shown in Equation (11):

$$\hat{y}_t = \sum_{i=1}^{m} \omega_i y_{i(t)} \tag{11}$$

$$\begin{cases} \omega_1 + \omega_2 + \cdots + \omega_m = 1 \\ \omega_i \geq 0 \end{cases} \tag{12}$$

where $\hat{y}_t$ is the combined forecast value at moment t; $y_{i(t)}$ is the forecast value of the ith forecast model at moment t; $W = (\omega_1, \omega_2, \cdots, \omega_m)^T$ is the weighting coefficient of the linear combination of m forecast models and satisfies the requirement, as shown in Equation (12).

The key to the linear combination prediction model is to determine a reasonable number of weights ω_i, based on the principle of the minimum sum of squares of errors (SSE) [48], which can make the prediction model more effective and accurate.

$$SSE = \sum_{t=1}^{n} e_t^2 = \sum_{t=1}^{n} \left(\sum_{i=1}^{m} \omega_i e_{it} \right)^2 = W^T E W \tag{13}$$

$$\begin{cases} \min SSE = W^T E W \\ s.t. R_m W = 1, W \geq 0 \end{cases} \tag{14}$$

$$W_0 = \frac{E^{-1} R_m T}{R_m E^{-1} R_m T} \tag{15}$$

where, $e_{it} = y_{(t)} - y_{i(t)}$ denotes the forecast error of the ith forecast model at moment t; $y_{(t)}$ is a sequence of actual values of a certain index of a forecast object; $e_t = y_{(t)} - \hat{y}_t$ denotes the forecast error of the linear combination model at moment t; $E = (e_{it})_{m \times n} (e_{it})^T{}_{m \times n}$ is the information error matrix; the optimal weighting coefficient W_0 is obtained by solving the optimal solution of the linear programming problem, where R_m is an m-dimensional row vector with all elements of 1, and the guaranteed non-negative optimal weighting coefficients enable the linear combinatorial model to effectively improve the prediction accuracy.

Our hybrid multivariate bike-sharing demand forecasting model based on the principle of minimum error sum of squares is shown in Figure 11.

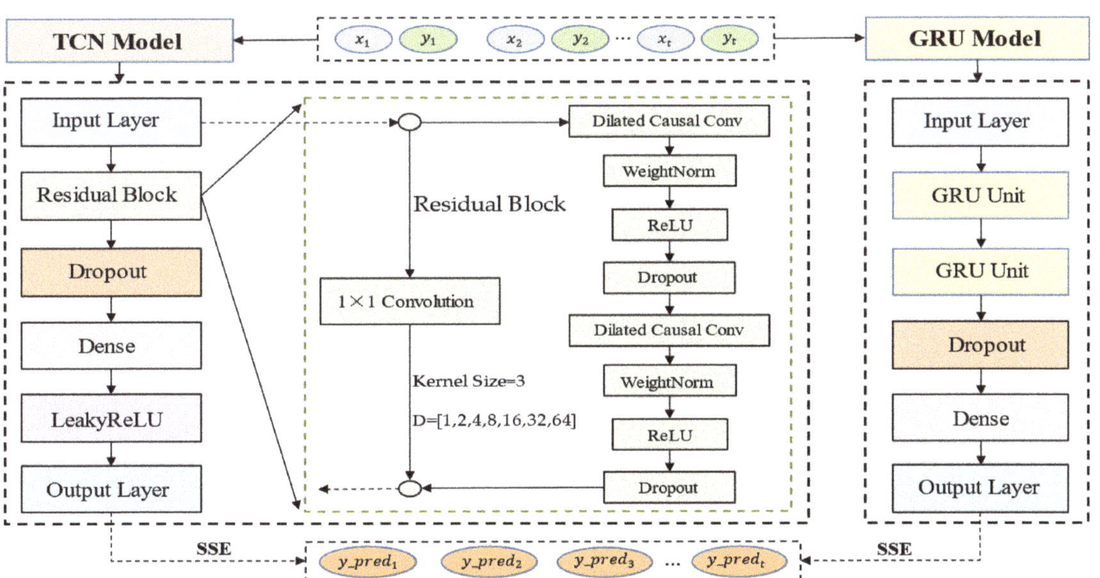

Figure 11. Basic structure of bike-sharing demand prediction combination model.

4.4. Variables Selection

The entropy of the variables in the data set will have a direct impact on the prediction model, and this paper uses the maximum information coefficient (MIC) [49] method based on entropy theory for variable selection. MIC is a combination of information theory and probability [50] based on mutual information, and is used to detect nonlinear correlations between different variables and eventually obtain a measure of the strength of dependencies between variables. The maximum information coefficient achieves universality and equilibrium, where universality, with the help of MIC, can discover functional and nonfunctional relationships between variables; equilibrium, with the help of MIC, can be used to compare the strength of relationships between different variables, both horizontally and vertically.

Suppose that, in the data set D, the sample size is s, where an explanatory variable $X = \{x_i, i = 1, 2, \cdots, s\}$ and the explanatory variable $Y = \{y_i, i = 1, 2, \cdots, s\}$; the $MIC(X, Y)$ between these two variables is calculated as follows.

(1) Calculate the mutual information $MI(X, Y)$ between the explanatory variable X and the explained variable Y:

$$MIC(X, Y) = \sum_{y_i \in Y} \sum_{x_i \in X} p(x_i, y_i) \log \frac{p(x_i, y_i)}{p(x_i)p(y_i)} \tag{16}$$

where $p(x_i, y_i)$ is the joint density function of the variables X and Y. $p(x_i)$ is the marginal probability density function of the explanatory variable X, and $p(y_i)$ is the marginal probability density function of the explanatory variable Y.

(2) The variables X and Y are divided into a grid of $m * n$ defined as $G = (m, n)$. To obtain the grid division that maximizes the MI, the value of MI is normalized. This normalized maximum MI can be expressed as follows:

$$MI_{D|G}(X, Y) = \frac{MI^*_{D|G}(X, Y)}{log_{min}\{m, n\}} \tag{17}$$

where $MI^*_{D|G}(X, Y)$ is the maximum MI of data set D under grid G.

(3) The MIC is defined as the maximum MI under all grids G, calculated as follows:

$$\begin{cases} MIC(X, Y) = \max_{m*n < B(s)} \left\{ MI_{D|G}(X, Y) \right\} \\ B(s) = s^{0.6} \end{cases} \tag{18}$$

where $B(s)$ is the maximum number of unit grids as a function of the number of samples.

The larger the value of $MIC(X, Y)$, the stronger the correlation between variables X and Y. Therefore, we calculate the MIC values between all explanatory and explained variables, and select the characteristics according to Equation (19):

$$MIC(X, Y) \geq \delta \tag{19}$$

where δ is the lowest variable selection threshold.

4.5. Model Evaluation Methods

To validate and compare the accuracy as well as the robustness of the models, we used R^2, $EVar$, MAE, $MedAE$, and $RMSE$ as evaluation metrics, respectively.

(1) Coefficient of determination (R2)

The coefficient of determination characterizes the extent to which the regression model explains the variation in the dependent variable, or the goodness of fit of the model to the observations.

$$R^2 = 1 - \frac{\sum_{i=1}^{N} (y_i - \hat{y}_i)^2}{\sum_{i=1}^{N} (y_i - \bar{y}_i)^2} \tag{20}$$

Here, y_i is the actual value of the ith data point; $\hat{y}_i$ is the corresponding predicted value; and $\bar{y}_i$ is the mean value of the time series. In general, the value of the coefficient of determination R^2 ranges from 0 to 1, where an R^2 equal to 0 means that the model cannot predict the target variable at all, and an R^2 equal to 1 means that the model can make a perfect prediction. R^2 can also have negative values, in which case the model's prediction ability is not as good as calculating the mean of the target variable directly.

(2)   Explainable Variance Score (EVar)

The explainable variance score measures the degree to which the dispersion of errors between all predicted and actual values is similar to the dispersion of the true values themselves.

$$EVar = 1 - \frac{Var(y - \hat{y})}{Var(y)} \tag{21}$$

A larger value of $EVar$ indicates the better prediction ability of the model, and the best possible value is 1.

(3)   Mean Absolute Error (MAE)

The mean absolute error is the expectation of the absolute value of the error between the predicted and actual values at each moment in time.

$$MAE = \frac{1}{N} \sum_{i=1}^{N} |y_i - \hat{y}_i| \tag{22}$$

(4)   Median Absolute Error (MedAE)

The median absolute error is the median of the absolute error of the predicted and actual values for all data points. The metric is robust to outliers.

$$MedAE = median(|y_1 - \hat{y}_i|, \cdots, |y_N - \hat{y}_i|) \tag{23}$$

(5)   Root Mean Square Error (RMSE)

The mean square error calculates the mean of the square of the error between the predicted and true values. The root mean square error, on the other hand, is the open square of the mean square error, which is consistent with the target variable in terms of magnitude.

$$RMSE = \sqrt{\frac{1}{N} \sum_{i=1}^{N} (y_i - \hat{y}_i)^2} \tag{24}$$

4.6. Verification Experiment and Result Analysis

To verify the validity of the proposed multivariate hybrid time series model, we conducted a validation experiment on the London area bike-sharing data set. The MIC method was first used for the variable selection part of this study, and the MIC values between the variables are shown in Figure 12.

The number of explanatory variables was studied in descending order according to the magnitude of MIC values between each explanatory variable and the dependent variable, and R2, EVar, MAE, and RMSE were used as measures.

It can be seen from Figure 13 that the model works best when the number of features is 5. That is, the lowest feature selection threshold $\delta = 0.07$ and the combination of explanatory variables chosen is {hour, hum, t1, is_weekend, day_of_week}. It can be seen that the set of selected explanatory variables includes not only hour, weekend and day of week, which closely correspond to the morning and evening peaks of people commuting to work, but also includes the weather characteristics t1 and hum obtained by using Granger causality tests.

We performed a parameter search with the goal of the optimization of the effect of the hybrid model. The parameter search results of the TCN and GRU models are shown in Tables 3 and 4.

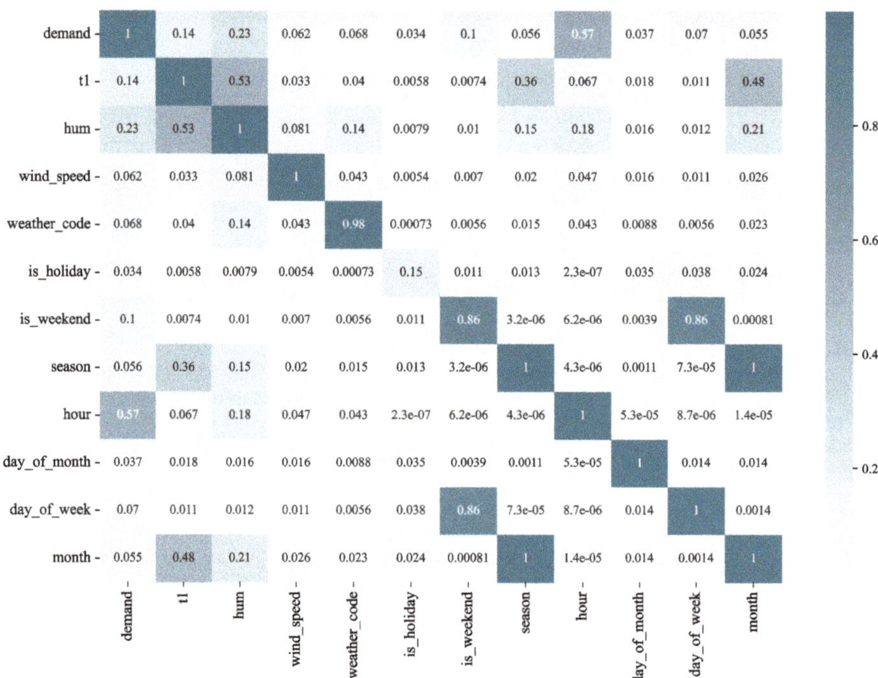

Figure 12. Heat map of MIC values between different variables.

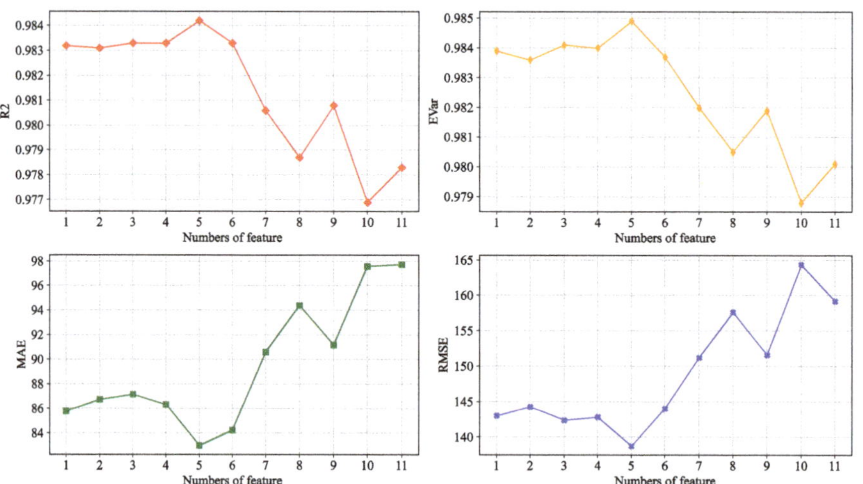

Figure 13. Comparison of the effects of models with different quantitative characteristics.

We conducted two experiments: univariate prediction of the demand for bike-sharing and multivariate prediction of the demand for bike-sharing, respectively. Univariate prediction refers to the demand for bike-sharing as the only input without considering other explanatory variables. Multivariate forecasting, on the other hand, considers the influence of other explanatory variables on demand with the demand of bike-sharing as input, and obtains the corresponding explanatory variables through variable selection methods, which are also used as inputs to the model.

Table 3. Parameter setting of the TCN model.

Parameter	Value
Time Steps	13
Nb_filters	64
Kernel_size	3
Nb_stacks	1
Epochs	80
Batch Size	32
Drop out	0.2
Dilations	[1, 2, 4, 8, 16, 32, 64]
Skip_connections	True
Kernel_initializer	he_normal
Optimizer	Adam
Activation Function	Rectified linear unit (ReLU)
Loss Function	Mean Squared Error (MSE)

Table 4. Parameters setting of GRU model.

Parameter	Value
Time Steps	13
Input Layer Units Number	100
Output Layer Units Number	1
Hide Layer Number	2
Hide Layer Units Number	100
Epochs	50
Batch Size	16
Learning Rate	0.001
Optimizer	Adam

The comparison models used for the experiments include:

(1) Support Vector Regression (SVR) [51] (kernel = 'rbf', C = 1.0, max_iter = −1);

(2) XGBoost [52] (max_depth = 6, learning_rate = 0.1, eta = 1);

(3) ARIMA [53] (autocorrelation order: $p = 9$, difference order: d = 1, moving average orders: q = 0);

(4) ARIMAX (autocorrelation order: p = 9, difference order: d = 1, moving average orders: q = 8, exogenous variables: hour, hum, t1, is_weekend, day_of_week);

(5) LSTM (input_size = 6, hidden_size = 100, num_layers = 2, batch_size = 64, dropout = 0.2);

(6) History Average Model (HA) (history time step = 13);

(7) Prophet [54] (growth = "linear", freq = "H", interval_width = 0.95);

(8) DeepAR [55] (input_size = 6, hidden_size = 64, num_layers = 3).

After averaging results over several iterations of the experiment, we determined the performance of each model on this dataset, and the specific evaluation metrics are shown in Table 5.

As can be seen from Table 5, the univariate model's predictions are less effective overall than the multivariate model's predictions, which indicates that the prediction performance of the model can be effectively improved with the inclusion of the selected explanatory variables; for example, the MAE and RMSE of the multivariate predictions are reduced by 7.0977 and 13.831, respectively, for the TCN-GRU model we used. In addition, some models such as DeepAR and Prophet may show non-adaptability to this dataset, and our experimental results are only better than those of the HA model. The hybrid model performs better than the single model in multivariate prediction, which proves that the hybrid model we use is more efficient and accurate based on the minimum sum of squares of errors.

The fit of our proposed multivariate TCN-GRU model to the actual values of bike-sharing demand for the last 480 data points (20 days) of the test set is shown in Figure 14.

Table 5. Prediction results of each model.

Model	Metrics	R2	EVar	MedAE	MAE	RMSE
Univariate	HA	0.4859	0.5234	457.8242	618.9324	864.8123
	Prophet	0.5971	0.6616	428.9174	504.2642	716.3489
	SVR	0.8287	0.8892	381.6209	308.4608	375.5922
	ARIMA	0.8379	0.8919	257.3966	297.9913	370.8495
	XGBoost	0.9657	0.9669	383.0468	111.5021	205.4212
	LSTM	0.9730	0.9748	315.4528	112.4182	178.9126
	GRU	0.9767	0.9769	312.3578	112.8749	171.3778
	TCN	0.9806	0.9813	288.7231	89.8644	154.1625
	TCN-LSTM	0.9808	0.9817	50.5265	90.0193	152.5853
	TCN-GRU	0.9819	0.9825	52.1868	90.0910	149.3043
Multivariate	DeepAR	0.7278	0.7861	401.2352	456.8923	613.7432
	ARIMAX	0.8529	0.8990	250.8287	285.9122	358.4603
	TCN	0.9829	0.9837	49.1962	86.2586	143.8991
	GRU	0.9817	0.9813	72.7963	104.2761	154.4806
	LSTM	0.9799	0.9807	61.567	98.7257	156.6573
	TCN-LSTM	0.9833	0.9841	48.1795	84.6395	142.0784
	TCN-GRU	0.9842	0.9849	47.7591	82.9933	138.7543

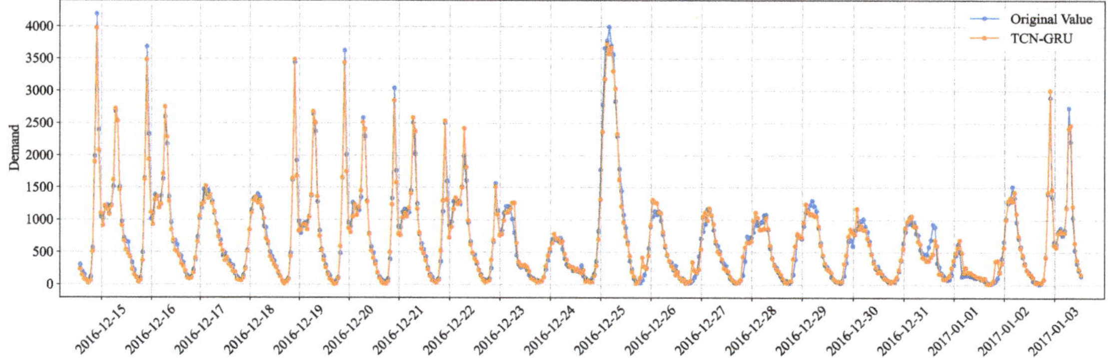

Figure 14. Fitting curve for bike-sharing demand data prediction.

5. Discussion

In recent years, bike-sharing has become an important way for people to travel in an environmentally conscious way. However, this free-form development mode has gradually revealed many problems, such as over-placement, the serious waste of public resources, and excessive growth, causing huge costs for urban management. The phenomenon of the indiscriminate parking of bike-sharing vehicles has led to a large number of public resources, such as subway station entrances, bus stops, bicycle lanes and pedestrian lanes, being occupied. The surge in the number of shared bicycles not only affects the cityscape, but also affects the safety of other public transportation. The uneven distribution of bicycles makes it difficult to meet the volatile users' travel demands. These problems are new challenges for urban transportation managers.

To address the above problems, we took advantage of the fitting of deep learning models on nonlinear and nonsmooth sample data, and we used TCN and GRU models for bike-sharing demand prediction on the data set, combining the models with the principle of the minimum error sum of squares. The hybrid model improved the prediction accuracy, reduced the error, and effectively avoided the overfitting phenomenon. The experiments also proved that the models were less effective than multivariate prediction in the univariate prediction of bike-sharing demand, which meant that adding explanatory variables such as time, humidity, and temperature to the model input could improve the prediction effect. The R2 and EVar of the proposed multivariate TCN-GRU model in this paper were improved by at least 0.0023 and 0.0024, respectively, and the MedAE, MAE, and RMSE decreased by at least 2.7674, 7.026, and 10.55, respectively, compared with univariate forecasting models. At the same time, the R2 and EVar values of this model improved by at least 0.0009 and 0.0008, respectively, and the MedAE, MAE, and RMSE decreased by at least 0.4204, 1.6462, and 3.3241, respectively, compared with other multivariate forecasting models. In order to achieve a more intuitive comparison of the prediction accuracy, we drew a scatter density plot of the prediction effect of the compared models, as shown in Figure 15. In the comparison, we can see that the density distribution of the predicted values of the univariate SVR model, as well as the multivariate ARIMAX model, are not uniform, the distribution is relatively more dispersed, and the prediction effect is average. Our proposed multivariate TCN-GRU model predicts the values, while converging towards the actual values, and the fitting effect is better. Thus, we have established an efficient and robust short-term hybrid prediction model for bike-sharing demand considering multiple variables.

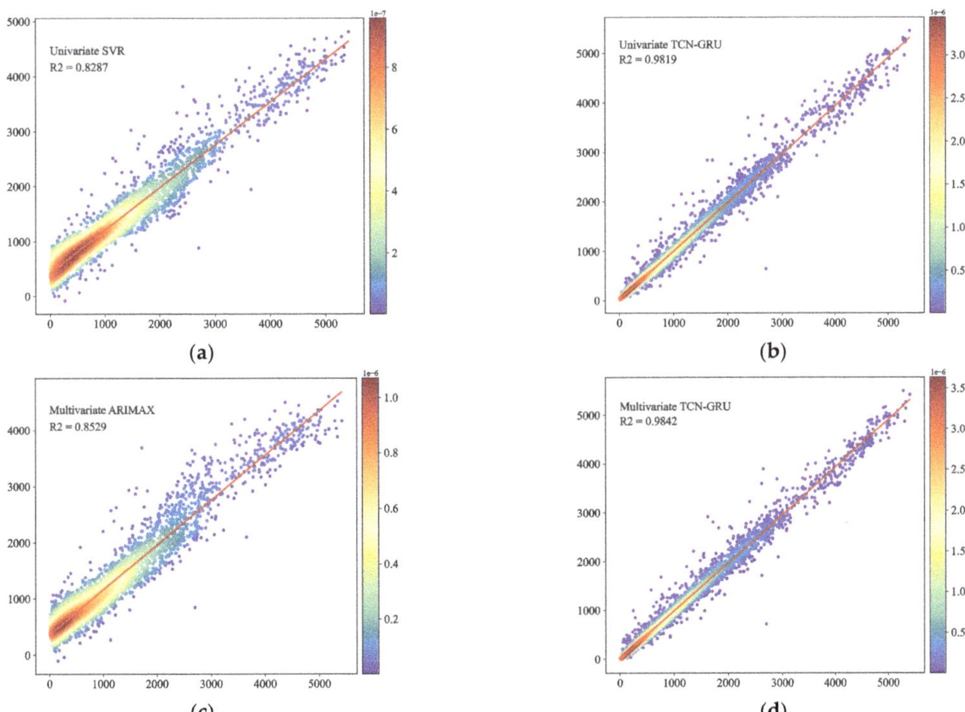

Figure 15. Model scatter density plot. (**a**) Univariate SVR scatter density plot. (**b**) Univariate TCN-GRU scatter density plot. (**c**) Multivariate ARIMAX scatter density plot. (**d**) Multivariate TCN-GRU scatter density plot.

There are still several areas for improvement in this study.

(1)   The combined model proposed in this paper showed good results in short-term bike-sharing demand prediction, and when we tried long-term prediction, the results were not satisfactory. Later, we will try to combine other models to improve performance in long-term prediction.

(2) In this study, we used a small-scale parameter-tuning method based on a grid search, and subsequently we considered other optimization algorithms for parameter searching which might improve the performance of the model.

(3) Due to limited data conditions, we were unable to obtain the main gathering locations of bike-sharing in the region, and thus could not extract spatial characteristics that could be used for further research following demand prediction.

6. Conclusions

In this paper, we built an accurate model that can be used for the short-term prediction of bike-sharing demand, using bike-sharing data from 2015 to 2017 in the London area. First, we analyzed multidimensional bike-sharing travel characteristics based on the explanatory variables such as weather, temperature, and humidity to understand the travel characteristics of local people, and thus facilitate traffic management and, to a certain extent, improve traffic congestion. Considering the nonlinear relationship between each explanatory variable and bike-sharing demand, we used the MIC method for variable selection, where variables were then used as part of the model input, and the experiments proved that adding explanatory variables could greatly improve the prediction performance of the model. In addition, considering the problems of over-fitting and poor stability that arise when using a single model on a small sample of data, we proposed a hybrid multivariate TCN-GRU model with the principle of the minimum error sum of squares, and the model showed strong efficiency and robustness. This can facilitate the accurate short-term prediction of bike-sharing demand in the region, which in turn provides decision support for intelligent dispatching and urban traffic safety improvements. It will also help to promote the development of green and low-carbon mobility in the future.

This study focuses on the possible prediction of factors affecting future bike-sharing in the London area by studying the time series data of bike-sharing traffic demand. Probably due to sensitivity issues, the data we obtained are limited, and we have been unable to obtain the actual locations of the main concentrations of shared bicycles, i.e., individual stations in the area. It would be useful to conduct a more in-depth study of intelligent scheduling, if the researchers can obtain the specific cluster locations of shared bikes in this area.

Author Contributions: Conceptualization, S.Z.; data curation, C.S. and T.W.; formal analysis, X.P.; funding acquisition, S.Z.; investigation, L.Y. and C.S.; methodology, S.Z. and W.C.; Project administration, W.C.; resources, X.P.; supervision, S.Z.; validation, L.Y. and C.S.; visualization, T.W. and C.S.; writing—original draft, C.S.; writing—review and editing, S.Z. and W.C. All authors have read and agreed to the published version of the manuscript.

Funding: This research was funded by the National Natural Science Foundation of China (Grants 71971013) and the Fundamental Research Funds for the Central Universities (YWF-22-L-943). The study was also sponsored by the New Engineering Disciplines research and practice project of Ministry of Education (Grant No. E-DSJ20201102), the key project of the production and education integration collaborative education series of the research branch of the Chapter of Industry-Education Integration Research, China Association of Higher Education (Grant No. CJRH1901) and the Graduate Student Education & Development Foundation of Beihang University.

Institutional Review Board Statement: Not applicable.

Informed Consent Statement: Not applicable.

Data Availability Statement: Publicly available dataset was analyzed in this study. It can be found here: https://www.kaggle.com/hmavrodiev/london-bike-sharing-dataset (accessed on 1 May 2022).

Acknowledgments: We acknowledge the Kaggle platform for providing the bike-sharing dataset in London.

Conflicts of Interest: The authors declare no conflict of interest.

References

1. Chen, X.Y.; Jiang, Y.K.; Li, M.Y.; Ke, X.W. A study of public bicycle single-site scheduling demand based on BP neural network. *Transp. Res.* **2016**, *3*, 30–35. [CrossRef]
2. Xu, C.; Yang, Y.; Jin, S.; Qu, Z.; Hou, L. Potential risk and its influencing factors for separated bicycle paths. *Accid. Anal. Prev.* **2016**, *87*, 59–67. [CrossRef]
3. Campbell, A.A.; Cherry, C.R.; Ryerson, M.S.; Yang, X. Factors influencing the choice of shared bicycles and shared electric bikes in Beijing. *Transp. Res. Part C Emerg. Technol.* **2016**, *67*, 399–414. [CrossRef]
4. Mattson, J.; Godavarthy, R. Bike share in Fargo, North Dakota: Keys to success and factors affecting ridership. *Sustain. Cities Soc.* **2017**, *34*, 174–182. [CrossRef]
5. Faghih-Imani, A.; Eluru, N.; El-Geneidy, A.M.; Rabbat, M.; Haq, U. How land-use and urban form impact bicycle flows: Evidence from the bicycle-sharing system (BIXI) in Montreal. *J. Transp. Geogr.* **2014**, *41*, 306–314. [CrossRef]
6. Nosal, T.; Miranda-Moreno, L.F. The effect of weather on the use of North American bicycle facilities: A multi-city analysis using automatic counts. *Transp. Res. Part A Policy Pract.* **2014**, *66*, 213–225. [CrossRef]
7. Gebhart, K.; Noland, R.B. The impact of weather conditions on bikeshare trips in Washington, DC. *Transportation* **2014**, *41*, 1205–1225. [CrossRef]
8. Faghih- Imani, A.; Hampshire, R.; Marla, L.; Eluru, N. An empirical analysis of bike sharing usage and rebalancing: Evidence from Barcelona and Seville. *Transp. Res. Part A Policy Pract.* **2017**, *97*, 177–191. [CrossRef]
9. Rixey, R.A. Station-Level Forecasting of Bikesharing Ridership: Station Network Effects in Three, U.S. Systems. *Transp. Res. Rec.* **2013**, *2387*, 46–55. [CrossRef]
10. Wang, X.; Lindsey, G.H.; Schoner, J.E.; Harrison, A. Modeling bike share station activity: Effects of nearby businesses and jobs on trips to and from stations. *J. Urban Plan. Dev.* **2016**, *142*, 04015001. [CrossRef]
11. El-Assi, W.; Mahmoud, M.S.; Habib, K.N. Effects of built environment and weather on bike sharing demand: A station level analysis of commercial bike sharing in Toronto. *Transportation* **2017**, *44*, 589–613. [CrossRef]
12. Fishman, E.; Washington, S.; Haworth, N.; Mazzei, A. Barriers to bike-sharing: An analysis from Melbourne and Brisbane. *J. Transp. Geogr.* **2014**, *41*, 325–337. [CrossRef]
13. Cock, J. Bike share in small and medium-sized cities. In Proceedings of the Presentation at 2016 Transportation Research Board Tools of the Trade Conference, Washington, DC, USA, 12–14 September 2016.
14. Li, Y.H.; Ma, Y. LSTM-based bicycle sharing demand prediction. *Smart City* **2019**, *5*, 1–4. [CrossRef]
15. Li, Y.; Zheng, Y.; Zhang, H.; Chen, L. Traffic prediction in a bike-sharing system. In Proceedings of the 23rd SIGSPATIAL International Conference on Advances in Geographic Information Systems (SIGSPATIAL '15), Seattle, WA, USA, 2–6 November 2015; Association for Computing Machinery: New York, NY, USA, 2015. Article 33. pp. 1–10. [CrossRef]
16. Chen, L.; Zhang, D.; Wang, L.; Yang, D.; Ma, X.; Li, S.; Wu, Z.; Pan, G.; Nguyen, T.-M.; Jakubowicz, J. Dynamic cluster-based over-demand prediction in bike sharing systems. In Proceedings of the 2016 ACM International Joint Conference on Pervasive and Ubiquitous Computing, Heidelberg, Germany, 12–16 September 2016; ACM: New York, NY, USA, 2016; pp. 841–852. [CrossRef]
17. Liu, X.N.; Wang, J.J.; Zhang, T.F. A Method of Bike Sharing Demand Forecasting. In *Applied Mechanics and Materials*; Trans Tech Publications Ltd.: Bach, Switzerland, 2014; Volume 587–589, pp. 1813–1816. [CrossRef]
18. Kaltenbrunner, A.; Meza, R.; Grivolla, J.; Codina, J.; Banchs, R. Urban cycles and mobility patterns: Exploring and predicting trends in a bicycle-based public transport system. *Pervasive Mob. Comput.* **2010**, *6*, 455–466. [CrossRef]
19. Yanping, L.; Wanfeng, D. Research on short-term forecasting method of urban public bicycle demand based on ARIMA model. *J. Nanjing Norm. Univ.* **2016**, *16*, 36–40.
20. Xia, Y. *Demand Prediction of Public Bicycle System Based on Station Clustering*; Dalian University of Technology: Dalian, China, 2018.
21. Zhou, Y.; Wang, L.; Zhong, R.; Tan, Y. A Markov Chain Based Demand Prediction Model for Stations in Bike Sharing Systems. *Math. Probl. Eng.* **2018**, *2018*, 1–8. [CrossRef]
22. Wang, W. Forecasting Bike Rental Demand Using New York Citi Bike Data. Master's Thesis, Dublin Institute of Technology, Dublin, Ireland, 2016.
23. Froehlich, J.E.; Neumann, J.; Oliver, N. Sensing and predicting the pulse of the city through shared bicycling. In Proceedings of the Twenty-First International Joint Conference on Artificial Intelligence, Los Angeles, CA, USA, 12–17 July 2009; pp. 1420–1426.
24. Hulot, P.; Aloise, D.; Jena, S.D. Towards Station-Level Demand Prediction for Effective Rebalancing in Bike-Sharing Systems. In Proceedings of the 24th ACM SIGKDD International Conference on Knowledge Discovery & Data Mining (KDD '18), London, UK, 19–23 August 2018; Association for Computing Machinery: New York, NY, USA, 2018; pp. 378–386. [CrossRef]
25. Liu, J.; Li, Q.; Qu, M.; Chen, W.; Yang, J.; Xiong, H.; Zhong, H.; Fu, Y. Station site optimization in bike sharing systems. In Proceedings of the 2015 IEEE International Conference on Data Mining, Atlantic City, NJ, USA, 14–17 November 2015; pp. 883–888. [CrossRef]
26. Wang, B.; Kim, I. Short-term prediction for bike-sharing service using machine learning. *Transp. Res. Procedia* **2018**, *34*, 171–178. [CrossRef]
27. Chen, P.; Hsieh, H.; Su, K.; Sigalingging, X.K.; Chen, Y.; Leu, J. Predicting station level demand in a bike-sharing system using recurrent neural networks. *IET Intell. Transp. Syst.* **2020**, *14*, 554–561. [CrossRef]

28. Zhang, C.; Zhang, L.; Liu, Y.; Yang, X. Short-term prediction of bike-sharing usage considering public transport: A LSTM approach. In Proceedings of the 2018 21st International Conference on Intelligent Transportation Systems (ITSC), Maui, HI, USA, 4–7 November 2018; pp. 1564–1571. [CrossRef]
29. He, M.; Xue, X.; Zhang, X.; Zhou, C. A Bike-sharing Demand Predicting Model with Integrating Temporal Convolutional Network and Self-Attention. In Proceedings of the 2021 International Conference on Electronic Information Engineering and Computer Science (EIECS), Changchun, China, 23–26 September 2021; pp. 278–281. [CrossRef]
30. Ma, X.; Yin, Y.; Jin, Y.; He, M.; Zhu, M. Short-Term Prediction of Bike-Sharing Demand Using Multi-Source Data: A Spatial-Temporal Graph Attentional LSTM Approach. *Appl. Sci.* **2022**, *12*, 1161. [CrossRef]
31. Dastjerdi, A.M.; Morency, C. Bike-Sharing Demand Prediction at Community Level under COVID-19 Using Deep Learning. *Sensors* **2022**, *22*, 1060. [CrossRef]
32. Chang, W.; Ji, X.; Wang, L.; Liu, H.; Zhang, Y.; Chen, B.; Zhou, S. A Machine-Learning Method of Predicting Vital Capacity Plateau Value for Ventilatory Pump Failure Based on Data Mining. *Healthcare* **2021**, *9*, 1306. [CrossRef]
33. Zhou, S.; Chen, B.; Liu, H.; Ji, X.; Wei, C.; Chang, W.; Xiao, Y. Travel Characteristics Analysis and Traffic Prediction Modeling Based on Online Car-Hailing Operational Data Sets. *Entropy* **2021**, *23*, 1305. [CrossRef]
34. Zhou, M.; Wang, D.; Li, Q.; Yue, Y.; Tu, W.; Cao, R. Impacts of weather on public transport ridership: Results from mining data from different sources. *Transp. Res. Part C Emerg. Technol.* **2017**, *75*, 17–29. [CrossRef]
35. Wu, J.; Liao, H. Weather, travel mode choice, and impacts on subway ridership in Beijing. *Transp. Res. Part A Policy Pract.* **2020**, *135*, 264–279. [CrossRef]
36. Zhao, P.; Li, S.; Li, P.; Liu, J.; Long, K. How does air pollution influence cycling behaviour? Evidence from Beijing. *Transp. Res. Part D Transp. Environ.* **2018**, *63*, 826–838. [CrossRef]
37. Ren, W.J.; Han, M. A review of research on multivariate time series causality analysis. *J. Autom.* **2021**, *47*, 64–78. [CrossRef]
38. Bai, S.; Kolter, J.Z.; Koltun, V. An empirical evaluation of generic convolutional and recurrent networks for sequence modeling [EB/OL]. *arXiv* **2018**, arXiv:1803.01271.
39. Zhou, D.; Li, Z.; Zhu, J.; Zhang, H.; Hou, L. State of health monitoring and remaining useful life prediction of lithium-ion batteries based on temporal convolutional network. *IEEE Access* **2020**, *8*, 53307–53320. [CrossRef]
40. Wu, P.; Sun, J.; Chang, X.; Zhang, W.; Arcucci, R.; Guo, Y.; Pain, C.C. Data-driven reduced order model with temporal convolutional neural network. *Comput. Methods Appl. Mech. Eng.* **2020**, *360*, 112766–112778. [CrossRef]
41. Oord, A.V.D.; Dieleman, S.; Zen, H.; Simonyan, K.; Vinyals, O.; Graves, A.; Kavukcuoglu, K. WaveNet: A generative model for raw audio [EB/OL]. *arXiv* **2016**, arXiv:1609.03499.
42. Wang, Y.; Yang, K.; Li, H. Industrial time-series modeling via adapted receptive field temporal convolution networks integrating regularly updated multi-region operations based on PCA. *Chem. Eng. Sci.* **2020**, *228*, 115956–115971. [CrossRef]
43. Richardson, R.R.; Osborne, M.A.; Howey, D.A. Gaussian process regression for forecasting battery state of health. *J. Power Sources* **2017**, *357*, 209–219. [CrossRef]
44. He, K.; Zhang, X.; Ren, S.; Sun, J. Deep residual learning for image recognition [EB/OL]. *arXiv* **2015**, arXiv:1512.03385.
45. Borovykh, A.; Bohte, S.; Oosterlee, C.W. Conditional time series forecasting with convolutional neural networks [EB/OL]. *arXiv* **2017**, arXiv:1703.04691.
46. Bengio, Y.; Simard, P.; Frasconi, P. Learning long-term dependencies with gradient descent is difficult. *IEEE Trans. Neural Netw.* **1994**, *5*, 157–166. [CrossRef]
47. Cho, K.; van Merrienboer, B.; Gülçehre, Ç.; Bahdanau, D.; Bougares, F.; Schwenk, H.; Bengio, Y. Learning Phrase Representations using RNN Encoder–Decoder for Statistical Machine Translation. *arXiv* **2014**, arXiv:1406.1078.
48. Gan, J.S.; Chen, G.L. Linear combinatorial prediction models and their applications. *Comput. Sci.* **2006**, *9*, 191–194.
49. Reshef, D.N.; Reshef, Y.A.; Finucane, H.K.; Grossman, S.R.; McVean, G.; Turnbaugh, P.J.; Lander, E.S.; Mitzenmacher, M.; Sabeti, P.C. Detecting novel associations in large data sets. *Science* **2011**, *334*, 1518–1524. [CrossRef]
50. Ham, T.J.; Wu, L.; Sundaram, N.; Satish, N.; Martonosi, M. Graphicionado: A high-performance and energy-efficient accelerator for graph analytics. In Proceedings of the IEEE ACM International Symposium on Microarchitecture, Taipei, Taiwan, 15–19 October 2016. [CrossRef]
51. Vladimir, N. Vapnik. *The Nature of Statistical Learning Theory*; Springer: New York, NY, USA, 2000; pp. 138–167. [CrossRef]
52. Chen, T.; Guestrin, C. Xgboost: A scalable tree boosting System. In Proceedings of the ACM SIGKDD International Conference on Knowledge Discovery and Data Mining, San Francisco, CA, USA, 13–17 August 2016; pp. 785–794. [CrossRef]
53. Box, G.E.P.; Jenkins, G.M. Time series analysis: Forecasting and control. *J. Time* **2010**, *31*, 303.
54. Taylor, S.J.; Letham, B. Forecasting at Scale. *Am. Stat.* **2018**, *72*, 37–45. [CrossRef]
55. Salinas, D.; Flunkert, V.; Gasthaus, J.; Januschowski, T. DeepAR: Probabilistic forecasting with autoregressive recurrent networks. *Int. J. Forecast.* **2020**, *36*, 1181–1191. [CrossRef]

Article

Selected Data Mining Tools for Data Analysis in Distributed Environment

Mikhail Moshkov [1], Beata Zielosko [2,*] and Evans Teiko Tetteh [3]

[1] Computer, Electrical and Mathematical Sciences and Engineering Division and Computational Bioscience Research Center, King Abdullah University of Science and Technology (KAUST), Thuwal 23955-6900, Saudi Arabia; mikhail.moshkov@kaust.edu.sa

[2] Institute of Computer Science, Faculty of Science and Technology, University of Silesia in Katowice, Będzińska 39, 41-200 Sosnowiec, Poland

[3] Doctoral School, University of Silesia in Katowice, Bankowa 14, 40-007 Katowice, Poland; evans.tetteh@us.edu.pl

* Correspondence: beata.zielosko@us.edu.pl

Abstract: In this paper, we deal with distributed data represented either as a finite set $\mathcal{T}$ of decision tables with equal sets of attributes or a finite set $\mathcal{I}$ of information systems with equal sets of attributes. In the former case, we discuss a way to the study decision trees common to all tables from the set $\mathcal{T}$: building a decision table in which the set of decision trees coincides with the set of decision trees common to all tables from $\mathcal{T}$. We show when we can build such a decision table and how to build it in a polynomial time. If we have such a table, we can apply various decision tree learning algorithms to it. We extend the considered approach to the study of test (reducts) and decision rules common to all tables from $\mathcal{T}$. In the latter case, we discuss a way to study the association rules common to all information systems from the set $\mathcal{I}$: building a joint information system for which the set of true association rules that are realizable for a given row ρ and have a given attribute a on the right-hand side coincides with the set of association rules that are true for all information systems from $\mathcal{I}$, have the attribute a on the right-hand side, and are realizable for the row ρ. We then show how to build a joint information system in a polynomial time. When we build such an information system, we can apply various association rule learning algorithms to it.

Keywords: distributed data; decision tables; information systems; decision trees; decision rules; tests; reducts; association rules

Citation: Moshkov, M.; Zielosko, B.; Tetteh, E.T. Selected Data Mining Tools for Data Analysis in Distributed Environment. *Entropy* **2022**, *24*, 1401. https://doi.org/10.3390/e24101401

Academic Editors: Przemysław Juszczuk and Jan Kozak

Received: 2 August 2022
Accepted: 14 September 2022
Published: 1 October 2022

Publisher's Note: MDPI stays neutral with regard to jurisdictional claims in published maps and institutional affiliations.

Copyright: © 2022 by the authors. Licensee MDPI, Basel, Switzerland. This article is an open access article distributed under the terms and conditions of the Creative Commons Attribution (CC BY) license (https://creativecommons.org/licenses/by/4.0/).

1. Introduction

Along with technological development, we are dealing with an increasing amount of data that must be processed and stored. The way they are processed depends on many factors, including the purpose of use and the type of data. One of the main goals is to extract knowledge from data, for example, by discovering patterns and relationships hidden in the data. Such knowledge can be presented by a set of decision rules, decision trees, or association rules. When a selection of features is required in order to find the most important and relevant ones, a test (reduct) is used. It is a (minimal) set of attributes that provides the same classification of objects as the whole input set of features.

An important element that influences the result of the chosen approach to extracting knowledge from data is their preparation. Pre-processing includes various algorithms, depending on the needs. These can be, for example, the imputation of missing attribute values, data normalization, or discretization. The type of method used depends on the goal and affects the subsequent stages of the data mining process. This phase is particularly difficult when we are dealing with distributed data that come from various data sources and appear in a different format, depending on the data owner [1].

One popular form of data representation is the tabular form, presented either as a decision table or as an information system. In the case of a distributed environment, such data can be represented as a finite set of decision tables with the same decision attribute [2,3]. Generally, these decision tables can have different sets of conditional attributes. However, the consideration of the sets of decision tables with equal sets of attributes is of particular interest. Data can also be represented by information systems [4,5]. As for the case of decision tables, considering the sets of information systems with equal sets of attributes is of most interest to us. This paper consists of the two parts. In the first one, we deal with dispersed data represented by a finite set of decision tables with equal sets of attributes. In the second part, we deal with dispersed data represented by a finite set of information systems with equal sets of attributes.

In the first part of the paper, we assume that we have a finite set $\mathcal{T} = \{T_1, \ldots, T_k\}$ of decision tables with equal sets of attributes. Our aim is to create tools for the work with decision trees, rules, and tests (reducts) [4–6] that are common to all decision tables from $\mathcal{T}$.

There are different algorithms for the construction and optimization of decision trees for single decision tables [7–10]. To apply these algorithms to the set of decision tables $\mathcal{T}$, we need to build a single decision table (called a joint decision table for $\mathcal{T}$) such that the set of decision trees for this table is equal to the set of common decision trees for all decision tables from $\mathcal{T}$. The situation is the same for decision rules and tests (reducts). In this paper, we show when we can build joint decision tables and how to build them in a polynomial time.

Note that in the case of dispersed decision tables with different sets of conditional attributes, instead of considering a joint decision table, we should study its lower and upper approximations, which leads to the investigation of NP-hard problems [2].

In the second part of the paper, we assume that we have a finite set $\mathcal{I} = \{I_1, \ldots, I_k\}$ of information systems, in which columns are labeled with the same attributes $a_1, \ldots, a_n$. We fix a row ρ from one of the information systems from $\mathcal{I}$ and an attribute $a_j \in \{a_1, \ldots, a_n\}$, and we consider the set $Arules(\mathcal{I}, \rho, a_j)$ of association rules of the form $(a_{i_1} = \sigma_1) \wedge \cdots \wedge (a_{i_m} = \sigma_m) \rightarrow (a_j = \sigma)$ that are true for each information system from $\mathcal{I}$ and are realizable for the row ρ (i.e., such rule covers the row ρ). Our aim is to create tools for the work with association rules from this set.

There are different algorithms for the construction and optimization of association rules for single information systems [11–16]. To apply these algorithms to the set of information systems $\mathcal{I}$, we need to build an information system J (called a joint information system for $\mathcal{I}$, ρ, and a_j) such that $Arules(\{J\}, \rho, a_j) = Arules(\mathcal{I}, \rho, a_j)$. In this paper, we show how to build joint information systems in a polynomial time.

The main contribution of this work is a proposed new methodology for working with distributed data, presented as a set of decision tables or a set of information systems. It is an interesting direction of research, especially in the areas of distributed data mining, data processing, and knowledge extraction from dispersed data sources. The proposed approach is different from the approaches described in the framework of distributed data mining (Section 2.1). Our methodology is based on the transformation of distributed data sources into the so-called joint tabular form of data, presented as a joint decision table or as a joint information system. An important element is that the obtained decision table or information system allows for the induction of decision rules, decision trees, reducts, or association rules common to the distributed data. Moreover, existing algorithms for their induction can be used.

The present paper is an extended version of two conference papers [17,18].

The rest of the paper is organized as follows. Section 2 presents some background information related to distributed data, decision trees and rules, tests, and reducts as well as association rules. In Section 3, we study distributed data represented as a finite set of decision tables, and in Section 4, we study distributed data represented as a finite set of information systems. Section 5 contains brief conclusions.

2. Background Information

In this section some basic information related to distributed data, decision trees and rules, tests, and reducts as well as association rules is presented.

2.1. Distributed Data

Technological development means that we are dealing with an increasing amount of data that can be heterogeneous, taking into account their format and location.

One of the popular solutions for processing and storing decentralized data are data warehouses [19,20]. They are used to store huge data sets. By using appropriate analytical tools that allow for the employment of data mining algorithms, it is possible to mine knowledge from data by analyzing trends, anomalies, or searching patterns. On this basis, business decisions are made regarding, for example, sales planning or marketing campaigns. In addition, data warehouses have ETL (Extraction, Trasformation, Loading) tools, which are designed to properly prepare data from heterogeneous sources and various locations.

Along with technological development and the necessity to process large amounts of distributed data, the field referred to as distributed data mining has been developing in recent years [21,22]. In this framework, different algorithms and approaches have been developed and proposed for classification, association mining, clustering, and other data mining tasks [23,24].

In this paper, a new methodology for working with distributed data is proposed. It is based on the idea of constructing one tabular form of data representation, i.e, a decision table or an information system for distributed sources, and then applying known algorithms for the induction of data mining tools, i.e., association and decision rules, decision trees, and reducts.

It should also be taken into account that distributed data mining techniques are more complex in comparison to centralized ones. The main issues which should be considered are: (i) heterogeneous data, i.e., local data sources can provide data with different formats and attributes with different domains; (ii) data fragmentation, i.e., local sources can be viewed as a horizontal or vertical fragmentation of the global data table, and therefore based on them, only part of the knowledge can be induced; (iii) data replication, i.e., replication provides better data availability, but on the other hand, it can make it difficult to ensure the consistency of distributed data; (iv) cost of communication in a distributed environment plays an important role; (v) security, privacy, and autonomy of local sources; (vi) integration results, i.e., discovered global interesting patterns and associations should be collected from local sources, and their utility should be verified globally.

Distributed data mining aims to analyze and process distributed data while taking into account resource constraints [25]. This task can be realized in the framework of a meta-learning, multi-agent system, or based on grid. The multi-agent data mining environment inherits properties of agents as interoperability and performance aspects. Interoperability concerns working collaboratively with other agents in the entire system. Performance measures can be improved or impaired by the data distribution at the local level. The meta-learning system constitutes a learning method at the local level. Learning at the meta level is based on accumulating experience on the performance of multiple applications of a learning system. Data mining based on grid aims to create a distributed computing environment in order to enable local data sources to use computing resources on demand.

2.2. Data Mining Tools

Data mining is a complex process that allows for the performance of analyses and the acquisition of knowledge from data by using different methods, depending on the aim and kind of data. Among data mining tools, decision rules, decision trees, reducts, and association rules can be used. They can be considered as algorithms for solving different problems and also as classifiers used in the area of machine learning [26]. A short description can be found in the sections below.

2.2.1. Decision Rules

Decision rules are popular and an often used form of knowledge representation. In general, decision rules can be presented in the following form:

$$IF \quad condition_1 \wedge \ldots \wedge condition_k \quad THEN \quad conclusion. \tag{1}$$

Conditions (pairs attribute = value) correspond to descriptors that are present in the premise part of the rule. Conclusion corresponds to the rule consequent part that present a class label. Rules presented in such a form can be considered as a compact form of knowledge representation. This form is simple and easily accessible from the point of view of understanding and interpreting knowledge represented by rules. Moreover, decision rules based on background knowledge can be employed in classification tasks, where a class label for a new object is assigned based on its conditions. Hence, decision rules can be applied in data mining tasks related to (i) knowledge representation and (ii) classification [27]. Taking into account these two perspectives, there are different measures used for rule evaluation and many different approaches for the induction of decision rules. The aim is to find patterns or regularities hidden in the data that are interesting and useful for users.

It should be noted that the minimization of length (number of conditions) and the maximization of support (which allows to discover major patterns in data) of decision rules are NP-hard problems [6,14]. The most part of approaches for construction of decision rules, with the exception of brute force, Boolean reasoning [28], and dynamic programming [6], cannot guarantee the construction of optimal rules, i.e., rules with minimum length or maximum support. Consequently, different heuristic approaches have been proposed in the literature [26,27,29,30]. Among them, greedy algorithms, genetic algorithms, ant colony optimization algorithms, approaches based on a sequential covering procedure, and many others can be mentioned.

2.2.2. Decision Trees

Decision trees are often used as classifiers, as a means of knowledge representation, and as algorithms. A decision tree learning algorithm approximates a target concept using a tree representation, where each internal node corresponds to an attribute, and each terminal node known as a leaf corresponds to a class label. The root node is at the top and leafs are at the bottom of a tree.

Most of the algorithms for decision tree induction use a greedy approach and a top-down, recursive, divide-and-conquer technique. In general, the algorithm for decision tree induction starts with the tree, which initially contains a single root node that is associated with the objects included in a data set. Then, the instances are recursively partitioned into smaller subsets according to a given splitting criterion. It indicates the attribute chosen as the test condition and how the instances should be distributed to the child nodes of the constructed tree. The creation and expansion of a node is finished when the stop criterion is satisfied, for example, when all the instances associated with the node in the divided data set have the same class label. However, there are also other criteria that allow for the expansion of a node to be stopped earlier even if corresponding assigned instances have different decisions.

An advantage of decision trees is that by reading a tree from root to leaves, a decision (class label) is proposed for a considered case (object); it is also possible to see the reasons for choosing a given decision. This feature is a very important element used in the domain of applications aimed at supporting decision making. In addition, based on the decision tree, decision rules can be obtained.

There are many algorithms for decision tree induction. The most popular are [8,9,31,32]: CART (Classification and Regression Trees), ID3 (Iterative Dichotomiser 3), C4.5 (improved version of the ID3 algorithm, where "C" shows that algorithm was written in C and the the 4.5 specifics version of this algorithm), Sprint (Scalable PaRallelizable INduction of decision Trees), Chaid (Chi-square automatic interaction detection), and their many modifications.

There are also a variety of approaches based on meta-heuristics [33] such as genetic algorithms, simulated annealing, ant colony optimization, and many others. An important element during decision tree induction is selecting the best split, which allows for the partitioning of instances into two or more subsets that are associated with the nodes of the decision tree. Among the popular ones, measures based on entropy and the Gini index used in CART can be distinguished.

2.2.3. Tests and Reducts

The construction of reducts and tests (super reducts) is closely connected with the feature selection area [34–36]. The aim of this domain is to select from the entire set of features only those attributes that are the most relevant while maintaining the descriptive and classification properties of the original feature space. Hence, this reduced set of attributes can be used instead of the entire attributes set for knowledge discovery. It is an important task, especially in areas where data sets contain a huge number of features, for example, in market basket analysis, stock trading, and sequence pattern discovery in bioinformatics.

Reduct is as an irreducible subset of features providing a satisfactory level of information about the considered target variable, which can be, for example, the accuracy of the classifier constructed based on the features contained in it. Therefore, from the classification point of view, a reduct can be interpreted as a minimal subset of attributes that has the same classification power as the entire set of features. Definitions for attribute reducts can be based on different criteria, for example, a reduct can also be considered as a minimal set of attributes that preserves the degree of dependency of the full set of attributes [37].

In the rough sets theory, where the construction of reducts constitutes one of the main research directions, decision super reduct (test) is defined as a subset of condition attributes that is sufficient for discerning any of the objects in a decision table with different class labels. A decision reduct is a test in the sense that each proper subset of this test is not a test for the considered problem.

Unfortunately, finding a reduct with minimum cardinality is an NP-hard problem. It is also known that the upper bound of a potential number of all reducts that can be found for a given dataset with k attributes is equal to $\binom{k}{\lfloor k/2 \rfloor}$. Taking into account that these issues represent high computational costs and complexity brought by the tasks of all reduct construction, different approaches and heuristics have been proposed for the construction of many reducts in some acceptable time. The popular ones are Boolean reasoning [28], genetic algorithms [38], greedy algorithms [39], fuzzy-rough approach, and others [14,40].

Based on the reduct constructed for a given decision table, decision rules can be induced from reduced sets of attributes. In this indirect method of rule induction, it is easy to see that the number of attributes which constitute a reduct is an important factor from the point of view of knowledge representation. Short reducts allow for the construction of short decision rules, which are more preferred from the point of view of understanding and interpretation by users.

2.2.4. Association Rules

Association rule mining is one of the key and interesting methods of data mining and knowledge discovery. It aims to extract co-occurrences of items as well as associations and patterns hidden in the data. One of the most popular applications of association rules is the market basket analysis, which finds associations between different items that customers place in their shopping baskets. Other areas include business fields involving decision making and effective marketing, medical diagnosis, stock trading, and others.

There are different types of association rules, for example: boolean association rules, which are used in market basket analysis; qualitative association rules [11], which are induced from business data; spatial association rules [41]; multilevel association rules [42], and others [29]. In general, association rules are presented in the following form:

$$X \to Y, \tag{2}$$

where X and Y are sets of items.

Two main quality measures of association rules are support and confidence [15]. Rules that satisfy minimum thresholds of these measures indicated by a user are called strong association rules.

It should be also noted that there are many algorithms for construction of association rules, however the process of mining of association rules consists of two main stages: (i) find all frequent itemsets, i.e., they occur at least as frequently as a predetermined minimum support threshold, and (ii) generate strong association rules from the frequent itemsets, i.e., rules that satisfy minimum support and minimum confidence thresholds. The most popular algorithm based on mining frequent itemset is Apriori [43]. However, many other approaches were proposed by researchers, for example, algorithms that use frequent pattern growth approach [44], vertical data format [45], hash based technique, partitioning the data and others [46].

One very important task in data mining is the classification process. In this framework, association rules also have an application. The associative classification task aims to find association rules that have only the class label in the consequent part of the rule and which satisfies the minimum support and the confidence thresholds, the so-called Class Association Rules. There are many methods for the construction of classifiers, which differ in the approaches used for mining association rules and their selection [47].

3. Sets of Decision Tables

In this section, we deal with dispersed data represented as a finite set of decision tables with equal sets of attributes.

3.1. Main Notions

A decision table T is a table filled with numbers from the set $\omega = \{0, 1, 2, \ldots\}$ of non-negative integers, in which columns are labeled with conditional attributes $a_1, \ldots, a_n$ and each row is labeled with a decision that is a number from ω (see Figure 1). We assume that equal rows in the table T are labeled with equal decisions, i.e., we consider only consistent decision tables. We associate the following problem with the table T: for a given row ρ of T, we should recognize the decision attached to ρ using values of the condition attributes from $\{a_1, \ldots, a_n\}$ in this row. To this end, we can use decision trees, rules, and test (reducts).

A decision tree Γ over T is a finite directed tree with a root, in which each internal node is labeled with an attribute from the set $\{a_1, \ldots, a_n\}$, edges leaving this node are labeled with pairwise different numbers from ω, and each leaf node is labeled with a decision from ω. For a given row $\rho = (\delta_1, \ldots, \delta_n)$, the tree Γ work starts in the root of Γ. If the node under consideration is a leaf, then the number attached to this node is the result of the Γ work. Let the node under consideration be an internal node with an attribute a_i attached to it. If there is an edge that leaves the considered node and is labeled with δ_i, then we pass along this edge. Otherwise, the decision tree Γ finishes its work without a result. We say that Γ is a decision tree for T if, for any row of T, the work of Γ finishes in a leaf that is labeled with the same decision as the considered row (see Figure 1). We denote with $Trees(T)$ the set of decision trees for T.

Any decision rule over T can be represented in the following form:

$$(a_{i_1} = \sigma_1) \wedge \cdots \wedge (a_{i_m} = \sigma_m) \to t \tag{3}$$

where $a_{i_1}, \ldots, a_{i_m} \in \{a_1, \ldots, a_n\}$ and $\sigma_1, \ldots, \sigma_m, t \in \omega$. This rule is called realizable for a row $\rho = (\delta_1, \ldots, \delta_n) \in \omega^n$ (it is possible that this row does not belong to T) if $\delta_{i_1} = \sigma_1, \ldots, \delta_{i_m} = \sigma_m$. This rule is called true for T if, for any row ρ' of T, such that rule (3) is realizable for ρ', the row ρ' is labeled with the decision t. We say that (3) is a rule for T and ρ if this rule is true for T and realizable for ρ (see Figure 1). We denote with $Rules(T,\rho)$ the set of decision rules for T and ρ. One can show that (3) is a rule for T and ρ if (i) ρ is labeled with the decision t if ρ belongs to T, and (ii) if each row ρ' of T, which is labeled with a decision different from t, is different from ρ on at least one attribute from the set $\{a_{i_1}, \ldots, a_{i_m}\}$.

A test for T is a subset of the set of conditional attributes $\{a_1, \ldots, a_n\}$, such that any two rows from T with different decisions are different on at least one attribute from this subset. A reduct for T is a test for T, for which each proper subset is not a test (see Figure 1). We denote with $Tests(T)$ the set of tests for T.

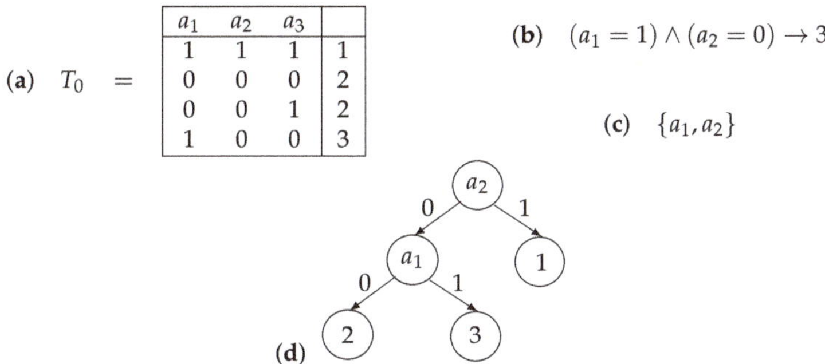

Figure 1. Considered objects: (**a**) decision table T_0, (**b**) decision rule for T_0 and row $(1,0,0)$, (**c**) reduct for T_0, (**d**) decision tree for T_0.

Let $\mathcal{T} = \{T_1, \ldots, T_k\}$ be a finite nonempty set of decision tables, in which columns are labeled with the same conditional attributes $a_1, \ldots, a_n$. Each decision table from this set is consistent, but different tables from $\mathcal{T}$ can contain equal rows labeled with different decisions. Let ρ be a row of a decision table from $\mathcal{T}$. We denote $Trees(\mathcal{T}) = \bigcap_{T_i \in \mathcal{T}} Trees(T_i)$, $Rules(\mathcal{T}, \rho) = \bigcap_{T_i \in \mathcal{T}} Rules(T_i, \rho)$, and $Tests(\mathcal{T}) = \bigcap_{T_i \in \mathcal{T}} Tests(T_i)$. In the next three sections, we will consider joint decision tables for these sets of common decision trees, rules, and tests (reducts) for $\mathcal{T}$.

3.2. Joint Decision Tables for Decision Trees

Let $\mathcal{T} = \{T_1, \ldots, T_k\}$ be a set of decision tables, in which the columns are labeled with the attributes $a_1, \ldots, a_n$. The set of decision tables $\mathcal{T}$ is called consistent if there are no two tables in $\mathcal{T}$ containing equal rows labeled with different decisions.

First, we show that if the set $\mathcal{T}$ is not consistent, then $Trees(\mathcal{T}) = \varnothing$. Since $\mathcal{T}$ is not consistent, there exist two tables T_i and T_j in $\mathcal{T}$ and a row ρ, such that ρ is a row of T_i labeled with a decision p, ρ is a row of T_j labeled with a decision q, and $p \neq q$. Let us assume that $Trees(\mathcal{T}) \neq \varnothing$ and $\Gamma \in Trees(\mathcal{T})$. Then, the output of Γ for the row ρ should be equal to p and to q at the same time, but this is impossible. Therefore, $Trees(\mathcal{T}) = \varnothing$.

Let us assume now that the set $\mathcal{T}$ is consistent. With $T^{trees}(\mathcal{T})$, we denote a decision table in which columns are labeled with attributes $a_1, \ldots, a_n$, and the set of rows coincides with the union of sets of rows of the tables $T_1, \ldots, T_k$. Each row belonging to $T^{trees}(\mathcal{T})$ is labeled with the decision attached to this row in the tables from $\mathcal{T}$ which this row belongs to (see Figure 2). Note that the table $T^{trees}(\mathcal{T})$ can be constructed in polynomial time.

We now show that $Trees(\mathcal{T}) = Trees(T^{trees}(\mathcal{T}))$. Let $\Gamma \in Trees(\mathcal{T})$. Then, for any $T_i \in \mathcal{T}$ and any row ρ belonging to T_i, Γ returns the decision attached to ρ in T_i. Therefore, for any row ρ of $T^{trees}(\mathcal{T})$, Γ returns the decision attached to ρ, i.e., $\Gamma \in Trees(T^{trees}(\mathcal{T}))$. Now, let $\Gamma \in Trees(T^{trees}(\mathcal{T}))$. Then, for any row ρ of $T^{trees}(\mathcal{T})$, Γ returns the decision attached to ρ. Therefore, for any table $T_i \in \mathcal{T}$ and any row ρ of T_i, Γ returns the decision attached to ρ in T_i, i.e., $\Gamma \in Trees(\mathcal{T})$.

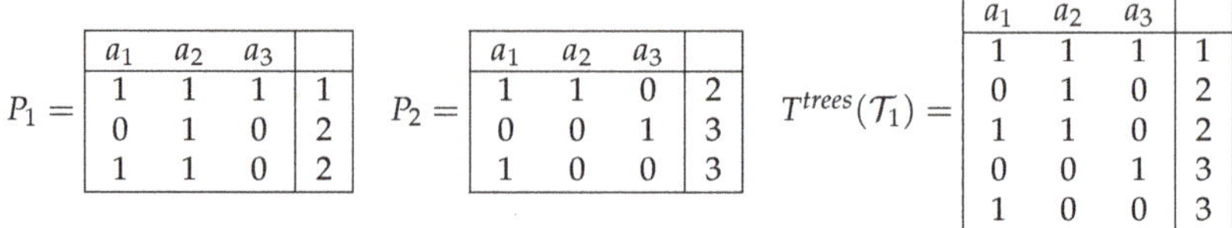

$$P_1 = \begin{array}{|ccc|c|} \hline a_1 & a_2 & a_3 & \\ \hline 1 & 1 & 1 & 1 \\ 0 & 1 & 0 & 2 \\ 1 & 1 & 0 & 2 \\ \hline \end{array} \qquad P_2 = \begin{array}{|ccc|c|} \hline a_1 & a_2 & a_3 & \\ \hline 1 & 1 & 0 & 2 \\ 0 & 0 & 1 & 3 \\ 1 & 0 & 0 & 3 \\ \hline \end{array} \qquad T^{trees}(\mathcal{T}_1) = \begin{array}{|ccc|c|} \hline a_1 & a_2 & a_3 & \\ \hline 1 & 1 & 1 & 1 \\ 0 & 1 & 0 & 2 \\ 1 & 1 & 0 & 2 \\ 0 & 0 & 1 & 3 \\ 1 & 0 & 0 & 3 \\ \hline \end{array}$$

Figure 2. Joint decision table $T^{trees}(\mathcal{T}_1)$ for the set of decision tables $\mathcal{T}_1 = \{P_1, P_2\}$.

3.3. Joint Decision Tables for Decision Rules

Let $\mathcal{T} = \{T_1, \ldots, T_k\}$ be a set of decision tables, in which columns are labeled with attributes $a_1, \ldots, a_n$. A row ρ of a decision table from the set $\mathcal{T}$ is called inconsistent if there are two tables in $\mathcal{T}$ that contain it and if the row ρ in these tables is labeled with different decisions. Otherwise, the row ρ is called consistent.

First, we show that if the row ρ is inconsistent, then $Rules(\mathcal{T}, \rho) = \varnothing$. Since ρ is inconsistent, there exist two tables T_i and T_j in $\mathcal{T}$, such that ρ is a row of T_i labeled with a decision p, ρ is a row of T_j labeled with a decision q, and $p \neq q$. Let us assume that $Rules(\mathcal{T}, \rho) \neq \varnothing$. Then, the right-hand side of each rule from $Rules(\mathcal{T}, \rho)$ should be equal to p and to q at the same time, but this is impossible. Therefore, $Rules(\mathcal{T}, \rho) = \varnothing$.

Let us assume now that the row ρ is consistent, and that it is labeled with the decision t. We denote with $T^{rules}(\mathcal{T}, \rho)$ a decision table in which columns are labeled with attributes $a_1, \ldots, a_n$, the first row is ρ, and the set of all other rows coincides with the union of the sets of rows of the tables $T_1, \ldots, T_k$, which are labeled with decisions different from t. The first row of $T^{rules}(\mathcal{T}, \rho)$ is labeled with the decision t, and all other rows are labeled with the decision $t + 1$ (see Figure 3). We cannot keep the initial decisions for rows that are now labeled with $t + 1$ since in this case, the table $T^{rules}(\mathcal{T}, \rho)$ can be inconsistent. Note that the table $T^{rules}(\mathcal{T}, \rho)$ can be constructed in polynomial time.

$$Q_1 = \begin{array}{|ccc|c|} \hline a_1 & a_2 & a_3 & \\ \hline 1 & 1 & 1 & 1 \\ 0 & 1 & 0 & 2 \\ 1 & 1 & 0 & 3 \\ \hline \end{array} \qquad Q_2 = \begin{array}{|ccc|c|} \hline a_1 & a_2 & a_3 & \\ \hline 1 & 1 & 1 & 2 \\ 0 & 1 & 1 & 3 \\ 1 & 0 & 0 & 3 \\ \hline \end{array} \qquad T^{rules}(\mathcal{T}_2, \rho) = \begin{array}{|ccc|c|} \hline a_1 & a_2 & a_3 & \\ \hline 1 & 0 & 0 & 3 \\ 1 & 1 & 1 & 4 \\ 0 & 1 & 0 & 4 \\ \hline \end{array}$$

Figure 3. Joint decision table $T^{rules}(\mathcal{T}_2, \rho)$ for the set of decision tables $\mathcal{T}_2 = \{Q_1, Q_2\}$ and row $\rho = (1, 0, 0)$.

We now show that $Rules(\mathcal{T}, \rho) = Rules(T^{rules}(\mathcal{T}, \rho), \rho)$. Let $\rho \in Rules(\mathcal{T}, \rho)$ and ρ be equal to (3). Then, for any table T_i from $\mathcal{T}$, any row of T_i labeled with a decision different from t is different from ρ on at least one attribute from the set $\{a_{i_1}, \ldots, a_{i_m}\}$. Therefore, any row of $T^{rules}(\mathcal{T}, \rho)$ labeled with the decision $t + 1$ is different from ρ on at least one attribute from the set $\{a_{i_1}, \ldots, a_{i_m}\}$, i.e., $\rho \in Rules(T^{rules}(\mathcal{T}, \rho), \rho)$. Now, let $\rho \in Rules(T^{rules}(\mathcal{T}, \rho), \rho)$. Then, any row of $T^{rules}(\mathcal{T}, \rho)$ labeled with the decision $t + 1$ is different from ρ on at least one attribute from the set $\{a_{i_1}, \ldots, a_{i_m}\}$. Therefore, for any table T_i from $\mathcal{T}$, any row of T_i labeled with a decision different from t is different from ρ on at least one attribute from the set $\{a_{i_1}, \ldots, a_{i_m}\}$, i.e., $\rho \in Rules(\mathcal{T}, \rho)$.

3.4. Joint Decision Tables for Tests (Reducts)

Let $\mathcal{T} = \{T_1, \ldots, T_k\}$ be a set of decision tables, in which columns are labeled with attributes $a_1, \ldots, a_n$. Each decision table from this set is consistent, but different tables from $\mathcal{T}$ can contain equal rows labeled with different decisions. It is clear that for each table T_i from $\mathcal{T}$, the set of attributes $\{a_1, \ldots, a_n\}$ is a test. Therefore, $Tests(\mathcal{T}) \neq \varnothing$.

We denote with $T^{tests}(\mathcal{T})$ a decision table in which columns are labeled with attributes $a_1, \ldots, a_n$, the first row is filled with zeros, and the set of all other rows is constructed in the following way. For any table T_i from $\mathcal{T}$ and any two rows ρ_1 and ρ_2 of T_i labeled with different decisions, we add to the table $T^{tests}(\mathcal{T})$ the row $c(\rho_1, \rho_2)$ filled with numbers from the set $\{0, 1\}$. For $i = 1, \ldots, n$, the row $c(\rho_1, \rho_2)$ has the number 1 in the ith position if and only if the rows ρ_1 and ρ_2 are different on the attribute a_j. The first row of the table $T^{tests}(\mathcal{T})$ is labeled with the decision 1. All other rows are labeled with the decision 2 (see Figure 4). It is clear that the rows ρ_1 and ρ_2 are different on an attribute a_j if and only if the first row of the table $T^{tests}(\mathcal{T})$ and the row $c(\rho_1, \rho_2)$ are different on the attribute a_j. Note that the table $T^{tests}(\mathcal{T})$ can be constructed in polynomial time.

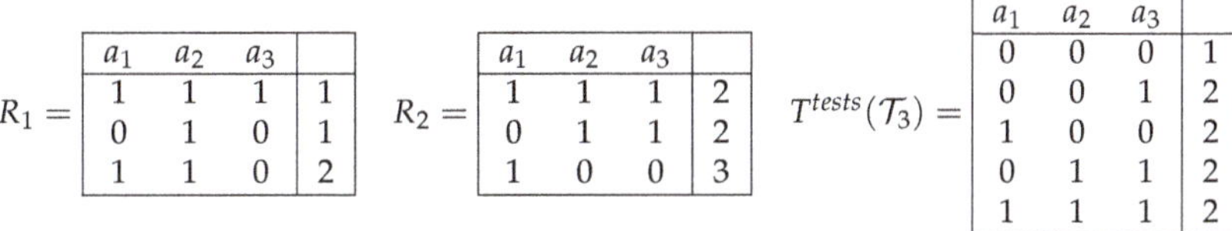

$$R_1 = \begin{array}{|ccc|c|} \hline a_1 & a_2 & a_3 & \\ \hline 1 & 1 & 1 & 1 \\ 0 & 1 & 0 & 1 \\ 1 & 1 & 0 & 2 \\ \hline \end{array} \qquad R_2 = \begin{array}{|ccc|c|} \hline a_1 & a_2 & a_3 & \\ \hline 1 & 1 & 1 & 2 \\ 0 & 1 & 1 & 2 \\ 1 & 0 & 0 & 3 \\ \hline \end{array} \qquad T^{tests}(\mathcal{T}_3) = \begin{array}{|ccc|c|} \hline a_1 & a_2 & a_3 & \\ \hline 0 & 0 & 0 & 1 \\ 0 & 0 & 1 & 2 \\ 1 & 0 & 0 & 2 \\ 0 & 1 & 1 & 2 \\ 1 & 1 & 1 & 2 \\ \hline \end{array}$$

Figure 4. Joint decision table $T^{tests}(\mathcal{T}_3)$ for the set of decision tables $\mathcal{T}_3 = \{R_1, R_2\}$.

We now show that $Tests(\mathcal{T}) = Tests(T^{tests}(\mathcal{T}))$. Let $B \in Tests(\mathcal{T})$. Then, for any table T_i from $\mathcal{T}$, any two rows from T_i with different decisions are different on at least one attribute from B. Therefore, the first row of the table $T^{tests}(\mathcal{T})$ is different from all other rows of the table $T^{tests}(\mathcal{T})$ on the attributes from B, i.e., $B \in Tests(T^{tests}(\mathcal{T}))$. Let $B \in Tests(T^{tests}(\mathcal{T}))$. Then, the first row of the table $T^{tests}(\mathcal{T})$ is different from all other rows of the table $T^{tests}(\mathcal{T})$ on the attributes from B. Therefore, for any table T_i from $\mathcal{T}$, any two rows from T_i with different decisions are different on at least one attribute from B, i.e., $B \in Tests(\mathcal{T})$.

4. Sets of Information Systems

In this section, we deal with dispersed data represented as a finite set of information systems with equal sets of attributes.

4.1. Main Notions

An information system I is a table filled with numbers from the set $\omega = \{0, 1, 2, \ldots\}$ of non-negative integers, in which columns are labeled with attributes $a_1, \ldots, a_n$. Each row ρ of the information system I is interpreted as an object, and the number in the intersection of the row ρ and the column a_i is interpreted as the value $a_i(\rho)$ of the attribute a_i for the object ρ.

Any association rule over the set of attributes $\{a_1, \ldots, a_n\}$ can be represented in the following form:

$$(a_{i_1} = \sigma_1) \wedge \cdots \wedge (a_{i_m} = \sigma_m) \rightarrow (a_j = \sigma), \tag{4}$$

where $a_j \in \{a_1, \ldots, a_n\}$, $a_{i_1}, \ldots, a_{i_m} \in \{a_1, \ldots, a_n\} \setminus \{a_j\}$, and $\sigma_1, \ldots, \sigma_m, \sigma \in \omega$. We will say that this rule is based on the attribute a_j. Rule (4) is called realizable for a row $\rho = (\delta_1, \ldots, \delta_n) \in \omega^n$ if $\delta_{i_1} = \sigma_1, \ldots, \delta_{i_m} = \sigma_m$. This rule is called true for the information system I if for any row ρ' of I such that rule (4) is realizable for ρ', $a_j(\rho') = \sigma$ (see Figure 5).

$$I_0 = \begin{array}{|cccc|} \hline a_1 & a_2 & a_3 & a_4 \\ \hline 1 & 0 & 1 & 0 \\ 0 & 0 & 0 & 1 \\ 0 & 0 & 1 & 1 \\ 0 & 1 & 0 & 0 \\ \hline \end{array} \qquad\qquad (a_1 = 0) \wedge (a_2 = 0) \rightarrow (a_4 = 1)$$

Figure 5. Information system I_0 and the association rule, which is based on the attribute a_4, true for the information system I_0, and realizable for the row $(0, 0, 0, 1)$.

4.2. Joint Information Systems for Association Rules

Let $\mathcal{I} = \{I_1, \ldots, I_k\}$ be a finite nonempty set of information systems, in which columns are labeled with the same attributes $a_1, \ldots, a_n$. Let $\rho = (\delta_1, \ldots, \delta_n)$ be a row of an information system from $\mathcal{I}$ and $a_j \in \{a_1, \ldots, a_n\}$. We denote with $Arules(\mathcal{I}, \rho, a_j)$ the set of association rules over the set of attributes $\{a_1, \ldots, a_n\}$, each of which is based on the attribute a_j, is realizable for the row ρ, and is true for each information system from $\mathcal{I}$.

Our aim is to construct a so-called joint information system J, for which

$$Arules(\{J\}, \rho, a_j) = Arules(\mathcal{I}, \rho, a_j). \tag{5}$$

In the information system $J = J(\mathcal{I}, \rho, a_j)$, columns are labeled with the attributes $a_1, \ldots, a_n$. This information system contains row ρ and all rows ρ' from the information systems $I_1, \ldots, I_k$, such that $a_j(\rho) \neq a_j(\rho')$ (we keep only one row from any group of equal rows) (see Figure 6). Note that the information system J can be constructed in polynomial time.

It is easy to show that the set of rules $Arules(\{J\}, \rho, a_j) \cup Arules(\mathcal{I}, \rho, a_j)$ is a subset of the set A of rules in the following form:

$$(a_{i_1} = \delta_{i_1}) \wedge \cdots \wedge (a_{i_m} = \delta_{i_m}) \to (a_j = \delta_j),$$

where $a_{i_1}, \ldots, a_{i_m} \in \{a_1, \ldots, a_n\} \setminus \{a_j\}$. To show that equality (5) holds, it is enough to prove that, for any rule $r \in A$, $r \notin Arules(\{J\}, \rho, a_j)$ if and only if $r \notin Arules(\mathcal{I}, \rho, a_j)$. It is clear that each rule from A is based on the attribute a_j and is realizable for the row ρ. Let $r \notin Arules(\{J\}, \rho, a_j)$. Then, the rule r is not true for J, and there exists a row ρ' from J such that r is realizable for ρ' and $a_j(\rho) \neq a_j(\rho')$. It is clear that ρ' is a row from an information system I_i from $\mathcal{I}$. Then, r is not true for I_i and $r \notin Arules(\mathcal{I}, \rho, a_j)$. Let $r \notin Arules(\mathcal{I}, \rho, a_j)$. Then, there exists an information system $I_i \in \mathcal{I}$ for which r is not true, and there exists a row ρ' from I_i such that r is realizable for ρ' and $a_j(\rho) \neq a_j(\rho')$. It is clear that ρ' is a row from the information system J. Then, r is not true for J, and $r \notin Arules(\{J\}, \rho, a_j)$. Thus, the equality (5) holds.

$$I_1 = \begin{array}{|ccc|} \hline a_1 & a_2 & a_3 \\ \hline 1 & 1 & 1 \\ 0 & 1 & 0 \\ 1 & 1 & 0 \\ \hline \end{array} \qquad I_2 = \begin{array}{|ccc|} \hline a_1 & a_2 & a_3 \\ \hline 1 & 1 & 1 \\ 0 & 1 & 1 \\ 1 & 0 & 0 \\ \hline \end{array} \qquad J(\mathcal{I}, \rho, a_3) = \begin{array}{|ccc|} \hline a_1 & a_2 & a_3 \\ \hline 1 & 0 & 0 \\ 1 & 1 & 1 \\ 0 & 1 & 1 \\ \hline \end{array}$$

Figure 6. Joint information system $J(\mathcal{I}, \rho, a_3)$ for the set of information systems $\mathcal{I} = \{I_1, I_2\}$, row $\rho = (1, 0, 0)$, and attribute a_3.

5. Conclusions

In this simple methodological paper, we have shown the problem of studying common decision trees for a dispersed set of decision tables with equal sets of attributes and how to reduce this to the study of decision trees for a single decision table. We accomplished the same for common decision rules and tests (reducts). The proposed approach allows us to generalize known methods in the study of single decision tables to the case of dispersed tables with equal sets of attributes.

We also showed the problem of studying common association rules for a dispersed set of information systems with equal sets of attributes and how to reduce this to the study of association rules for a single information system. The proposed approach allows us to generalize known methods in the study of association rules for single information systems to the case of dispersed information systems with equal sets of attributes.

The presented idea is different from the methods offered in the framework of distributed data mining or data warehouses. In our approach, the cost of communication in a distributed environment is limited to the construction of a joint tabular form. Then, depending on the aim of the data analysis, different existing algorithms for the induction of decision trees, rules, reducts, or association rules can be used. In the case of data warehouses, the main application is the use of OLAP tools for supporting business decisions. In the case of distributed data mining, collaboration among agents in the entire system and learning at the local level are important factors that are omitted in the proposed approach.

Future research will be connected with developing an algorithm for the induction of decision rules from distributed data. The proposed idea will be different from the one presented in this paper, since decision rules will be induced from a set of decision tables without the process of transforming the distributed data into a joint tabular form.

Author Contributions: Conceptualization, all authors; methodology, all authors; formal analysis, all authors; investigation, all authors; writing—original draft preparation, M.M. and B.Z.; supervision, M.M.; project administration, M.M.; funding acquisition, M.M. All authors have read and agreed to the published version of the manuscript.

Funding: Research funded by King Abdullah University of Science and Technology.

Data Availability Statement: Not applicable.

Acknowledgments: The research work reported in this publication was supported by King Abdullah University of Science and Technology (KAUST).

Conflicts of Interest: The authors declare no conflict of interest.

References

1. Fu, Y. Distributed data mining: An overview. *Newsl. IEEE Tech. Comm. Distrib. Process.* **2001**, *4*, 5–9.
2. Moshkov, M. Decision trees and reducts for distributed decision tables. In *Monitoring, Security, and Rescue Techniques in Multiagent Systems, MSRAS 2004, Plock, Poland, 7–9 June 2004*; Advances in Soft Computing; Dunin-Keplicz, B., Jankowski, A., Skowron, A., Szczuka, M.S., Eds.; Springer: Berlin/Heidelberg, Germany, 2004; Volume 28, pp. 239–248.
3. Ślęzak, D. Decision value oriented decomposition of data tables. In *Foundations of Intelligent Systems, 10th International Symposium, ISMIS '97, Charlotte, NC, USA, 15–18 October 1997, Proceedings*; Lecture Notes in Computer Science; Ras, Z.W., Skowron, A., Eds.; Springer: Berlin/Heidelberg, Germany, 1997; Volume 1325, pp. 487–496.
4. Pawlak, Z. *Rough Sets-Theoretical Aspects of Reasoning about Data*; Theory and Decision Library: Series D; Kluwer: Alphen aan den Rijn, The Netherlands, 1991; Volume 9.
5. Pawlak, Z.; Skowron, A. Rudiments of rough sets. *Inf. Sci.* **2007**, *177*, 3–27. [CrossRef]
6. Moshkov, M.; Zielosko, B. *Combinatorial Machine Learning—A Rough Set Approach*; Studies in Computational Intelligence; Springer: Berlin/Heidelberg, Germany, 2011; Volume 360.
7. AbouEisha, H.; Amin, T.; Chikalov, I.; Hussain, S.; Moshkov, M. *Extensions of Dynamic Programming for Combinatorial Optimization and Data Mining*; Intelligent Systems Reference Library; Springer: Berlin/Heidelberg, Germany, 2019; Volume 146.
8. Breiman, L.; Friedman, J.H.; Olshen, R.A.; Stone, C.J. *Classification and Regression Trees*; Chapman and Hall/CRC: Boca Raton, FL, USA, 1984.
9. Moshkov, M. Time complexity of decision trees. In *Trans. Rough Sets III*; Lecture Notes in Computer Science; Peters, J.F., Skowron, A., Eds.; Springer: Berlin/Heidelberg, Germany, 2005; Volume 3400, pp. 244–459.
10. Rokach, L.; Maimon, O. *Data Mining with Decision Trees-Theory and Applications*; Series in Machine Perception and Artificial Intelligence; World Scientific: Singapore, 2007; Volume 69.
11. Agrawal, R.; Srikant, R. Fast algorithms for mining association rules in large databases. In *VLDB*; Bocca, J.B., Jarke, M., Zaniolo, C., Eds.; Morgan Kaufmann: San Francisco, CA, USA, 1994; pp. 487–499.
12. Alsolami, F.; Amin, T.; Moshkov, M.; Zielosko, B.; Żabiński, K. Comparison of heuristics for optimization of association rules. *Fundam. Inform.* **2019**, *166*, 1–14. [CrossRef]
13. Moshkov, M.; Piliszczuk, M.; Zielosko, B. Greedy algorithm for construction of partial association rules. *Fundam. Informaticae* **2009**, *92*, 259–277. [CrossRef]
14. Nguyen, H.S.; Ślęzak, D. Approximate reducts and association rules-correspondence and complexity results. In *RSFDGrC*; Lecture Notes in Computer Science; Zhong, N., Skowron, A., Ohsuga, S., Eds.; Springer: Berlin/Heidelberg, Germany, 1999; Volume 1711, pp. 137–145.
15. Wieczorek, A.; Słowiński, R. Generating a set of association and decision rules with statistically representative support and anti-support. *Inf. Sci.* **2014**, *277*, 56–70. [CrossRef]
16. Zielosko, B. Application of dynamic programming approach to optimization of association rules relative to coverage and length. *Fundam. Inform.* **2016**, *148*, 87–105. [CrossRef]
17. Moshkov, M. Common decision trees, rules, and tests (reducts) for dispersed decision tables (to appear). In Proceedings of the 26th International Conference on Knowledge-Based and Intelligent Information & Engineering Systems (KES 2022), Verona, Italy, 7–9 September 2022.
18. Moshkov, M.; Zielosko, B.; Tetteh, E.T. Common association rules for dispersed information systems (to appear). In Proceedings of the 26th International Conference on Knowledge-Based and Intelligent Information & Engineering Systems (KES 2022), Verona, Italy, 7–9 September 2022.

19. Amuthabala, P.; Santhosh, R. Robust analysis and optimization of a novel efficient quality assurance model in data warehousing. *Comput. Electr. Eng.* **2019**, *74*, 233–244. [CrossRef]
20. Theodorou, V.; Jovanovic, P.; Abelló, A.; Nakuçi, E. Data generator for evaluating ETL process quality. *Inf. Syst.* **2017**, *63*, 80–100. [CrossRef]
21. Cuzzocrea, A. Editorial: Models and algorithms for high-performance distributed data mining. *J. Parallel Distrib. Comput.* **2013**, *73*, 281–283. [CrossRef]
22. Lin, K.W.; Chung, S.H. A fast and resource efficient mining algorithm for discovering frequent patterns in distributed computing environments. *Future Gener. Comput. Syst.* **2015**, *52*, 49–58. [CrossRef]
23. Kargupta, H.; Kamath, C.; Chan, P. Distributed and parallel data mining: Emergence, growth, and future directions. In *Advances in Distributed and Parallel Knowledge Discovery*; AAAI/MIT Press: Cambridge, MA, USA, 2000; pp. 409–416.
24. Urmela, S.; Nandhini, M. A framework for distributed data mining heterogeneous classifier. *Comput. Commun.* **2019**, *147*, 58–75. [CrossRef]
25. Vilalta, R.; Giraud-Carrier, C.; Brazdil, P. Meta-learning-concepts and techniques. In *Data Mining and Knowledge Discovery Handbook*; Springer: Boston, MA, USA, 2010; pp. 717–731.
26. Chikalov, I.; Lozin, V.V.; Lozina, I.; Moshkov, M.; Nguyen, H.S.; Skowron, A.; Zielosko, B. *Three Approaches to Data Analysis-Test Theory, Rough Sets and Logical Analysis of Data*; Intelligent Systems Reference Library; Springer: Berlin/Heidelberg, Germany, 2013; Volume 41.
27. Stefanowski, J.; Vanderpooten, D. Induction of decision rules in classification and discovery-oriented perspectives. *Int. J. Intell. Syst.* **2001**, *16*, 13–27. [CrossRef]
28. Pawlak, Z.; Skowron, A. Rough sets and Boolean reasoning. *Inf. Sci.* **2007**, *177*, 41–73. [CrossRef]
29. Han, J.; Kamber, M. *Data Mining: Concepts and Techniques*; Morgan Kaufmann: San Francisco, CA, USA, 2000.
30. Żabiński, K.; Zielosko, B. Decision rules construction: Algorithm based on EAV model. *Entropy* **2021**, *23*, 14. [CrossRef] [PubMed]
31. Kotsiantis, S.B. Decision trees: A recent overview. *Artif. Intell. Rev.* **2013**, *39*, 261–283. [CrossRef]
32. Quinlan, J.R. *C4.5: Programs for Machine Learning*; Morgan Kaufmann Publishers Inc.: San Francisco, CA, USA, 1993.
33. Rivera-Lopez, R.; Canul-Reich, J.; Mezura-Montes, E.; Cruz-Chávez, M.A. Induction of decision trees as classification models through metaheuristics. *Swarm Evol. Comput.* **2022**, *69*, 101006. [CrossRef]
34. Guyon, I.; Gunn, S.; Nikravesh, M.; Zadeh, L. (Eds.) *Feature Extraction: Foundations and Applications*; Studies in Fuzziness and Soft Computing; Springer: Berlin/Heidelberg, Germany, 2006; Volume 207.
35. Liu, H.; Motoda, H. *Computational Methods of Feature Selection (Chapman & Hall/Crc Data Mining and Knowledge Discovery Series)*; Chapman & Hall/CRC: Boca Raton, FL, USA, 2007.
36. Stańczyk, U.; Zielosko, B.; Żabiński, K. Application of greedy heuristics for feature characterisation and selection: A case study in stylometric domain. In Proceedings of the Rough Sets-International Joint Conference, IJCRS 2018, Quy Nhon, Vietnam, 20–24 August 2018; Lecture Notes in Computer Science; Springer: Berlin/Heidelberg, Germany, 2018; Volume 11103, pp. 350–362.
37. Jia, X.; Shang, L.; Zhou, B.; Yao, Y. Generalized attribute reduct in rough set theory. *Knowl.-Based Syst.* **2016**, *91*, 204–218. [CrossRef]
38. Wróblewski, J. Theoretical foundations of order-based genetic algorithms. *Fundam. Inform.* **1996**, *28*, 423–430. [CrossRef]
39. Zielosko, B.; Piliszczuk, M. Greedy algorithm for attribute reduction. *Fundam. Inform.* **2008**, *85*, 549–561.
40. Grzegorowski, M.; Ślęzak, D. On resilient feature selection: Computational foundations of r-C-reducts. *Inf. Sci.* **2019**, *499*, 25 – 44. [CrossRef]
41. Lee, A.J.T.; Hong, R.W.; Ko, W.M.; Tsao, W.K.; Lin, H.H. Mining spatial association rules in image databases. *Inf. Sci.* **2007**, *177*, 1593–1608. [CrossRef]
42. Han, J.; Fu, Y. Discovery of multiple-level association rules from large databases. In *VLDB*; Dayal, U., Gray, P.M.D., Nishio, S., Eds.; Morgan Kaufmann: San Francisco, CA, USA, 1995; pp. 420–431.
43. Agrawal, R.; Imieliński, T.; Swami, A. Mining association rules between sets of items in large databases. In Proceedings of the 1993 ACM SIGMOD International Conference on Management of Data, Washington, DC, USA, 25–28 May 1993; pp. 207–216.
44. Han, J.; Pei, J.; Yin, Y.; Mao, R. Mining frequent patterns without candidate generation: A frequent-pattern tree approach. *Data Min. Knowl. Discov.* **2004**, *8*, 53–87. [CrossRef]
45. Borgelt, C. Simple algorithms for frequent item set mining. In *Advances in Machine Learning II*; Studies in Computational Intelligence; Koronacki, J., Raś, Z.W., Wierzchoń, S.T., Kacprzyk, J., Eds.; Springer: Berlin/Heidelberg, Germany, 2010; Volume 263, pp. 351–369.
46. Herawan, T.; Deris, M.M. A soft set approach for association rules mining. *Knowl.-Based Syst.* **2011**, *24*, 186–195. [CrossRef]
47. Mattiev, J.; Kavsek, B. Coverage-based classification using association rule mining. *Appl. Sci.* **2020**, *10*, 7013. [CrossRef]

Article

New Classification Method for Independent Data Sources Using Pawlak Conflict Model and Decision Trees

Małgorzata Przybyła-Kasperek * and Katarzyna Kusztal

Institute of Computer Science, University of Silesia in Katowice, Będzińska 39, 41-200 Sosnowiec, Poland
* Correspondence: malgorzata.przybyla-kasperek@us.edu.pl; Tel.: +48-32-269-17-56

Abstract: The research concerns data collected in independent sets—more specifically, in local decision tables. A possible approach to managing these data is to build local classifiers based on each table individually. In the literature, many approaches toward combining the final prediction results of independent classifiers can be found, but insufficient efforts have been made on the study of tables' cooperation and coalitions' formation. The importance of such an approach was expected on two levels. First, the impact on the quality of classification—the ability to build combined classifiers for coalitions of tables should allow for the learning of more generalized concepts. In turn, this should have an impact on the quality of classification of new objects. Second, combining tables into coalitions will result in reduced computational complexity—a reduced number of classifiers will be built. The paper proposes a new method for creating coalitions of local tables and generating an aggregated classifier for each coalition. Coalitions are generated by determining certain characteristics of attribute values occurring in local tables and applying the Pawlak conflict analysis model. In the study, the classification and regression trees with Gini index are built based on the aggregated table for one coalition. The system bears a hierarchical structure, as in the next stage the decisions generated by the classifiers for coalitions are aggregated using majority voting. The classification quality of the proposed system was compared with an approach that does not use local data cooperation and coalition creation. The structure of the system is parallel and decision trees are built independently for local tables. In the paper, it was shown that the proposed approach provides a significant improvement in classification quality and execution time. The Wilcoxon test confirmed that differences in accuracy rate of the results obtained for the proposed method and results obtained without coalitions are significant, with a p level = 0.005. The average accuracy rate values obtained for the proposed approach and the approach without coalitions are, respectively: 0.847 and 0.812; so the difference is quite large. Moreover, the algorithm implementing the proposed approach performed up to 21-times faster than the algorithm implementing the approach without using coalitions.

Keywords: Pawlak conflict analysis model; independent data sources; coalitions; decision trees; dispersed data

Citation: Przybyła-Kasperek, M.; Kusztal, K. New Classification Method for Independent Data Sources Using Pawlak Conflict Model and Decision Trees. *Entropy* **2022**, *24*, 1604. https://doi.org/10.3390/e24111604

Academic Editors: Przemysław Juszczuk and Jan Kozak

Received: 24 October 2022
Accepted: 1 November 2022
Published: 4 November 2022

Publisher's Note: MDPI stays neutral with regard to jurisdictional claims in published maps and institutional affiliations.

Copyright: © 2022 by the authors. Licensee MDPI, Basel, Switzerland. This article is an open access article distributed under the terms and conditions of the Creative Commons Attribution (CC BY) license (https://creativecommons.org/licenses/by/4.0/).

1. Introduction

In today's world, data are often collected in a decentralized and dispersed manner. There are many examples that illustrate this process: hospitals that separately collect data on the same issue/disease; banks that store data on their clients; applications on mobile devices that collect various data. These data are collected independently and in separate data storage.

It is crucial to use these data sets simultaneously to construct a classification of new objects. Of course, a very significant consideration is to guarantee high efficiency in the classification process based on dispersed data.

The issues of dispersed data are mainly considered in distributed learning approaches [1,2]. The distributed models process all or part of the data at different nodes [3,4]. A solution in which all the data are simultaneously aggregated and stored in a single set is both inefficient and often impossible to apply [5]. Therefore, most research papers have proposed a collaborative solution without data aggregation. In federated learning [6,7], nodes perform multiple rounds with local data and send the local model to the central server for aggregation into new global models. The main idea here is to guarantee data protection and privacy. Moreover, models are much shorter than raw data, so the exchange of data is faster and less complex. In the distributed learning approach, methods can be found in which local models are built independently, and the final decision is simply generated by applying fusion methods. Various models have been proposed, both parallel [8] and hierarchical [9,10]. The concept of agent collaboration is also key here [11]; however, we do not build aggregated tables as a result of this collaboration. In the literature, examples of classifier ensembles in which feature subsets are considered can be found [12–14]. There are also ensembles of classifiers built based on subsets of objects [15,16]. In the paper [17], an approach that considers missing values in the context of ensembles is considered. A crucial matter that affects the quality of classification is diversity among the base classifiers [18,19]. The method for generating the final decision also has a significant impact on the efficiency of ensembles [20,21]. Approaches recognizing relations between local data are considered in the literature. In the paper [22], a hierarchical federated learning approach was proposed. On the other hand, the paper [23] proposed a hierarchical approach in classifier ensembles. Mainly in the literature, distributed learning is considered in terms of the following issues [2,24]: data division—horizontal or vertical fragmentation; type of base classifiers— can be homogeneous or heterogeneous; type and cost of communication—data or models may be shared; privacy and data security—whether raw data exchange is allowed; fusion methods—if local models are built (global model is not created) then fusion of predictions is necessary to generate global decisions; data consistency—it can be assumed that objects are shared between local tables and are consistent, or data can be independently created and inconsistent. However, proposed approaches do not analyze the contents of local tables and the relationships between them. In addition, the aggregation of local tables is seldom considered in the literature.

Therefore, in this paper we fill this gap and propose a solution that performs a complex analysis of tables' content. The proposed approach aims to identify conflicts of local tables. The term conflict used here refers to significant differences in the values of conditional attributes occurring in local tables. We analyze relations and create coalitions of local tables containing similar data. Based on the aggregated tables, a model is built. It is expected that in this way we achieve better classification accuracy because models created via this approach have a better ability to generalize concepts compared to approaches that use a single model created based on a single table.

In the literature, conflict analysis is widely considered and various models are proposed. Group decision-making represents an approach that solves the situation in which each individual has their own private perspective [24]. In [25], a model is proposed for distributed group-decision support system that is suitable for use over the Internet. The theory of negotiation and coalition formation presents an important issue regarding social interaction and is also studied in computer science in the context of distributed systems [26,27]. Pawlak's conflict analysis model [28,29] is yet another approach to conflict recognition that provides excellent solutions in a variety of applications [30,31]. Pawlak conflict analysis model was also considered in the context of dispersed data in the papers [32–34]. This application shows that the Pawlak model provides excellent results for dispersed data when tables are aggregated within coalitions. However, the approach discussed in their study is completely different from the one proposed in this paper. Here, the compatibility of tables is examined in terms of the information stored in them—the values on the attributes. In contrast, the papers [32–34] consider compatibility in terms of predictions generated by the base models created based on the tables. Another difference is that in this paper we assume that in local tables the same attributes are present, while in the papers [32–34] there was no such assumption. Furthermore, in this paper, the system is static, whereas previously it was dynamic. However, the success of the previous model provides the inspiration for proposing a new approach in this paper. The main differences between these approaches are listed in Table 1.

Table 1. Comparison of the new approach with the approach proposed in the papers [32–34].

	New Proposed Approach	**Approach Proposed in the Papers [32–34]**
System's Structure	**Static**	**Dynamic**
Changeability of coalitions	Coalitions of local tables determined only once regardless of the object that is being classified.	Coalitions of local tables determined for each classified object from scratch.
Basis for coalitions designation	Information system in Pawlak model created based on characteristics of values stored in local tables. So coalitions are created based on conditional attributes' values occurring in local tables.	Information system in Pawlak model created based on prediction vectors generated for the classified object.
Definition of aggregated table for one coalition	Aggregated table is defined by a sum of objects.	Aggregated table is defined by the approximated method for the aggregation of decision tables—computationally complex.
Base classifiers	Decision tree, CART	k–nearest neighbor classifier
Constraints on local tables	The same conditional attributes in all local tables.	None

This paper proposes the use of the Pawlak conflict analysis method to generate coalitions of decision tables, in which there are similar values on a set of conditional attributes. The goal is to achieve a better quality of classification by ensuring that similar units work together. Formally, this approach requires that data are collected in a set of decision tables (that were collected independently) in which the names of the conditional attributes are identical (but the values on the objects may differ). Thus, coalitions of tables containing similar values will be created. The tables in one coalition are then aggregated and a common model is determined based on the aggregated table. This approach seems natural, since in everyday life we also notice that similar entities join forces to form better decisions or to guarantee better management. This paper describes the process of using characteristics of attribute values stored in decision tables in the Pawlak conflict analysis model. The paper proposes a static and hierarchical classification model. The model is static because coalitions—the model's structure—are determined only once. Hierarchy of the model results from the fact that tables in coalitions are aggregated and then models are built based on them and these models perform classification. In this paper, decision trees are used as base models. Specifically, classification and regression trees with Gini index (CART) [35] are applied. The final classification of new objects is determined using majority voting based on the predictions generated by the decision trees.

The paper also considers a parallel approach in which conflict analysis is not considered. In this approach, the CART trees are also employed as base models, but the cooperation of tables is not implemented, and the final decisions are made by majority voting of decision trees generated independently based on tables.

The main objective in this study is to analyze how building coalitions of tables using the Pawlak conflict analysis model affects the quality of classification and the running time of the model. The two research hypotheses are verified in the paper. The first is that applying the proposed model with Pawlak analysis and coalitions provides better classification quality than an approach in which coalitions are not used (in both models the same base classifiers are used—the CART trees). The second research hypothesis is that the algorithm implementing the proposed model has a lower time complexity than the algorithm implementing the approach in which decision trees are built based on each local table separately.

Herein, it is shown that combining local tables into aggregated tables significantly improves classification quality. In addition, it reduces the number of generated trees and thus reduces the time complexity of the method.

The main contributions of the paper are:

- proposing a new classification model using cooperation and coalitions of local tables (tables contain the same attributes),
- proposing a new method for creating coalitions of tables using the Pawlak conflict analysis model,
- developing a hierarchical system with CART trees for classification based on dispersed data.

The structure of the paper is organized as follows. Section 2 presents the proposed model. The method of defining the coalitions and steps in building the model are described there. Section 3 is dedicated to presenting the experimental results. The data, the measures used and the methodology of the experiments are described in this section, and the results obtained are also provided in tables. Section 4 contains the discussion and comparisons of the obtained results. Section 5 gives conclusions and future research plans.

2. Materials and Methods

This section describes a new proposed hierarchical system for classification based on dispersed data. In this research, we assume that the sets of attributes appearing in local tables are equal. Stages of system construction are described in the following subsections. The first step involves creating the system's structure—generating coalitions of local tables. This stage is implemented only once. Our goal here is the cooperation of tables that store similar conditional attribute values. This concept detailing the cooperation of units that share similar views with each other—have compatible values in this case—represents a natural behavior that we can observe in everyday life and nature. For this purpose, characteristics of conditional attributes' values are calculated. In the next step, coalitions are created based on these characteristics using the Pawlak conflict analysis model. The final step is the aggregation of tables from one coalition. Based on such aggregated coalition's data, a classifier is built. In this study, we use a decision tree model. The final classification model is a set of such decision trees generated for coalitions. The classification of an object is conducted by the majority voting of these trees. Figure 1 illustrates the workflow of the proposed model.

2.1. Basic Concepts and Method of Defining Characteristics of Conditional Attributes

We assume that a set of decision tables is given. The tables were collected independently by separate units, but it is required that the same attributes are stored in all tables. We do not impose any restrictions on the objects contained within the tables. We assume that we do not know which objects are shared between local tables.

Formally, we assume that a set of decision tables $D_i = (U_i, A, d), i \in \{1, \ldots, n\}$ from one discipline is available, where U_i is the universe, a set of objects; A is a set of conditional attributes; d is a decision attribute. As can be seen the sets of objects are different between local tables. The names of attributes that occur in local tables, both conditional and decision, are the same. Therefore, the conditional attributes A and decision attribute d in all local tables are denoted in the same way. Clearly, from a formal point of view, the attribute $a \in A$ in the decision table D_i is a function $a : U_i \rightarrow V^a$, where V^a is the set of values of the attribute a. Thus, the domains of the functions between local tables are different. However, for the sake of simplicity, the same designations for attributes were adopted in all local tables, and the domain of the function will be directly derived from the attribute's membership in the decision table. Aggregation for these tables is a difficult process and can generate inconsistencies. Another aspect that should be taken into account is data protection and privacy. In addition, the process of aggregating all local tables is highly complex. Thus, in the literature, rather, methods are proposed for partial aggregation of tables or even building separate models based on each local tables, and then aggregating these models or the predictions generated by the models [7,21,36].

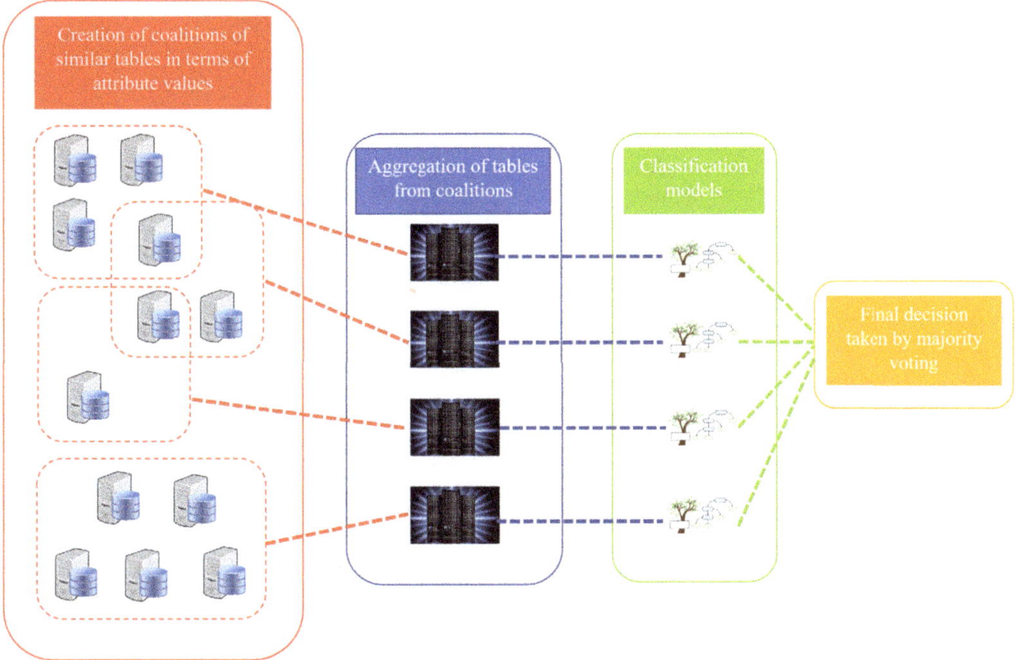

Figure 1. The overall workflow of the proposed model.

In this paper, a new approach is proposed in which we aggregate tables that contain similar values on conditional attributes. For this purpose, for each local table and for each attribute, some characteristics of the attribute's values occurring in the table are generated. Suppose that in each local table we have m attributes $card\{A\} = m$ (*card* denotes the number of elements in the set). Let us assume that we have m_1 quantitative attributes and m_2 qualitative attributes, so $m_1 + m_2 = m$.

For each quantitative attribute $a_{quan} \in A$, we determine the average of all attribute's values present in local table D_i, for each $i \in \{1, \ldots, n\}$. Let us denote this value as $\overline{Val}^i_{a_{quan}}$. We also calculate the global average and the global standard deviation. Let us denote them as $\overline{Val}_{a_{quan}}$ and $SD_{a_{quan}}$. These values are determined based on the averages calculated for the local decision tables according to the following formulas:

$$\overline{Val}_{a_{quan}} = \frac{1}{n} \sum_{i=1}^{n} \overline{Val}^i_{a_{quan}} \tag{1}$$

$$SD_{a_{quan}} = \sqrt{\frac{1}{n} \sum_{i=1}^{n} \left(\overline{Val}_{a_{quan}} - \overline{Val}^i_{a_{quan}} \right)^2} \tag{2}$$

These characteristics for quantitative attributes will be used in the coalitions generation process.

For each qualitative attribute $a_{qual} \in A$, we determine a vector over the values of that attribute. Suppose attribute a_{qual} has c values $val_1, \ldots, val_c$. The vector $Val^i_{a_{qual}} = (n^i_1, \ldots, n^i_c)$

represents the number of occurrences of each of these values in the decision table D_i. More precisely, the coordinate n_j represents the number of objects in table D_i that have value val_j on attribute a_{qual}. This vector is normalized. This is done to ensure that in further analysis the percentage of occurrences of a given value in the table matters rather than the number of objects in the table.

The Pawlak conflict analysis model is employed to determine coalitions of local tables that store similar attribute values. The next section presents the method to create an information system with a description of the conflict situation and how coalitions are generated with the use of the Pawlak model.

2.2. Pawlak Conflict Analysis Model and Creation of Coalitions

The Pawlak conflict analysis model is a very simple yet effective approach for recognizing coalitions of units involved in a conflicting situation [28,29]. In this model, an information system is defined in which the views of agents—units involved in a conflict situation—on the issues that are the matter of the conflict are stored. In the considered approach, the agents are local tables while the issues are conditional attributes stored in these tables. Formally, an information system is defined $S = (U, A)$, where U is a set of local decision tables $U = \{D_1, \ldots, D_n\}$ and A is a set of conditional attributes (qualitative and quantitative) occurring in local tables, which was defined in the previous section. In the Pawlak model, opinions of agents on issues are expressed by using three values. Value 1 means an agent is in favor of an issue, value 0 means an agent is neutral to an issue, while value -1 means an agent is against an issue. The original interpretation differs from that used herein. In this paper, the values refer rather to the differences in values of a given attribute appearing in the local decision table. Depending on the type of attribute (qualitative or quantitative), a different method of determining these values is used.

For the quantitative attribute $a_{quan} \in A$ a function $a_{quan} : U \to \{-1, 0, 1\}$ is defined

$$
a_{quan}(D_i) = \begin{cases} 1 & \text{if } \overline{Val}_{a_{quan}} + SD_{a_{quan}} < \overline{Val}^i_{a_{quan}} \\ 0 & \text{if } \overline{Val}_{a_{quan}} - SD_{a_{quan}} \leq \overline{Val}^i_{a_{quan}} \leq \overline{Val}_{a_{quan}} + SD_{a_{quan}} \\ -1 & \text{if } \overline{Val}^i_{a_{quan}} < \overline{Val}_{a_{quan}} - SD_{a_{quan}} \end{cases} \tag{3}
$$

The motivation for proposing this function originates from the method of estimating typical values of normal distribution. It is known that about 68% of the typical values from the normal distribution fall within the range: average $\pm$ standard deviation. Thus, we assign the value 0 on attribute a_{quan} to decision tables D_i when the average of the attribute's values occurring in the table falls in the $SD_{a_{quan}}$-neighborhood of the global average $\overline{Val}_{a_{quan}}$.

This means that the values of the attribute occurring in the decision table are typical.

In contrast, the value 1 means that the average of the conditional attribute values in the decision table is above the global average more than $SD_{a_{quan}}$ value; it deviates more than the value of the standard deviation. Similarly, the value -1 indicates an atypical—lower—average value of the conditional attribute in the decision table compared to the global average value.

As mentioned above, the vectors that determine the distribution of values occurring in the decision tables are generated for qualitative attributes. For an attribute $a_{qual} \in A$ we have the vectors $Val^i_{a_{qual}} = (n^i_1, \ldots, n^i_c), i \in \{1, \ldots, n\}$. In order to define three groups of decision tables with similar distribution of the attribute's a_{qual} values, we group these vectors with the k–means clustering algorithm, fixed number of groups $k = 3$ and the Euclidean distance. We then place in descending order the centroids obtained for groups. Ordering with respect to the value of the first centroid coordinate was applied. Let us denote the groups of decision tables obtained from the k–means algorithm and indexed in relation to the centroids' order as G_1, G_2, G_3. For the qualitative attribute $a_{qual} \in A$ a function $a_{qual} : U \to \{-1, 0, 1\}$ is defined

$$a_{qual}(D_i) = \begin{cases} 1 & \text{if } D_i \in G_1 \\ 0 & \text{if } D_i \in G_2 \\ -1 & \text{if } D_i \in G_3 \end{cases} \qquad (4)$$

The function above assigns values on a qualitative attribute to local tables that reflect the consistency of the characteristics of this attribute appearing in the table. Thus, decision tables that contain similar distribution of values of the qualitative attribute will have the same value assigned in the information system S.

In this way, the information system S is defined that stores information about the compatibility of values of conditional attributes occurring in local tables. Based on this system, we calculate the general similarity of values of all attributes for each pair of tables. For this purpose, a conflict function is used that was proposed by Pawlak in their conflict analysis model [28]. The conflict function $\rho : U \times U \to [0,1]$ is defined as follows

$$\rho(D_i, D_j) = \frac{card\{a \in A : a(D_i) \neq a(D_j)\}}{card\{A\}}. \qquad (5)$$

A pair of decision tables $D_i, D_j \in U$ is said to be [28]:

- allied, if $\rho(D_i, D_j) < 0.5$,
- in conflict, if $\rho(D_i, D_j) > 0.5$,
- neutral, if $\rho(D_i, D_j) = 0.5$.

Set $X \subseteq U$ is a coalition if for every $D_i, D_j \in X$ decision tables are allied $\rho(D_i, D_j) < 0.5$.

By applying the Pawlak conflict analysis model, we obtain coalitions of local tables that share similar values of conditional attributes. It should be noted that coalitions do not have to be disjointed—one local table can be included in several coalitions. In fact, this is a quite common case, as will be shown in the experimental section.

The pseudo-code of the algorithm that generates the coalitions of local tables is given in Algorithm 1.

Algorithm 1 Pseudo-code of algorithm generating coalitions of local tables

---

Input: A set of local decision tables $D_i = (U_i, A, d), i \in \{1, \ldots, n\}$.
Output: A set of coalitions of local tables $X_1, \ldots, X_k$.
Construction of an information system $S = (U, A)$, where $U = \{D_1, \ldots, D_n\}$ and A is a set of conditional attributes
for each $a \in A$:

    if a is a quantitative attribute then

        Use Equation (3) to define the function a

    else

        Use Equation (4) to define the function a

Conflict function values
for each pair $D_i, D_j \in U$:

    Use Equation (5) to calculate the value $\rho(D_i, D_j)$

Creation of coalitions
$X_1 = U, i = 1, j = 1$
while $i \leq j$:

    Repeat until there is a pair of tables $D_l, D_k \in X_i$ so that $\rho(D_l, D_k) \geq 0.5$:

        $j = j + 1$
        $X_j = X_i \setminus \{D_l\}, X_i = X_i \setminus \{D_k\}$

    $i = i + 1$

Return only the largest sets, due to the inclusion relation, from the sets $X_i, i = 1, \ldots, j$

The computational complexity of the algorithm is exponential due to the number of local tables. The greatest complexity is noted when there exists no pair of local tables similar enough to satisfy the conditions of being allied. Subsequently, all subsets of the set of local tables will eventually be checked. However, in most applications, the number of local tables is not so large. In the experimental section, the application of the proposed model is checked for dispersed data containing up to eleven local tables. The obtained times in the worst cases are expressed in minutes.

2.3. Aggregation of Tables from Coalitions and Final Classification

An aggregated decision table is defined for each coalition of local tables generated in the previous step. Suppose we have coalitions of tables $X_1, \ldots, X_k$. The aggregated decision table for the coalition X_j is denoted as $D_j^{aggr} = (U_j^{aggr}, A, d)$, where $U_j^{aggr} = \bigcup_{D_i \in X_j} U_i$ and the names of attributes in the aggregated table are the same as those in local tables. The attribute a from the aggregated table is a function defined on U_j^{aggr} that takes values in V^a. The attribute a from the aggregated table has the same value, on object $x \in U_i$, as the corresponding attribute a from the local table D_i on that object. Thus, an aggregated table is defined by summing objects from local tables in the coalition without recognizing whether there are common objects in the local tables (based on the assumptions, we do not possess this possibility). In the aggregated table, the values assigned to objects on the attributes are taken from local tables.

Based on aggregated tables, models are generated. In this paper, the classification and regression tree algorithm is used with Gini index [35]. It should be noted that prepruning and postpruning were not used for this tree. An implementation available in Python language was used for this purpose [37]. Specifically, *DecisionTreeClassifier(criterion = "gini")* function was used. The tree is built independently for each aggregated table, thus we obtain k models $M_1, \ldots, M_k$.

The classification of a new object x is realized by each model separately. The final decision—the global decision, which we denote as $\hat{d}(x)$—is made by majority voting. This means that there may be a tie, which we do not resolve in any way. Thus, $\hat{d}(x)$ is the set of decisions that were most frequently indicated by models $M_1, \ldots, M_k$. In the experimental part, the relevant measures for evaluating the quality of classification, which takes into account the possibility of draws, were used.

In the section below, an illustrative example of the proposed approach is provided for clarification.

2.4. Baseline Model without the Use of Coalitions

The results obtained using the proposed method are compared with the results generated by an approach without any conflict analysis. In the baseline approach, a model is built based on each local table. In order to perform a fair comparison of the impact of the proposed novelty on the results obtained, the same classification model was used—for each local table the CART tree is used. Classification of a new object is realized by applying the majority voting method to the classification results obtained using these decision trees. Ties can occur, but as stated before, we do not resolve them in any way. The adequate measures were used in the experimental part.

2.5. Example of Use of the Proposed Approach

Let us consider an example that uses the proposed approach. Suppose we have a set of four local tables $D_i = (U_i, A, d), i \in \{1, \ldots, 4\}$. Each of them contains a set of five conditional attributes $A = \{a_1, \ldots, a_5\}$ and a decision attribute d. We assume that $V^{a_i} = \{0, 1, 2\}, i \in \{1, \ldots, 5\}$, and $V^d = \{d_1, d_2\}$ for each of the tables. For the purposes of this example, the conditional attributes in the tables are quantitative. The local tables defined above are given in Table 2.

Table 2. Local tables used in the example.

U_1	a_1	a_2	a_3	a_4	a_5	d
x_1	1	0	2	0	0	d_2
x_2	2	1	0	1	0	d_2
x_3	0	0	1	2	2	d_1
x_4	2	1	1	1	1	d_1
x_5	1	2	0	1	2	d_2

U_2	a_1	a_2	a_3	a_4	a_5	d
x_1	0	2	1	0	0	d_2
x_2	2	1	2	1	2	d_1
x_3	2	0	0	2	1	d_2
x_4	1	1	2	0	0	d_2
x_5	2	0	2	1	1	d_1

U_3	a_1	a_2	a_3	a_4	a_5	d
x_1	1	1	0	2	2	d_1
x_2	1	1	2	0	1	d_1
x_3	2	0	1	2	1	d_2
x_4	0	2	0	2	0	d_2
x_5	2	0	2	1	2	d_2

U_4	a_1	a_2	a_3	a_4	a_5	d
x_1	1	0	0	2	2	d_1
x_2	2	1	0	1	0	d_2
x_3	0	2	1	2	2	d_2
x_4	2	0	2	1	1	d_1
x_5	1	2	0	1	1	d_2

Based on the attribute values in the local tables (Table 2), the information system is generated as described in Section 2.2. In the first step, the average of all attribute's values occurring in the local table for each attribute and each table is calculated. These values are denoted as $\overline{Val}^{i}_{a_j}$, $i \in \{1,\dots,4\}, j \in \{1,\dots,5\}$ and are given in Table 3. Furthermore, the global average and the global standard deviation for each attribute are calculated, the values are also shown in Table 3.

Table 3. Averages $\overline{Val}^{i}_{a_j}$, $i \in \{1,\dots,4\}, j \in \{1,\dots,5\}$.

Local Table	a_1	a_2	a_3	a_4	a_5
D_1	$\overline{Val}^{1}_{a_1} = 1.2$	$\overline{Val}^{1}_{a_2} = 0.8$	$\overline{Val}^{1}_{a_3} = 0.8$	$\overline{Val}^{1}_{a_4} = 1$	$\overline{Val}^{1}_{a_5} = 1$
D_2	$\overline{Val}^{2}_{a_1} = 1.4$	$\overline{Val}^{2}_{a_2} = 0.8$	$\overline{Val}^{2}_{a_3} = 1.4$	$\overline{Val}^{2}_{a_4} = 0.8$	$\overline{Val}^{2}_{a_5} = 0.8$
D_3	$\overline{Val}^{3}_{a_1} = 1.2$	$\overline{Val}^{3}_{a_2} = 0.8$	$\overline{Val}^{3}_{a_3} = 1$	$\overline{Val}^{3}_{a_4} = 1.4$	$\overline{Val}^{3}_{a_5} = 1.2$
D_4	$\overline{Val}^{4}_{a_1} = 1.2$	$\overline{Val}^{4}_{a_2} = 1$	$\overline{Val}^{4}_{a_3} = 0.6$	$\overline{Val}^{4}_{a_4} = 1.4$	$\overline{Val}^{4}_{a_5} = 1.2$
Global metrics	$\overline{Val}_{a_1} = 1.25$ $SD_{a_1} = 0.087$	$\overline{Val}_{a_2} = 0.85$ $SD_{a_2} = 0.087$	$\overline{Val}_{a_3} = 0.95$ $SD_{a_3} = 0.296$	$\overline{Val}_{a_4} = 1.15$ $SD_{a_4} = 0.260$	$\overline{Val}_{a_5} = 1.05$ $SD_{a_5} = 0.166$

Thus, according to Equation (3), the values in the information system for attribute a_1 are assigned as follows

$$a_1(D_i) = \begin{cases} 1 & \text{if } 1.337 < \overline{Val}^{i}_{a_1} \\ 0 & \text{if } 1.163 \leq \overline{Val}^{i}_{a_1} \leq 1.337 \\ -1 & \text{if } \overline{Val}^{i}_{a_1} < 1.163 \end{cases} \tag{6}$$

which means that $a_1(D_1) = 0, a_1(D_2) = 1, a_1(D_3) = 0, a_1(D_4) = 0, a_1(D_5) = 0$. For other attributes, the values in the information system are determined similarly. The obtained information system is shown in Table 4.

Table 4. Information system.

U	a_1	a_2	a_3	a_4	a_5
D_1	0	0	0	0	0
D_2	1	0	1	−1	−1
D_3	0	0	0	0	0
D_4	0	1	−1	0	0

In the next step, the values of conflict function for the local tables are determined according to Equation (5). For example, for the pair (D_1, D_2) of local tables, the value is calculated as follows

$$\rho(D_1, D_2) = \frac{card\{a \in A : a(D_1) \neq a(D_2)\}}{card\{A\}} = \frac{4}{5}. \tag{7}$$

The values of the conflict function for the above information system are presented in Table 5.

Table 5. Function values.

	D_1	D_2	D_3	D_4
D_1				
D_2	0.8			
D_3	0	0.8		
D_4	0.4	1.0	0.4	

Figure 2 shows a graphical representation of the conflict situation. When agents (local tables) are allied $(\rho(D_i, D_j) < 0.5)$, the circles representing the agents are linked. In order to find coalitions, all cliques should be identified in the graph. In this example, there are two coalitions: $\{D_1, D_3, D_4\}$ and $\{D_2\}$.

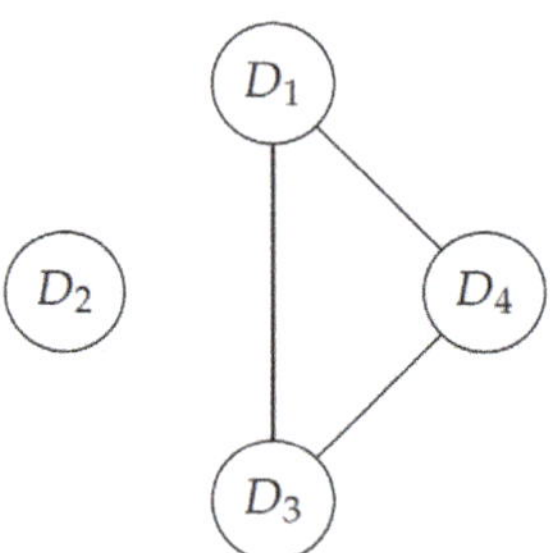

Figure 2. A graphical representation of the conflict situation example.

An aggregated decision table is generated for each coalition. The aggregated tables are presented in Table 6.

Now, a decision tree is built for each aggregated table. This is done using the function implemented in the Scikit-learn library *tree.DecisionTreeClassifier(criterion = "gini")*. The built decision trees are presented in Figure 3. Test objects are classified based on these models using the simple voting method.

Table 6. Aggregated local tables.

$U_1{}^{aggr}$	a_1	a_2	a_3	a_4	a_5	d
x_1^{aggr}	1	0	2	0	0	d_2
x_2^{aggr}	2	1	0	1	0	d_2
x_3^{aggr}	0	0	1	2	2	d_1
x_4^{aggr}	2	1	1	1	1	d_1
x_5^{aggr}	1	2	0	1	2	d_2
x_6^{aggr}	1	1	0	2	2	d_1
x_7^{aggr}	1	1	2	0	1	d_1
x_8^{aggr}	2	0	1	2	1	d_2
x_9^{aggr}	0	2	0	2	0	d_2
x_{10}^{aggr}	2	0	2	1	2	d_2
x_{11}^{aggr}	1	0	0	2	2	d_1
x_{12}^{aggr}	2	1	0	1	0	d_2
x_{13}^{aggr}	0	2	1	2	2	d_2
x_{14}^{aggr}	2	0	2	1	1	d_1
x_{15}^{aggr}	1	2	0	1	1	d_2

$U_2{}^{aggr}$	a_1	a_2	a_3	a_4	a_5	d
x_1^{aggr}	0	2	1	0	0	d_2
x_2^{aggr}	2	1	2	1	2	d_1
x_3^{aggr}	2	0	0	2	1	d_2
x_4^{aggr}	1	1	2	0	0	d_2
x_5^{aggr}	2	0	2	1	1	d_1

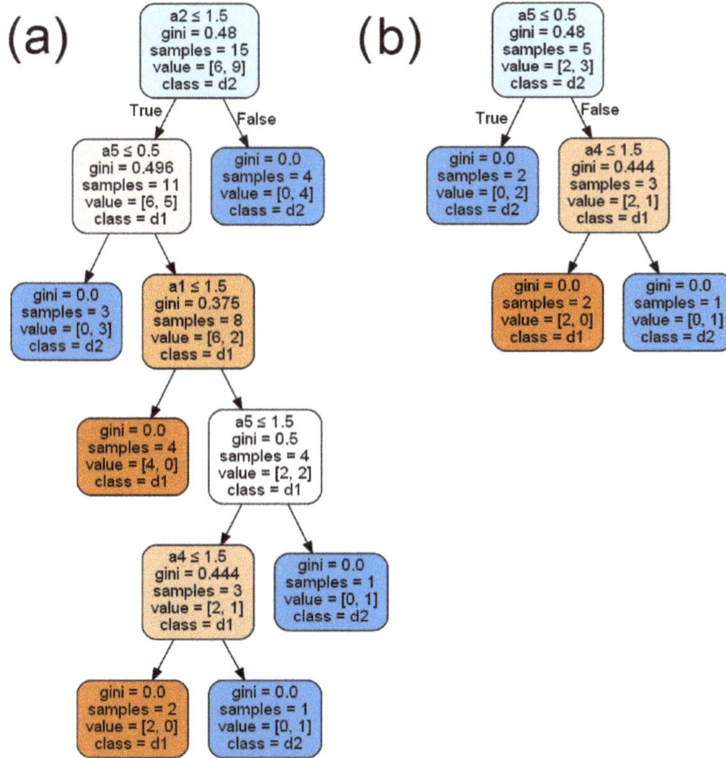

Figure 3. Decision trees created for aggregated decision tables. (**a**) The aggregated table $D_1{}^{aggr}$ (**b**) The aggregated table $D_2{}^{aggr}$.

Since local table D_2 is left in a coalition containing only one element, the second aggregated table is the same as the local table D_2, therefore, the trees generated based on them are also the same. So we should mainly focus on the tree generated based on the first aggregated table and the three trees generated from local tables D_1, D_3 and D_4. As we can see, they are quite different. For example, in the tree generated based on the aggregated table there is a condition $a_2 \leq 1.5$ the root, which does not correspond to the conditions occurring in the trees in Figure 4a,c,d. In addition, in the aggregated tree, there is the attribute a_5 in two internal nodes and the attribute a_4 in one internal node. These attributes are not included at all in the trees generated from local tables D_1, D_3 and D_4.

Since tables are combined into coalitions in terms of similarity of conditional attributes' values, trees generated based on aggregated tables should not be very altered compared to trees generated from local tables. In general, trees generated from a larger number of training objects are expected to be more accurate and have better classification quality.

For comparison, let us also consider the baseline model, in which coalitions are not generated. In this case, the decision trees are generated directly based on local tables. Thus, we obtain four decision trees generated from the tables given in Table 2, which are presented in Figure 4.

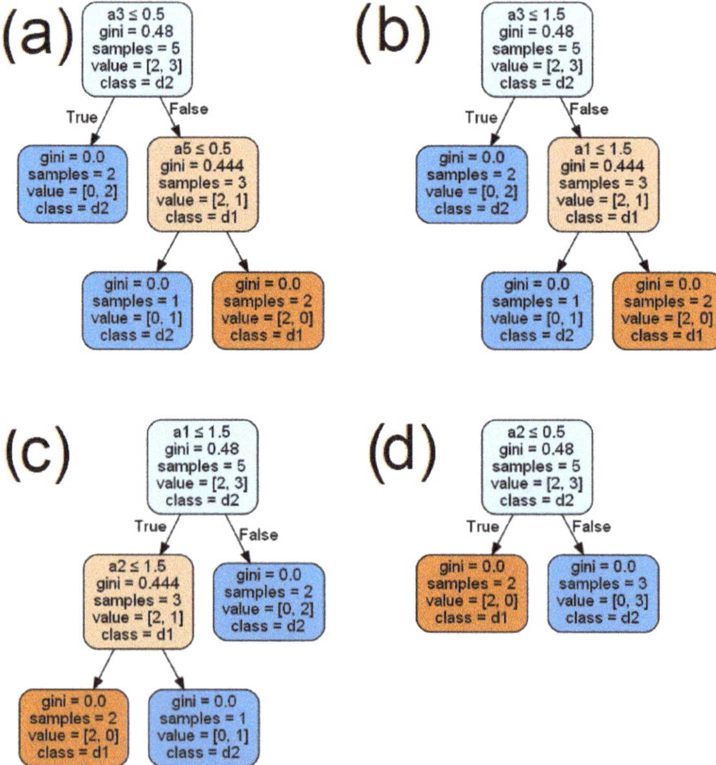

Figure 4. Decision trees created for local decision tables, (**a**) for the local table D_1, (**b**) for the local table D_2, (**c**) for the local table D_3, (**d**) for the local table D_4.

3. Results

The experiments were carried out using the data available from the UC Irvine Machine Learning Repository [38]. A total of three data sets were selected for the analysis—the Vehicle Silhouettes, the Landsat Satellite and the Soybean (Large) data sets. Regarding the Landsat Satellite and Soybean data sets, the training and test sets are located in the repository.

The Vehicle data set was randomly split into two disjoint subsets, the training set (70% of objects) and the test set (30% of objects). Data characteristics are given in Table 7.

Table 7. Data set characteristics.

Data Set	# The Training Set	# The Test Set	# Conditional Attributes	# Decision Classes
Vehicle Silhouettes	592	254	18	4
Landsat Satellite	4435	2000	36	7
Soybean	307	376	35	19

The training sets of the above data sets were dispersed. A total of 5 different dispersed versions with 3, 5, 7, 9 and 11 local tables were prepared to check for different degrees of dispersion for each data set. This was done using a stratified mode. Each local table contained the full set of attributes, and a subset of the set of objects.

The quality of classification was evaluated based on the test set. The following measures were used:

- the classification accuracy

$$acc = \frac{1}{card\{U_{test}\}} \sum_{x \in U_{test}} I(d(x) \in \hat{d}(x)),$$

where $I(d(x) \in \hat{d}(x)) = 1$, when $d(x_i) \in \hat{d}(x)$ and $I(d(x) \in \hat{d}(x)) = 0$, when $d(x) \notin \hat{d}(x)$; $\hat{d}(x)$ is a set of global decisions generated by the system for the test object x from the test set U_{test}

- the classification ambiguity accuracy

$$acc_{ONE} = \frac{1}{card\{U_{test}\}} \sum_{x \in U_{test}} I(d(x) = \hat{d}(x)),$$

where $I(d(x) = \hat{d}(x)) = 1$, when $\{d(x)\} = \hat{d}(x)$ and $I(d(x) = \hat{d}(x)) = 0$, when $\{d(x)\} \neq \hat{d}(x)$

- the average size of the global decision sets

$$\bar{d} = \frac{1}{card\{U_{test}\}} \sum_{x \in U_{test}} card\{\hat{d}(x)\}.$$

The classification accuracy refers to the ratio of correctly classified objects from the test set to their total number in this set. When the correct decision class of an object is contained within the generated decision set, the object is considered to be correctly classified. The classification ambiguity accuracy also describes the ratio of correctly classified objects from the test set to their total number in this set. With the difference being that this time when only one correct decision class is generated, the object is considered to be correctly classified. The third measure allows us to assess the frequency and number of draws generated by the classification model.

The experiments were conducted according to the following scheme:

- Generating coalitions of local tables using the Pawlak conflict analysis model. Detailed information on the coalitions that were generated is shown in Table 8. In cases where no coalitions were generated for a set of local tables then the dispersed set was not considered for further analysis. The reason for this is that the data in the tables are so different that they should not be combined and the proposed model does not bring any changes compared to the baseline approach.
- Defining aggregated tables for coalitions and generating decision tree models based on them. The classifier is a set of decision trees generated based on the aggregated tables for coalitions. Evaluating the proposed model using a test set.

- Analysis of the baseline approach. Generating decision trees based on the local tables (without any conflict analysis or coalitions). The final decision is made by simple voting. Evaluating the baseline approach using a test set.

As mentioned above, Table 8 shows the coalitions generated during construction of the proposed model. As can be seen, in two cases no coalitions were generated—for the Satellite and Soybean data sets with three local tables. In most cases, coalitions were created and, as can be seen, they are not disjoint sets. This means that some local tables were involved in the creation of several aggregated tables. The reason for this is that a given local table is partially similar to different sets of local tables and provides additional knowledge to the construction of trees representing different concepts.

Table 8. Coalitions generated using the Pawlak conflict analysis model for dispersed data. LT denotes local table.

Data Set	No. of Local Tables	Coalitions
Vehicle	3	$\{LT1, LT3\}, \{LT2\}$
	5	$\{LT2, LT3, LT4\}, \{LT4, LT5\}, \{LT1\}$
	7	$\{LT1, LT3, LT5, LT6, LT7\}, \{LT2\}, \{LT4\}$
	9	$\{LT1, LT3, LT4, LT9\}, \{LT3, LT4, LT5, LT6\}, \{LT3, LT4, LT5, LT9\},$ $\{LT2, LT3, LT4, LT9\}, \{LT7, LT8\}$
	11	$\{LT2, LT4, LT5, LT8\}, \{LT2, LT5, LT7, LT8\},$ $\{LT2, LT5, LT6, LT8\}, \{LT1, LT9\}, \{LT8, LT9\}, \{LT3, LT10\}, \{LT11\}$
Satellite	3	NO COALITIONS
	5	$\{LT1, LT4\}, \{LT2\}, \{LT3\}, \{LT5\}$
	7	$\{LT1, LT4, LT6, LT7\}, \{LT3, LT6\}, \{LT2\}, \{LT5\}$
	9	$\{LT1, LT4, LT5, LT6, LT9\}, \{LT3, LT4, LT5\}, \{LT2\}, \{LT7\}, \{LT8\}$
	11	$\{LT1, LT2, LT7, LT10\}, \{LT1, LT2, LT7, LT11\}, \{LT2, LT6, LT7, LT10\},$ $\{LT2, LT3, LT7, LT9\}, \{LT2, LT4, LT7\},$ $\{LT5, LT9\}, \{LT5, LT11\}, \{LT8\}$
Soybean	3	NO COALITIONS
	5	$\{LT2, LT4\}, \{LT1\}, \{LT5\}, \{LT3\}$
	7	$\{LT2, LT3, LT5\}, \{LT1, LT3\}, \{LT5, LT7\}, \{LT2, LT4\}, \{LT6\}$
	9	$\{LT1, LT2, LT4\}, \{LT1, LT2, LT5\}, \{LT1, LT5, LT6\}, \{LT1, LT3, LT5\},$ $\{LT1, LT9\}, \{LT8, LT9\}, \{LT7\}$
	11	$\{LT1, LT4, LT6, LT7, LT8, LT9\}, \{LT1, LT4, LT6, LT7, LT9, LT10\},$ $\{LT1, LT4, LT7, LT8, LT9, LT11\}, \{LT1, LT4, LT7, LT9, LT10, LT11\},$ $\{LT4, LT5, LT6, LT7, LT9, LT10\}, \{LT2\}, \{LT3\}$

Table 9 presents the classification accuracy *acc* values, the classification ambiguity accuracy acc_{ONE} values and the average number of generated decisions set $\bar{d}$ obtained for all dispersed data sets. The table shows the results obtained for both the proposed approach and the baseline approach. For each data set, the better result is indicated in bold. As can be seen, in the vast majority of cases better results are generated by the proposed model with creation of coalitions and recognition of similarity of data stored in local tables.

To better visualize the differences in the results generated by the models, Figure 5 was prepared with the classification accuracy marked for each data set. As can be seen, the most significant improvement in classification quality using the proposed approach was observed for the Soybean data set. Here, the improvement is around 0.1. For the Vehicle Silhouettes data set, the improvement in most cases is around 0.03 (even greater in certain scenarios). Furthermore, for the Landsat Satellite data set, the improvement in results was also noticed, but smaller at around 0.015. However, for all data sets, there is a noticeable and seemingly significant improvement obtained using the proposed approach compared to the baseline approach.

Table 9. Results of classification accuracy *acc*, classification ambiguity accuracy acc_{ONE} and the average number of generated decisions set $\bar{d}$ for all dispersed data sets.

Data Set	No. of Local Tables	Baseline Approach $acc/acc_{ONE}/\bar{d}$	Proposed Approach $acc/acc_{ONE}/\bar{d}$
Vehicle	3	0.803/0.673/1.268	**0.831**/0.496/1.409
	5	0.756/0.677/1.094	**0.791**/0.709/1.173
	7	0.752/0.681/1.114	**0.780**/0.669/1.228
	9	**0.760**/0.693/1.098	0.740/0.685/1.075
	11	0.740/0.673/1.087	**0.776**/0.728/1.051
Satellite	5	0.875/0.839/1.053	**0.893**/0.820/1.099
	7	0.870/0.841/1.040	**0.888**/0.822/1.093
	9	**0.874**/0.847/1.035	0.873/0.841/1.045
	11	0.877/0.850/1.034	**0.892**/0.857/1.042
Soybean	5	0.858/0.784/1.142	**0.868**/0.791/1.132
	7	0.807/0.716/1.135	**0.899**/0.834/1.074
	9	0.794/0.703/1.152	**0.905**/0.875/1.037
	11	0.787/0.723/1.108	**0.878**/0.855/1.064
Average		0.812/0.746/1.105	**0.847**/0.768/1.117

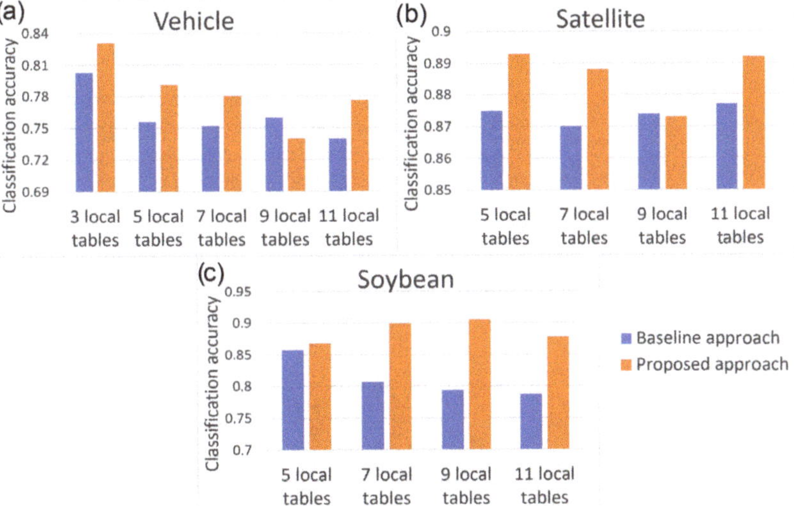

Figure 5. Comparison of classification accuracy (*acc*) of the baseline approach versus the proposed approach: (**a**) the Vehicle data set (**b**) the Landsat Satellite data set (**c**) the Soybean data set.

In order to investigate the significance in differences of accuracy rate obtained for the proposed model and the baseline approach, the results from Table 9 were used. Two dependent samples were created—one containing the results for the proposed model and one containing the results for the baseline approach. Each sample had a cardinality equal to 13 observations—results obtained for different data sets and number of local tables. The Wilcoxon test confirmed that differences in the accuracy rate between these two groups are significant, with $p = 0.005$.

Additionally, a comparative box-plot chart for the accuracy rate values was created (Figure 6). We can observe an increase in accuracy rate when the proposed model is used. Both the box alignment and the median itself are significantly higher when the proposed model is employed.

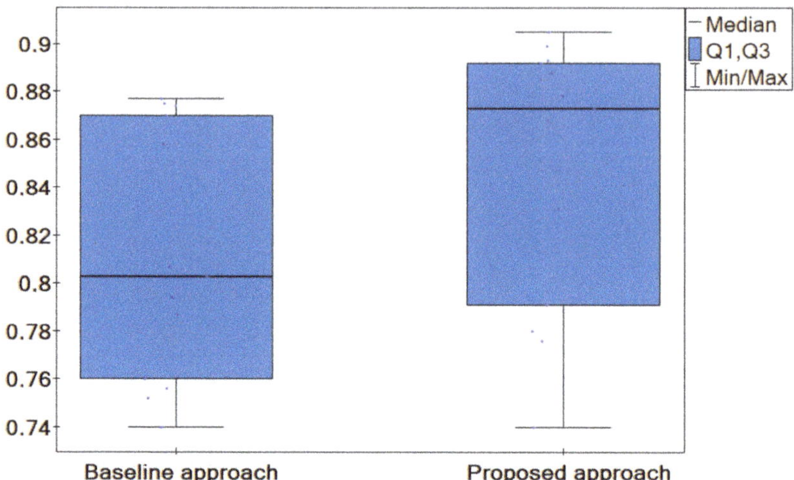

Figure 6. Box-plot chart with (median, the first quartile—Q1, the third quartile—Q3) the value of accuracy rate *acc* for the proposed model and the baseline approach.

Furthermore, we also analyzed the time needed to generate decision trees in both approaches. In the baseline method, the time needed to generate trees directly from local tables was investigated, and in the proposed approach the time required to generate trees from aggregated tables was considered. Table 10 shows the execution times of the decision tree generation algorithms in the baseline approach and with coalitions.

Table 10. Execution times of the decision tree generation algorithms in the base approach and with coalitions.

Data Set	No. of Local Tables	Baseline Approach Time [s]	Proposed Approach Time [s]	Ratio $\frac{Baseline}{Proposed}$
Vehicle	3	41.258	3.423	12.05
	5	46.694	4.332	10.78
	7	52.810	4.294	12.30
	9	61.634	6.704	9.19
	11	68.064	7.760	8.77
Satellite	5	3044.087	139.973	21.75
	7	3228.569	160.59	20.10
	9	3497.267	175.614	19.91
	11	3658.961	288.654	12.68
Soybean	5	58.542	4.538	12.90
	7	63.733	5.610	11.36
	9	72.051	7.714	9.34
	11	82.072	8.560	9.59

The differences in execution times are notably significant. The proposed model has significantly lower time complexity. This is due to the fact that with the proposed approach—coalitions creation—a smaller number of trees is created than when decision trees are generated based on each local table separately. This results in the significantly reduced execution time of making a final decision based on dispersed data.

Figure 7 illustrates the ratio of execution times of the baseline approach to the proposed approach. As can be seen for the Satellite data set, in some cases, the proposed approach exhibits an execution time more than 20-fold faster than the baseline approach. In general, it can be seen that for the largest data set (Satellite) the execution acceleration is the most significant.

In addition, for a smaller degree of dispersion—smaller number of local tables—the reduction in execution time using the proposed approach is greater than for data with a larger degree of dispersion—greater number of local tables. This is due to the fact that for a larger degree of dispersion, there is also a greater number of coalitions generated using the Pawlak analysis model (as can be seen in Table 8).

Figure 7. Ratio of execution times of the algorithms implementing the baseline approach and the approach with coalitions.

All experiments were performed on a portable computer with the following technical specifications:

- AMD Ryzen 54,600 h CPU,
- 32 GB RAM Memory,
- Microsoft Windows 11 Operating System.

The code used for the analyzed approaches has been implemented in Python and all data-related calculations have been saved in a text document. Decision trees were built using the function implemented in the Scikit-learn library *tree.DecisionTreeClassifier(criterion = "gini")*. In all cases, the Gini index was used. The postpruning and prepruning methods were intentionally not applied, since the main goal of this study focused on analyzing how building coalitions of tables using the Pawlak conflict analysis model affects classification quality and model running time. Combining local tables into aggregated tables was shown to significantly improve classification quality. In addition, it also reduces the number of generated trees and thus reduces the time complexity of the method.

4. Discussion

The paper proposes a new method for classification based on dispersed data. This method is used when the same set of conditional attributes occurs in all local tables. It should be noted that the conditional attributes can be of different types—both qualitative and quantitative. Sets of objects in local tables can be diversified. Indeed, we do not consider the possibility of examining whether identical objects occur in different local tables. The main idea behind this method is the aggregation of tables that store similar values on conditional attributes. In order to determine which tables should be aggregated, a new method for generating characteristics of values stored in tables and a new method for using the Pawlak conflict analysis model are proposed. Next, a method for defining aggregated tables and a method for final decision-making are defined. It was shown that the proposed method brings a significant improvement in the quality of classification obtained based on dispersed data compared to the approach when aggregation of tables and formation of coalitions are not considered.

The main advantages of the proposed approach are:

- The proposed method guarantees higher quality of classification in comparison with cases where conflict analysis and creation of coalitions are not used.
- The proposed method has less time complexity than methods where coalitions are not considered.
- Combining several similar tables—aggregation of tables into one—increases readability of the model. One decision tree generated based on an aggregated table provides better readability and possibility to interpret the described concepts than several trees generated independently from local tables.

The main limitations of the proposed approach are:

- The proposed model in the current stage of development is dedicated only to a set of local tables with the same sets of conditional attributes.
- Although with the proposed model, the readability of the system is increased by aggregating local tables, we still have not achieved full interpretability of the results. The final classifier consists of a set of decision trees.
- In the proposed approach, it is necessary to exchange data and make them available. The proposed model will not be suitable for dispersed data in which data protection and privacy is a priority.

There are practically no parameters in the proposed model, since the Pawlak model has no parameters, and the decision trees were built without prepruning or postpruning (this will be implemented in the next stage of the future work). The only parameter we can consider is the degree of data dispersion. The decision tables were dispersed to varying degrees into 3, 5, 7, 9 and 11 decision tables. The dispersion was performed in relation to the objects in stratified mode and ensuring the number of objects in the local tables remains equal. Figure 8 shows the function of classification accuracy values in relation to the number of local tables.

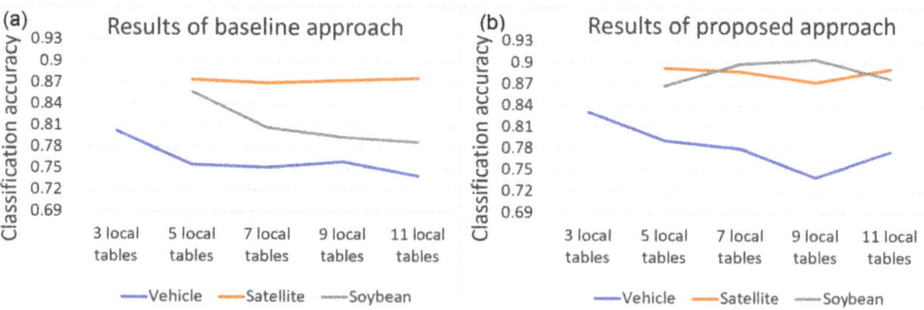

Figure 8. Classification of accuracy values in relation to the number of local tables: (**a**) for the baseline approach (**b**) for the approach with coalitions.

In the case of the baseline method for both the Soybean and the Vehicle data sets, an increase in the degree of data dispersion results in a deterioration of classification accuracy. For the Landsat Satellite data set, this relation is not observed. For the proposed approach, only for the Vehicle set can it be stated that an increase in the degree of dispersion affects the deterioration of classification accuracy. For the Soybean data set, the proposed method eliminates the negative effect of high dispersion on classification accuracy. Thus, it can be concluded that the use of the proposed approach allows improvement in the quality of classification, especially in the case of high dispersion where many local tables occur. In other words, the proposed model generally improves the quality of classification, but is particularly useful for data dispersed over a large number of local tables.

5. Conclusions

A new classification approach based on dispersed data was proposed in this paper. The main innovation lies in the proposal of a method that combines local decision tables into an aggregated table. For this purpose, a method based on the Pawlak conflict analysis model was proposed. The new approach was shown to improve both the quality of classification and the running time.

In future work, we plan to:

- use other classification models different from decision tree to build classifiers based on aggregated tables,
- conduct research on the impact of tree optimization—prepruning and postpruning—on the classification quality of the model,
- extend the proposed model to cases where only parts of the conditional attributes are shared between local tables.

Author Contributions: Conceptualization, M.P.-K.; methodology, M.P.-K., K.K.; software, K.K.; validation, M.P.-K., K.K.; formal analysis, M.P.-K., K.K.; investigation, M.P.-K., K.K.; resources, M.P.-K.; writing—original draft preparation, M.P.-K.; writing—review and editing, M.P.-K., K.K.; visualization, M.P.-K., K.K.; supervision, M.P.-K. All authors have read and agreed to the published version of the manuscript.

Funding: This research received no external funding.

Institutional Review Board Statement: Not applicable.

Data Availability Statement: Publicly available data sets were analyzed in this study. These data can be found here: [38]. One data set has been artificially generated and a description of the process behind the artifical generation is presented in the paper.

Conflicts of Interest: The authors declare no conflict of interest.

References

1. Czarnowski, I.; Jędrzejowicz, P. Ensemble online classifier based on the one-class base classifiers for mining data streams. *Cybern. Syst.* **2015**, *46*, 51–68. [CrossRef]
2. Verbraeken, J.; Wolting, M.; Katzy, J.; Kloppenburg, J.; Verbelen, T.; Rellermeyer, J.S. A survey on distributed machine learning. *ACM Comput. Surv.* **2020**, *53*, 1–33. [CrossRef]
3. Guo, Y.; Zhao, R.; Lai, S.; Fan, L.; Lei, X.; Karagiannidis, G.K. Distributed machine learning for multiuser mobile edge computing systems. *IEEE J. Sel. Top. Signal Process.* **2022**, *16*, 460–473. [CrossRef]
4. Ma, C.; Li, J.; Shi, L.; Ding, M.; Wang, T.; Han, Z.; Poor, H.V. When federated learning meets blockchain: A new distributed learning paradigm. *IEEE Comput. Intell. Mag.* **2022**, *17*, 26–33. [CrossRef]
5. Xiao, M.; Skoglund, M. Coding for Large-Scale Distributed Machine Learning. *Entropy* **2022**, *24*, 1284. [CrossRef] [PubMed]
6. Rodríguez-Barroso, N.; Stipcich, G.; Jiménez-López, D.; Ruiz-Millán, J.A.; Martínez-Cámara, E.; González-Seco, G.; Luzóna, M.V.; Veganzones M.A.; Herrera, F. Federated learning and differential privacy: Software tools analysis, the sherpa. ai fl framework and methodological guidelines for preserving data privacy. *Inf. Fusion* **2020**, *64*, 270–292. [CrossRef]
7. Yang, Q.; Liu, Y.; Chen, T.; Tong, Y. Federated machine learning: Concept and applications. *ACM Trans. Intell. Syst. Technol. (TIST)* **2019**, *10*, 1–19. [CrossRef]
8. Ng, W.W.; Zhang, J.; Lai, C.S.; Pedrycz, W.; Lai, L.L.; Wang, X. Cost-sensitive weighting and imbalance-reversed bagging for streaming imbalanced and concept drifting in electricity pricing classification. *IEEE Trans. Ind. Inform.* **2018**, *15*, 1588–1597. [CrossRef]
9. Czarnowski, I. Weighted Ensemble with one-class Classification and Over-sampling and Instance selection (WECOI): An approach for learning from imbalanced data streams. *J. Comput. Sci.* **2022**, *61*, 101614. [CrossRef]
10. Pławiak, P.; Abdar, M.; Pławiak, J.; Makarenkov, V.; Acharya, U.R. DGHNL: A new deep genetic hierarchical network of learners for prediction of credit scoring. *Inf. Sci.* **2020**, *516*, 401–418. [CrossRef]
11. Gupta, O.; Raskar, R. Distributed learning of deep neural network over multiple agents. *J. Netw. Comput. Appl.* **2018**, *116*, 1–8. [CrossRef]
12. Alsahaf, A.; Petkov, N.; Shenoy, V.; Azzopardi, G. A framework for feature selection through boosting. *Expert Syst. Appl.* **2022**, *187*, 115895. [CrossRef]

13. Hashemi, A.; Dowlatshahi, M.B.; Nezamabadi-Pour, H. Ensemble of feature selection algorithms: A multi-criteria decision-making approach. *Int. J. Mach. Learn. Cybern.* **2022**, *13*, 49–69. [CrossRef]
14. Ślęzak, D.; Janusz, A. Ensembles of bireducts: Towards robust classification and simple representation. In Proceedings of the International Conference on Future Generation of Information Technology (FGIT), Gangneug, Korea, 16–19 December 2011; Springer: Berlin, Germany, 2011; Volume 7105, pp. 64–77.
15. Kozak, J. *Decision Tree and Ensemble Learning Based on Ant Colony Optimization*; Springer International Publishing: Berlin/Heidelberg, Germany, 2019.
16. Tüysüzoğlu, G.Ö.K.S.U.; Birant, D. Enhanced bagging (eBagging): A novel approach for ensemble learning. *Int. Arab. J. Inf. Technol.* **2020**, *17*, 515–528.
17. Batra, S.; Khurana, R.; Khan, M.Z.; Boulila, W.; Koubaa, A.; Srivastava, P. A Pragmatic Ensemble Strategy for Missing Values Imputation in Health Records. *Entropy* **2022**, *24*, 533. [CrossRef]
18. Nam, G.; Yoon, J.; Lee, Y.; Lee, J. Diversity matters when learning from ensembles. *Adv. Neural Inf. Process. Syst.* **2021**, *34*, 8367–8377.
19. Ortega, L.A.; Cabañas, R.; Masegosa, A. Diversity and generalization in neural network ensembles. In Proceedings of the International Conference on Artificial Intelligence and Statistics, Valencia, Spain, 28–30 March 2022; pp. 11720–11743.
20. Kashinath, S.A.; Mostafa, S.A.; Mustapha, A.; Mahdin, H.; Lim, D.; Mahmoud, M.A.; Mohammed, M.A.; Al-Rimy, B.A.S.; Fudzee, M.F.; Yang, T.J. Review of data fusion methods for real-time and multi-sensor traffic flow analysis. *IEEE Access* **2021**, *9*, 51258–51276. [CrossRef]
21. Kuncheva, L.I. *Combining Pattern Classifiers: Methods and Algorithms*; John Wiley & Sons: Hoboken, NJ, USA, 2014.
22. Liu, L.; Zhang, J.; Song, S.H.; Letaief, K.B. Client-edge-cloud hierarchical federated learning. In Proceedings of the ICC 2020-2020 IEEE International Conference on Communications (ICC), Dublin, Ireland, 7–11 June 2020; pp. 1–6.
23. Zhou, C.; Zhang, H.; Valdebenito, M.A.; Zhao, H. A general hierarchical ensemble-learning framework for structural reliability analysis. *Reliab. Eng. Syst. Saf.* **2022**, *225*, 108605. [CrossRef]
24. Gholizadeh, N.; Musilek, P. Distributed Learning Applications in Power Systems: A Review of Methods, Gaps, and Challenges. *Energies* **2021**, *14*, 3654. [CrossRef]
25. Tang, M.; Liao, H.; Mi, X.; Lev, B.; Pedrycz, W. A hierarchical consensus reaching process for group decision making with noncooperative behaviors. *Eur. J. Oper. Res.* **2021**, *293*, 632–642. [CrossRef]
26. Dai, T.; Sycara, K.; Zheng, R. Agent reasoning in AI-powered negotiation. In *Handbook of Group Decision and Negotiation*; Springer: Berlin/Heidelberg, Germany, 2021; pp. 1187–1211.
27. Wyai, L.C.; WaiShiang, C.; Lu, M.V.A. Agent negotiation patterns for multi agent negotiation system. *Adv. Sci. Lett.* **2018**, *24*, 1464–1469. [CrossRef]
28. Pawlak, Z. Some remarks on conflict analysis. *Eur. J. Oper. Res.* **2005**, *166*, 649–654. [CrossRef]
29. Pawlak, Z. Conflict analysis. In Proceedings of the Fifth European Congress on Intelligent Techniques and Soft Computing (EUFIT'97), Aachen, Germany, 8–12 September 1997; pp. 1589–1591.
30. Tong, S.; Sun, B.; Chu, X.; Zhang, X.; Wang, T.; Jiang, C. Trust recommendation mechanism-based consensus model for Pawlak conflict analysis decision making. *Int. J. Approx. Reason.* **2021**, *135*, 91–109. [CrossRef]
31. Yao, Y. Three-way conflict analysis: reformulations and extensions of the Pawlak model. *Knowl. Based Syst.* **2019**, *180*, 26–37. [CrossRef]
32. Przybyła-Kasperek, M. Study of selected methods for balancing independent data sets in k-nearest neighbors classifiers with Pawlak conflict analysis. *Appl. Soft Comput.* **2022**, *129*, 109612. [CrossRef]
33. Przybyła-Kasperek, M. Coalitions' Weights in a Dispersed System with Pawlak Conflict Model. *Group Decis. Negot.* **2020**, *29*, 549–591. [CrossRef]
34. Przybyła-Kasperek, M. Three conflict methods in multiple classifiers that use dispersed knowledge. *Int. J. Inf. Technol. Decis. Mak.* **2019**, *18*, 555–599. [CrossRef]
35. Breiman, L.; Friedman, J.H.; Olshen, R.A.; Stone, C.J. *Classification and Regression Trees*; Routledge: Abingdon, UK, 2017.
36. Przybyła-Kasperek, M.; Wakulicz-Deja, A. Global decision-making system with dynamically generated clusters. *Inform. Sci.* **2014**, *270*, 172–191. [CrossRef]
37. Lamrini, B. Contribution to Decision Tree Induction with Python: A Review. In *Data Mining—Methods, Applications and Systems*; IntechOpen: London, UK, 2020. doi: 10.5772/intechopen.92438. [CrossRef]
38. Asuncion, A.; Newman, D.J. *UCI Machine Learning Repository*; University of Massachusetts: Amherst, MA, USA, 2007. Available online: https://archive.ics.uci.edu (accessed on 19 September 2022).

Article

Improved EAV-Based Algorithm for Decision Rules Construction

Krzysztof Żabiński [†] and Beata Zielosko [*,†]

Institute of Computer Science, Faculty of Science and Technology, University of Silesia in Katowice, Będzińska 39, 41-200 Sosnowiec, Poland

* Correspondence: beata.zielosko@us.edu.pl
† These authors contributed equally to this work.

Abstract: In this article, we present a modification of the algorithm based on EAV (entity–attribute–value) model, for induction of decision rules, utilizing novel approach for attribute ranking. The selection of attributes used as premises of decision rules, is an important stage of the process of rules induction. In the presented approach, this task is realized using ranking of attributes based on standard deviation of attributes' values per decision classes, which is considered as a distinguishability level. The presented approach allows to work not only with numerical values of attributes but also with categorical ones. For this purpose, an additional step of data transformation into a matrix format has been proposed. It allows to transform data table into a binary one with proper equivalents of categorical values of attributes and ensures independence of the influence of the attribute selection function from the data type of variables. The motivation for the proposed method is the development of an algorithm which allows to construct rules close to optimal ones in terms of length, while maintaining enough good classification quality. The experiments presented in the paper have been performed on data sets from UCI ML Repository, comparing results of the proposed approach with three selected greedy heuristics for induction of decision rules, taking into consideration classification accuracy and length and support of constructed rules. The obtained results show that for the most part of datasests, the average length of rules obtained for 80% of best attributes from the ranking is very close to values obtained for the whole set of attributes. In case of classification accuracy, for 50% of considered datasets, results obtained for 80% of best attributes from the ranking are higher or the same as results obtained for the whole set of attributes.

Keywords: decision rules; length; support; greedy heuristics; feature selection; rough sets

Citation: Żabiński, K.; Zielosko, B. Improved EAV-Based Algorithm for Decision Rules Construction. *Entropy* **2023**, *25*, 91. https://doi.org/10.3390/e25010091

Academic Editors: Jan Kozak, Przemysław Juszczuk and Geert Verdoolaege

Received: 1 December 2022
Revised: 21 December 2022
Accepted: 28 December 2022
Published: 2 January 2023

Copyright: © 2023 by the authors. Licensee MDPI, Basel, Switzerland. This article is an open access article distributed under the terms and conditions of the Creative Commons Attribution (CC BY) license (https://creativecommons.org/licenses/by/4.0/).

1. Introduction

Decision rules are one of popular and well-known form of data representation. They are also often used in the classifier building process. Generally, it can be said that the process of induction of decision rules may have two perspectives [1]: knowledge representation and classification.

One of the main purposes of knowledge representation is to discover patterns or anomalies hidden in the data. The patterns are presented in the form of decision rules that map dependencies between the values of conditional attributes and the label of the decision class. Taking into account this perspective of rule induction, there exists variety of rules' quality measures that are related to human perception. These are, among others number of induced rules, their length and support [2,3].

The purpose of rule-based classifier is to assign a decision class label to a new object based on the attributes values' describing that object. One of the popular measure of rule quality from this perspective belonging to the domain of supervised learning, is the classification error. It is a percentage of the number of incorrectly classified examples.

There are different approaches for construction of decision rules. It is known that the form of obtained rules, for example, their number, length, depend on the algorithm used for their induction. Moreover, the set of rules which consist of a classifier ensuring a low classification error, is not always easy to understand and interpret from the point of view of knowledge representation. On the other hand, a small number of induced rules that are short and only reflect general patterns from the data, will not always ensure a good classification quality. These discrepancies mean that different rule induction approaches may be proposed, depending on the purpose of their application and mentioned two perspectives of rule induction, i.e., classification and knowledge representation, which do not coincide often.

In the paper, an approach that allows induction of decision rules, taking into account both the knowledge representation and classification perspective is presented. The proposed algorithm is based on the idea of an extension of the dynamic programming approach for optimization of decision rules relative to length and partitioning table into subtables.

Unfortunately, for large data sets, i.e., with a large number of attributes with many different values, the time for obtaining an optimal solution may be relatively long, which motivated authors to develop the presented method. Moreover, the problem of minimization of length of decision rules is NP-hard [4,5] and the most part of approaches for decision rules construction, with the exception of brute force, Boolean reasoning, extension of dynamic programming, Apriori algorithm, cannot guarantee the construction of optimal rules, i.e., rules with minimum length. Exact algorithms for construction of decision rules with minimum length have very often exponential computational complexity. Thus, for large datasets the rule generation time can be significant. However, often results close to optimal ones are enough for given application. Taking into account above facts, some heuristic which allows to obtain rules close to optimal from the point of view of length and with relatively good accuracy of classification was presented. The proposed algorithm is an extension and modification of the approach presented in [6]. To ensure the possibility of working with categorical values of attributes, and the independence of the attribute selection function from the data type, the data preparation stage was introduced. It consist of transforming data set into a matrix form and allows to work with binary data table where each attribute value has the same weight and numerical values are assigned automatically. This step is important from the point of view of attribute selection process performed during rule construction phase. An other element of the proposed approach is transformation data table into EAV (attribute–entity–value) form which is convenient for processing large amounts of data.

The methods and approaches for choosing of the attributes that consist of rules' premises can be wrapped in the rule induction algorithms or can be performed immediately preceding the rule induction step. An example of the latter approach is rule construction based on reducts [7]. However, in both cases different measures, such as based on similarity, entropy, dependency, distance or statistical characteristics are employed and used for attributes evaluation. It is also possible that based on selected set of features their ranking is constructed. It allows to indicate importance of variables. In the paper, the method for selection of attributes directly precedes the rule induction step. It takes into account an influence of features' values into class labels and it is based on standard deviation of attributes values per decision classes. Obtained values of standard deviation function are used for creation of ranking of variables and user decides what percentage of attributes with highest position in the ranking is taken into account during rule construction phase.

Decision rules induced by presented algorithm were compared with three selected heuristics. The choice of these heuristics follows from the fact that they allow to obtain rules close to optimal ones in terms of length and support. In [8] the experimental results showed that the average relative difference between length of rules constructed by the best heuristic and minimum length of rules is at most 4%, similar situation was observed in case of support.

The paper consists of five sections. Section 2 is devoted to approaches and methods for attribute selection during process of induction of decision rules. The main stages of the proposed algorithm are presented in Section 3. Section 4 contains short description of three selected heuristics for induction of decision rules. Experimental results concerning analysis of obtained sets of rules from the point of view of knowledge representation and classification, and comparison with selected heuristics are included in Section 5. Conclusions and future plans are given in Section 6.

2. Selection of Attributes for Rule Construction

The attribute selection process, in general, leads to the selection of a certain subset of originally available features in order to accomplish a specific task, which is, e.g., creation a model for classification purposes [9]. It also allows for removal redundant or irrelevant variables from a set of all attributes. The feature selection stage is not only an important element of data preprocessing, it plays a key role during induction of decision rules. The obtained results impact on the knowledge representation perspective. A smaller set of attributes is easier to check, understand and visualize, it has lower storage requirements and from the classification point of view it allows to avoid overfitting [10]. Selection of features can lead to the creation of their ranking. This approach is called feature ranking and allows to estimate relevance of attributes based on some adopted threshold. As a result, the most important variables have assigned the highest positions in the ranking, and the least relevant—the lowest positions.

There are many algorithms for selecting features. The most popular is a division of methods into filters, wrappers, and embedded [11]. Filter methods can be considered as data preprocessing tasks that are independent on the classification systems. Therefore, their advantage is speed and main drawback is what makes them fast and easily applicable in almost all kinds of problems, i.e., neglecting the real-time influence on the classification system. Wrapper methods, as opposed to filters, can be treated as feedback-based systems by examining the influence of the choice of subsets of features on the classification result. The last group, embedded methods contain a feature subset evaluation mechanism built directly into the learning algorithm. As a result, they can provide good quality solutions for specific applications where knowledge about characteristics of learning algorithm is necessary.

A decision rule can be viewed as a hypothesis that maps to a pattern in the data or a function that predicts a decision class label for a given object. From this perspective, selection of attributes is one of element of decision rule construction process. It is often performed during the rule induction algorithm work and it is an iterative step in which the attributes are selected sequentially if adopted criterion is met. It is also possible to construct rules using filter approach, e.g., based on reducts. In both cases, the chosen attribute together with the corresponding value form a rule descriptor (attribute = value pair) which constitutes a rule premise part. The attributes contained in rules determine their quality, therefore the process of variable selection and the adopted criterion plays an important role.

In the framework of rough sets theory there are many algorithms for induction of decision rules [12]. During process of rules construction different evaluation measures are used and they are based on discernibility relation, upper and lower approximations, dependency degree concept, discernibility function and prime implicants and many others [13,14]. Reduct is a popular notion in the rough sets theory [15] and is interpreted as such minimal subset of attributes that is sufficient to discern any pairs of objects with different class labels. Based on the attributes which constitutes reduct, decision rules are constructed, so they are induced from the reduced set of attributes [16,17]. The popular measures for selection of attributes during reduct construction are based on, for example, discernibility matrix [18], positive region-based dependency [19], neighbourhood information granules [20], entropy and many others [21].

Another group of methods related to algorithms for induction of decision rules is based on sequential covering approach [22,23], e.g., family of AQ algorithms, CN2, Ripper. In this framework, candidates for the elementary conditions of a rule are evaluated taking into account, for example, maximization of the number of positive examples covered by the conjunction of elementary conditions in premise part of a rule, maximization of the ratio of covered positive examples to the total number of covered examples, minimization of a rule length and others [24,25].

It should be also noted that there are many heuristics algorithms which uses different criteria based on entropy, Gini index, information gain, statististical characteristics and different their modifications [26–32].

In the proposed approach, selection of attributes is based on standard deviation of attributes values in the framework of decision classes, described in Section 3.

3. Decision Rules Construction Approach

In this section, an algorithm for decision rules induction is presented. This algorithm can be considered as an extension and improvement of the algorithm based on EAV model presented in [6]. One of the important element of the considered approach is selection of attributes based on standard deviation of their values in the framework of decision classes. In order to calculate standard deviation of attributes values, categorical ones should be transformed to numerical. The modification proposed in this paper provides independence of the attribute selection function from the data type of variables and automatic assignment of numerical equivalents to categorical values, so each attribute has the same weight. This stage of the algorithm is considered as data preparation step which concerns transformation data table into matrix form [33]. Then, based on numerical form of data, EAV table [34] is created which allows to use the relational database engine to determine the standard deviation of attributes within decision classes. This step of proposed approach is presented as data transformation block on Figure 1. Employing selection of attributes based on standard deviation approach results ranking of features that indicates order and importance of attributes which are considered during process of rule construction. This stage of the approach is presented as attribute selection block on Figure 1. The third phase is indicated on Figure 1 as construction of decision rules block. The general idea of the proposed approach, expressed in the form of an activity diagram, is presented on the Figure 1 and described in detail in the following sections.

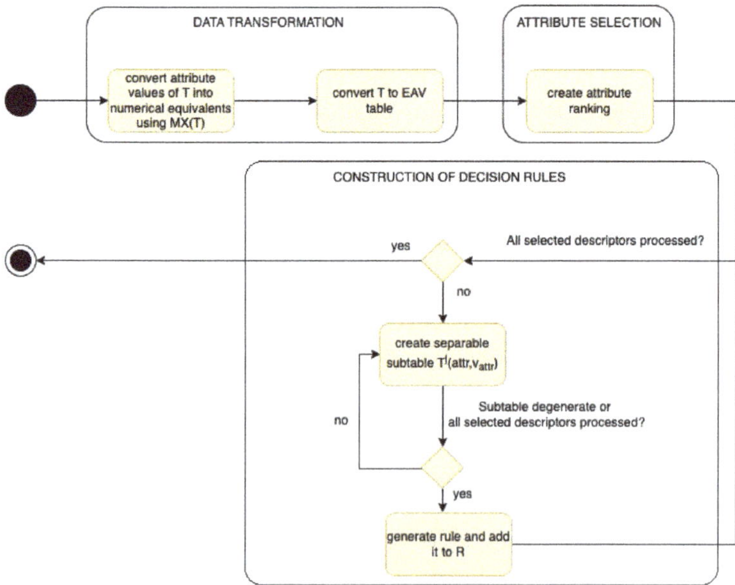

Figure 1. General idea of the approach for decision rules construction.

3.1. Data Transformation and Attribute Selection

Popular form of data representation is tabular form defined as a decision table T [15], $T = (U, A \cup \{d\})$, where U is a nonempty, finite set of objects (rows), $A = \{attr_1, \ldots, attr_n\}$ is nonempty, finite set of condition attributes, $attr : U \to V_{attr}$ is a function, for any $attr \in A$, V_{attr} is the set of values of an attribute $attr$. $d \notin A$ is a distinguished attribute called a decision attribute with values $V_d = \{d_1, \ldots, d_{|V_d|}\}$.

Data transformation stage consists of data transformation into matrix form and construction of EAV table. The first one is applied in order to facilitate statistical analysis if the attributes' values are categorical. Such way of data preparation is known from CART (ang. classification and regression trees) approach [35] and also used for induction of binary association rules [36]. It is a tabular form where each attribute and its value from T is represented as a single table column. Matrix data format incorporates two attribute values only: 0 or 1. 1 represents the situation where a given attribute with its value occurs for the given object, 0 represents the situation where a given attribute with its value does not occur for the given row of T. Algorithm 1 presents conversion of symbolic values of attributes from data table T into matrix form $MX(T)$.

Algorithm 1 Algorithm for conversion of symbolic values of attributes into numerical equivalents.

Input: decision table T with condition attributes $attr_1, \ldots, attr_n$, row $r = (v_{attr_1}, \ldots, v_{attr_n})$

Output: $MX(T)$-matrix data form of T

$AV \leftarrow \varnothing;$ // AV is a set of unique pairs $(attr, v_{attr})$ from T

for each r of T **do**

    add descriptor $(attr, v_{attr})$ to AV;

end for

for each descriptor $(attr, v_{attr})$ from AV **do**

    add column to $MX(T)$, named av_i, filled with 0's;

end for

for each r of T **do**

    set value to 1 for column named av_i where $a = attr$ and $v_i = v_{attr}$;

end for

An example of data table transformed into the matrix form is presented in Figure 2.

data table T				matrix_form						
f1	f2	f3	decision	f1_low	f1_high	f2_bad	f2_good	f3_small	f3_big	decision
high	bad	small	1	0	1	1	0	1	0	1
high	good	big	2	0	1	0	1	0	1	2
high	bad	big	3	0	1	1	0	0	1	3
low	bad	big	1	1	0	1	0	0	1	1
low	good	big	2	1	0	0	1	0	1	2
low	bad	small	3	1	0	1	0	1	0	3

Figure 2. Data table T transformed to matrix form.

Based on data presented in the matrix form, average values of each column of $MX(T)$ are obtained and used for replacement of symbolic values of attributes by their numerical equivalents in the table T.

The next stage of data transformation concerns conversion of a decision table with numerical equivalents into EAV form. It is a tabular form where each row contains an attribute, its corresponding value, class label and the ordinal number of object to which the given attribute is assigned. The main advantage of this approach is the possibility of using a relational database engine to analyze large data sets, as it was shown in case of induction of association and decision rules [37,38].

Then, calculation of standard deviation of attributes values per decision class is performed and ranking of attributes is obtained (see Figure 3).

EAV with average values from matrix format				Standard deviation of average values	
attribute	value	decision	row	attribute	value
f1	0.50	1	1	f2	0.19
f2	0.67	1	1	f3	0.10
f3	0.33	1	1	f1	0.00
f1	0.50	2	2		
f2	0.33	2	2		
f3	0.67	2	2		
f1	0.50	3	3		
f2	0.67	3	3		
f3	0.67	3	3		
f1	0.50	1	4		
f2	0.67	1	4		
f3	0.67	1	4		
f1	0.50	2	5		
f2	0.33	2	5		
f3	0.00	2	5		
f1	0.50	3	6		
f2	0.67	3	6		
f3	0.33	3	6		

Figure 3. EAV table and ranking of attributes for data presented in Figure 2.

The standard deviation of average values of attributes per decision classes has been chosen as a distinguishability level, following the intuitive idea that there is a correlation between average attribute value in a given class and the class itself. The relation is directly proportional, meaning that the highest the average standard deviation of the attribute, the biggest impact on the decision class. This intuitive approach follows the ideas of Bayesian analysis of data using Rough Bayesian model, which has been introduced in [39]. There was shown a correspondence between the main concepts of rough set theory and statistics where a hypothesis (target concept X_1) can be verified positively, negatively (in favour of the null hypothesis, which is a complement concept X_0) or undecided, under the given evidence E. The Rough Bayesian model is based on the idea of inverse probability analysis and Bayes factor B_0^1, defined as follows [39]:

$$B_0^1 = \frac{Pr(E|X_1)}{Pr(E|X_0)}.$$

Posterior probabilities can correspond to the accuracy factor in the machine learning domain [40]. Comparison of prior and posterior knowledge allows seeing if new evidence (satisfaction of attributes' values of objects) decreases or increases the belief in a given event, i.e., membership to a given decision class.

Let us assume that X_k are events, then $Pr(X_k)$ is the prior probability, $\sum_{l=0}^{|V_d|-1} Pr(X_l) = 1$. It is possible that X_k will occur, but there is no certainty for that. $Pr(X_k|E)$ is the posterior probability meaning X_k can occur when the evidence associated with E appears, $\sum_{l=0}^{|V_d|-1} Pr(X_l|E) = 1$. E can be considered in the framework of indiscernibility relation $E \in U/B$, $B \in A$, which provides a partition of objects U from decision table T into groups having the same values of B. The above-mentioned probabilities can be estimated as follows:

$$Pr(X_k) = \frac{|X_k|}{|U|}, Pr(X_k|E) = \frac{|X_k \cap E|}{|E|}.$$

Obviously, the bigger value of $Pr(X_k|E)$ is, the higher correlation between X_k and E exists. Then, using the probability density function, it is possible to visualize the influence of the posterior probability on the density range of E. This range can be approximated using the standard deviation of the attribute values within a given decision class. Such an approach was used in the feature selection process [41] and induction of decision rules [6,34,37].

3.2. Construction of Decision Rules

Based on the created ranking of attributes, it is possible to proceed to rules generation stage. In the proposed approach, user can indicate a specified number of best attributes which will be taken into consideration during the process of rules induction. On this basis, descriptors from set AV, which is a set of unique pairs $(attr, v_{attr})$ from T, are selected. Starting with the highest ranked attribute, a separable subtable is created. It is a subtable of the table T that contains only rows that have values $v_{attr_1}, \ldots, v_{attr_m}$ at the intersection with columns $attr_{i_1}, \ldots, attr_{i_m}$ and is denoted by $T' = T(attr_{i_1}, v_{attr_1}) \ldots (attr_{i_m}, v_{attr_m})$. The process of the partitioning of the table T into separable subtables is stopped when the considered subtable is degenerate, i.e., the same decision values are assigned to all rows or when all descriptors from AV based on the selected attributes were used. Pairs $(attr = v_a ttr)$ that form separable subtables T' at the bottom level corresponds to descriptors included in the premise part of decision rules. $mcd(T')$ denotes the most common decision for rows of T'. Algorithm 2 presents the algorithm for decision rules construction.

Algorithm 2 Algorithm for induction of decision rules.

Input: decision table T with numerical values of attributes, number p of best attributes to

be taken into consideration

Output: set of unique rules R

  $j \leftarrow \emptyset$;

  $Q \leftarrow \emptyset$;

  convert T into EAV table;

  $\forall_{attr \in A}$ calculate STD_{attr} grouped by V_d and create a ranking;

  select p attributes from the ranking and select descriptors from AV containing selected

  attributes;

  while all selected descriptors are not processed **do**

    create separable subtable $T^j(attr, v_{attr})$;

    $Q \leftarrow Q \cup \{attr = v_{attr}\}$;

    if $T^j(attr, v_{attr})$ is degenerate OR $j = p$ **then**

      $R \leftarrow R \cup \forall_{attr=v_{attr} \in Q}(attr_i = v_{attr_i}) \rightarrow mcd(T^j)$, where $mcd(T^j)$ is the most common

      decision for T^j;

    else

      $j = j + 1$;

    end if

  end while

The time and space complexity of the Algorithm 2 has been discussed in details in the previous authors' publication [6]. The mean computational complexity is linear and only decision table specificity can lead to square complexity in the worst case scenario. Algorithm 1 is part of the whole approach for decision rule construction with minor influence on the whole complexity itself.

4. Selected Greedy Heuristics

Greedy algorithms are often used to solve optimization problems. This approach, in order to determine the solution at each step, makes a greedy, i.e. the most promising partial solution at a given moment.

In the paper, three greedy heuristics are presented. They are called M, RM and log and used for rule induction. Detailed description of these heuristics can be found in [8]. The research has shown that on average the results of the greedy algorithms, in terms of length and support of induced rules, are close to optimal ones obtained by extensions of dynamic programming approach.

In general, the pseudocode of greedy heuristics is presented by Algorithm 3. Each heuristic (M, RM or log) constructs a decision rule for the table T and a given row r with assigned decision d_k, $k \in \{1, \ldots, |V_d|\}$. It is applied sequentially, for each row r of T and in each iteration selects an attribute $attr_i \in \{attr_1, \ldots, attr_n\}$ with a minimum index, fulfilling the given criterion.

Algorithm 3 Heuristic (M, RM or log) for induction of decision rules.

Input: Decision table T with condition attributes and row r

Output: Decision rule rul for T and given row r

$Q \leftarrow \varnothing$;

$T^0(attr, v_{attr}) \leftarrow T$;

while $T^j(attr, v_{attr})$ is not degenerate **do**

    select attribute $attr_i$ as follows:

- heuristic M selects $attr_i$ which minimizes the value $M(attr_i, r, d_k)$;
- heuristic RM selects $attr_i$ which minimizes the value $RM(attr_i, r, d_k)$;
- heuristic log selects $attr_i$ which maximizes the value $\frac{\beta(attr_i, r, d_k)}{\log_2(\alpha(attr_i, r, d_k) + 2)}$;

    $Q \leftarrow Q \cup \{attr\}$;

    $T^{(j+1)} \leftarrow T^j(attr, v_{attr})$;

    $j = j + 1$;

end while

$rul \leftarrow \forall_{attr \in Q}(attr_i = v_{attr_i}) \rightarrow d_k$;

During the heuristics work, the following notation was used: $N(T)$-number of rows in the table T, $N(T, d_k)$-number of rows from T with a given decision.

- $M(attr_i, r, d_k) = M(T^j, d_k) = N(T^{j+1}) - N(T^{j+1}, d_k)$,
- $RM(attr_i, r, d_k) = (N(T^{j+1}) - N(T^{j+1}, d_k))/N(T^{j+1})$,
- $\alpha(attr_i, r, d_k) = N(T^j, d_k) - N(T^{j+1}, d_k)$ and $\beta(attr_i, r, d_k) = M(T^j, d_k) - M(T^{j+1}, d_k)$.

Figure 4 presents separable subtables created based on the values of attributes assigned to the second row of data table T.

data table T			
f1	f2	f3	decision
high	bad	small	1
high	good	big	2
high	bad	big	3
low	bad	big	1
low	good	big	2
low	bad	small	3

T(f1,high)			
f1	f2	f3	decision
high	bad	small	1
high	good	big	2
high	bad	big	3

T(f2,good)			
f1	f2	f3	decision
high	good	big	2
low	good	big	2

T(f3,big)			
f1	f2	f3	decision
high	good	big	2
high	bad	big	3
low	bad	big	1
low	good	big	2

Figure 4. Separable subtables $T(f_1, high), T(f_2, good), T(f_3, big)$ of decision table T.

The selected heuristics work as follows:

- $M(f_1, r_2, 2) = 2, M(f_2, r_2, 2) = 0, M(f_3, r_2, 2) = 2,$
  $f_2 = good \rightarrow 2;$
- $RM(f_1, r_2, 2) = \frac{1}{3}, RM(f_2, r_2, 2) = 0, RM(f_3, r_2, 2) = \frac{1}{2},$
  $f_2 = good \rightarrow 2;$
- $\alpha(f_1, r_2, 2) = 1, \alpha(f_2, r_2, 2) = 0, \alpha(f_3, r_2, 2) = 0,$
  $\beta(f_1, r_2, 2) = 2, \beta(f_2, r_2, 2) = 4, \beta(f_3, r_2, 2) = 0,$
  $f_2 = good \rightarrow 2;$

Decision rules constructed by these heuristics for the second row from T are the same.

5. Experimental Results

Experiments have been executed on datasets from UCI Machine Learning Repository [42]. Unique valued attributes have been eliminated. Any missing values have been filled by the most common value for the given attribute. The sets taken into consideration are the following:

- balance-scale,
- breast-cancer,
- cars,
- flags,
- hayes-roth-data,
- house-votes,
- lymphography,
- tic-tac-toe.

The aim of the experiments is to compare the proposed algorithm with the selected heuristics. The study was performed from the point of view of knowledge representation taking into account length and support of constructed rules and from the point of view of classification accuracy. Length of the rule is defined as number of descriptors in the premise part of the rule. Support of the rule is the number of rows from T which matching conditions and the decision of a given rule. Classification accuracy is defined as the number of properly classified rows from the test part of T, divided by the number of all rows from the test part of T.

The algorithms have been implemented in Java 17 and Spring Boot framework and experiments have been executed with Macbook Pro: Intel i7-9750H CPU, 16 GB of RAM memory, macOS Monterey 12.2.1 operating system.

5.1. Comparison from the Point of Data Representation

From the point of view of data representation, two quality measures have been compared: rule length and rule support. Tables 1–3 present minimal, average and maximal length and support of rules obtained by proposed algorithm taking into account 100%, 80% and 60% of best attributes from the ranking.

Table 1. Values on minimum, average and maximum length and support of rules generated by proposed algorithm taking into account the whole set of attributes in data table.

Data Set	Number of		Length			Support		
	Rows	Attributes	Min	Avg	Max	Min	Avg	Max
balance-scale	625	4	3	3.64	4	1	2.44	5
breast-cancer	266	9	1	5.61	9	1	2.61	11
cars	128	6	2	3.90	6	1	79.31	192
flags	194	26	2	8.88	20	1	1.78	6
hayes-roth-data	69	5	1	2.64	4	1	3.81	12
house-votes	279	16	3	6.14	16	1	31.21	81
lymphography	148	18	1	8.40	16	1	2.85	6
tic-tac-toe	958	9	3	5.71	8	1	6.43	38

Table 2. Values on minimum, average and maximum length and support of rules generated by proposed algorithm taking into account 80% of best attributes from the ranking.

Data Set	Number of		Length			Support		
	Rows	Attributes	Min	Avg	Max	Min	Avg	Max
balance-scale	625	4	3	3.64	4	1	2.44	5
breast-cancer	266	9	1	5.58	8	1	2.61	11
cars	128	6	2	3.52	5	2	79.88	192
flags	194	26	2	8.88	20	1	1.78	6
hayes-roth-data	69	5	1	2.64	4	1	3.81	12
house-votes	279	16	3	6.09	13	1	31.23	81
lymphography	148	18	1	8.39	15	1	2.85	6
tic-tac-toe	958	9	3	5.71	8	1	6.43	38

Table 3. Values on minimum, average and maximum length and support of rules generated by proposed algorithm taking into account 60% of best attributes from the ranking.

Data Set	Number of		Length			Support		
	Rows	Attributes	Min	Avg	Max	Min	Avg	Max
balance-scale	625	4	3	3.00	3	2	3.94	5
breast-cancer	266	9	1	5.28	6	1	3.01	11
cars	128	6	2	3.11	4	6	82.78	192
flags	194	26	2	8.79	16	1	1.78	6
hayes-roth-data	69	5	1	2.51	3	1	3.94	12
house-votes	279	16	3	5.94	10	1	31.46	81
lymphography	148	18	1	8.17	11	1	3.01	6
tic-tac-toe	958	9	3	5.29	6	1	6.73	38

Figure 5 presents, the average length of rules relative to number of attributes, obtained for 100%, 80% and 60% of best attributes from the ranking, for considered datasets. It is possible to see that for most of the datasets, with the exceptions of breast-cancer and cars, the average length of rules obtained for 80% of best attributes from the ranking is very close to results obtained for the whole set of attributes. In case of average support the best results, were obtained for datasets cars and house-votes. The function that determines the choice of attributes during decision rule construction is the standard deviation of attribute values within decision classes. Thus, the distribution of such values has an impact on the obtained results.

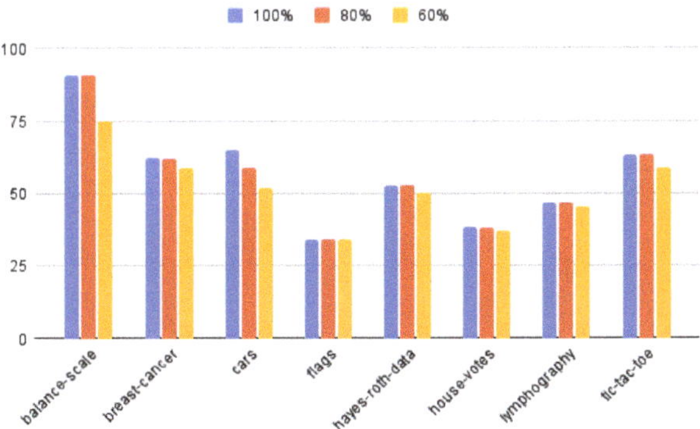

Figure 5. The average length of rules relative to number of attributes in given dataset, obtained for 100%, 80% and 60% of best attributes from the ranking.

Tables 4–6 present minimal, average and maximal length and support of rules obtained by heuristics M, RM and *log*.

Table 4. Values on minimum, average and maximum length and support of rules generated by means of M heuristic.

Data Set	Number of		Length			Support		
	Rows	Attributes	Min	Avg	Max	Min	Avg	Max
balance-scale	625	4	3	3.41	4	1	3.38	5
breast-cancer	266	9	1	2.97	6	1	2.81	24
cars	128	6	1	5.57	6	1	6.69	576
flags	194	26	1	2.04	4	1	2.04	18
hayes-roth-data	69	5	1	2.88	4	1	2.33	12
house-votes	279	16	2	3.17	6	1	22.86	95
lymphography	148	18	1	2.32	4	1	5.34	32
tic-tac-toe	958	9	3	4.12	5	1	7.32	90

Table 5. Values on minimum, average and maximum length and support of rules generated by means of RM heuristic.

Data Set	Number of		Length			Support		
	Rows	Attributes	Min	Avg	Max	Min	Avg	Max
balance-scale	625	4	3	3.41	4	1	3.38	5
breast-cancer	266	9	1	3.52	8	1	3.25	24
cars	128	6	1	5.44	6	1	8.14	576
flags	194	26	1	2.23	9	1	2.59	18
hayes-roth-data	69	5	1	2.92	4	1	2.56	12
house-votes	279	16	2	3.29	5	1	32.22	95
lymphography	148	18	1	2.56	5	1	7.70	32
tic-tac-toe	958	9	3	4.32	7	1	13.21	90

Table 6. Values on minimum, average and maximum length and support of rules generated by means of *log* heuristic.

Data Set	Number of		Length			Support		
	Rows	Attributes	Min	Avg	Max	Min	Avg	Max
balance-scale	625	4	3	3.41	4	1	3.38	5
breast-cancer	266	9	1	3.29	6	1	4.10	25
cars	128	6	1	5.45	6	1	8.11	576
flags	194	26	1	3.26	6	1	5.68	22
hayes-roth-data	69	5	1	2.90	4	1	2.87	12
house-votes	279	16	2	3.56	7	2	40.02	95
lymphography	148	18	1	2.85	5	1	10.83	32
tic-tac-toe	958	9	3	4.20	6	2	13.04	90

The statistical analysis by means of the Wilcoxon two-tailed test has been performed, to verify the null hypothesis that there are no significant differences in the assessment of rule from the point of view of length and support, average values of these measures have been taken into consideration. The results of rule length comparison have been gathered in the Figure 6.

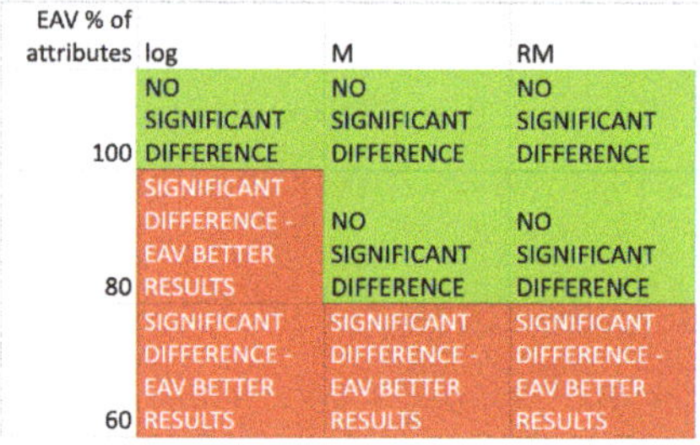

Figure 6. Wilcoxon test results-comparison of the average rules length.

The results of rule support comparison have been gathered in the Figure 7.

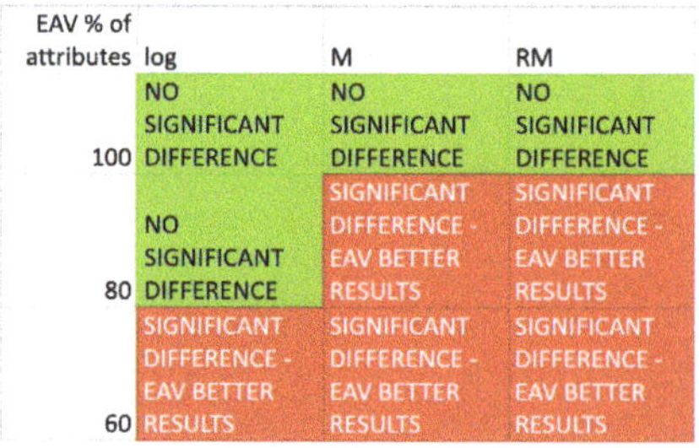

Figure 7. Wilcoxon test results-comparison of the average rules support.

The results show that the values of supports are comparable for all heuristics and 100% of attributes for presented algorithm. For 80% and 60% of selected best attributes, the supports results are noticeably better for the proposed approach. As for rule lengths, values are also comparable for all heuristics and 100% of attributes for the presented approach. Taking into account 80% and 60% of selected best attributes, it is possible to see that the length vales are noticeable smaller for the presented algorithm.

5.2. Comparison From The Point Of Data Classification

From the point of view of classification, accuracy has been compared (see Tables 7 and 8). 10-fold cross validation has been performed. Column std in presented tables denotes standard deviation of obtained results.

Table 7. Average classification accuracies of rules generated by means of the proposed algorithm.

Data Set	100%	Std	80%	Std	60%	Std
balance-scale	0.89	0.05	0.93	0.05	0.73	0.05
breast-cancer	0.92	0.03	0.94	0.03	0.88	0.03
cars	0.95	0.06	0.86	0.04	0.84	0.07
flags	0.93	0.05	0.92	0.05	0.91	0.05
hayes-roth-data	0.92	0.07	0.88	0.07	0.90	0.05
house-votes	0.98	0.03	0.98	0.03	0.97	0.04
lymphography	0.95	0.11	0.94	0.11	0.92	0.04
tic-tac-toe	0.95	0.06	0.95	0.06	0.88	0.06

Table 8. Average classification accuracies of rules generated by means of M, RM and log heuristics.

Data Set	M	Std	RM	Std	Log	Std
balance-scale	0.94	0.06	0.95	0.05	0.95	0.05
breast-cancer	0.94	0.03	0.95	0.03	0.95	0.03
cars	0.97	0.11	0.97	0.11	0.97	0.11
flags	0.97	0.08	0.99	0.08	0.99	0.08
hayes-roth-data	0.94	0.07	0.94	0.07	0.94	0.07
house-votes	0.99	0.11	0.99	0.11	0.99	0.11
lymphography	0.94	0.05	0.98	0.06	0.98	0.06
tic-tac-toe	0.97	0.04	0.98	0.05	0.98	0.05

Figure 8 presents, the average accuracy of classification, obtained for 100%, 80% and 60% of best attributes from the ranking, for considered datasets. For four datasets, i.e., balance-scale, breast-cancer, house-votes and tic-tac-toe, the classification accuracy obtained for 80% of best attributes from the ranking is higher or the same as results obtained for the whole set of attributes.

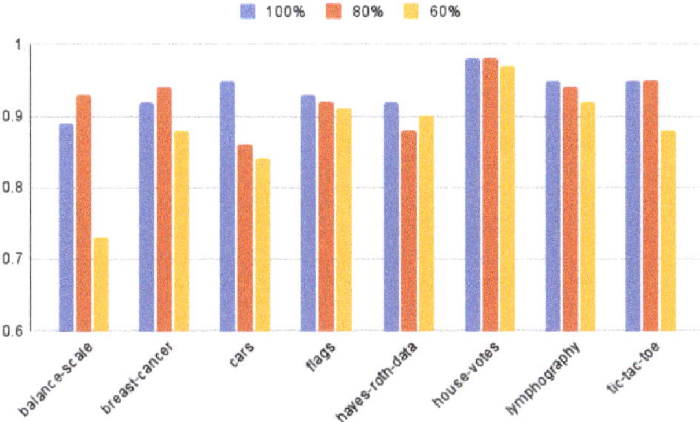

Figure 8. The average accuracy of classification, obtained for 100%, 80% and 60% of best attributes from the ranking.

The classification accuracy results once again have been compared by means of two-tailed Wilcoxon test, average values have been taken into this comparison, to verify the null hypothesis that there are no significant differences in the assessment of rule from the point of view of classification accuracy. The results are shown in the Figure 9.

EAV % of attributes	log	M	RM
100	NO SIGNIFICANT DIFFERENCE	NO SIGNIFICANT DIFFERENCE	NO SIGNIFICANT DIFFERENCE
80	NO SIGNIFICANT DIFFERENCE	NO SIGNIFICANT DIFFERENCE	NO SIGNIFICANT DIFFERENCE
60	SIGNIFICANT DIFFERENCE - EAV WORSE RESULTS	SIGNIFICANT DIFFERENCE - EAV WORSE RESULTS	SIGNIFICANT DIFFERENCE - EAV WORSE RESULTS

Figure 9. Wilcoxon test results-comparison of the average classification accuracy.

The results show that the classification accuracies are comparable for all heuristics and 100% as well as 80% of selected best attributes for proposed algorithm. For 60% of selected best attributes the classification results are noticeably worse, for the proposed approach. Such a situation is opposite to results obtained from knowledge representation point of view.

6. Conclusions

Taking into account results obtained by the experiments performed, it is possible to say that the proposed algorithm allows to obtain rules enough good from both perspectives: data representation and classification. The described approach is a heuristic one, and it has been compared with M, RM and log heuristics, which are good from the point of view of knowledge representation. The obtained result show that the presented approach allows to construct rules which are comparable with the heuristics in terms of classification accuracy (except for 60% of selected best attributes). As for rule support and rule length it was shown that the proposed algorithm allows to construct enough short rules with sufficiently good support.

Unfortunately, the proposed algorithm does not allow to automatically perform the feature selection stage. This issue will be considered as the next step on algorithm's improvement. Additionally, the possibility of working with missing values of attributes will be studied. Future works will also concentrate on comparison with algorithms for induction of decision rules based on sequential covering approach.

Author Contributions: Conceptualization B.Z. and K.Ż.; Methodology B.Z. and K.Ż.; Investigation B.Z. and K.Ż.; Software K.Ż.; Validation B.Z. and K.Ż.; Writing–original draft preparation B.Z. and K.Ż.; Writing review and editing B.Z. and K.Ż.; All authors have read and agreed to the published version of the manuscript.

Funding: This research received no external funding.

Institutional Review Board Statement: Ethical review and approval were waived for this study due to usage of publicly available datasets. Therefore, there was no any research done on humans or animals.

Data Availability Statement: Publicly available datasets were analyzed in this study. This data can be found here: http://archive.ics.uci.edu/ml (accessed on 23 March 2022).

Conflicts of Interest: The authors declare no conflict of interest.

References

1. Stefanowski, J.; Vanderpooten, D. Induction of decision rules in classification and discovery-oriented perspectives. *Int. J. Intell. Syst.* **2001**, *16*, 13–27. [CrossRef]
2. An, A.; Cercone, N. Rule Quality Measures Improve the Accuracy of Rule Induction: An Experimental Approach. In *International Symposium on Methodologies for Intelligent Systems*; Raś, Z.W., Ohsuga, S., Eds.; Springer: Berlin/Heidelberg, Germany, 2000; Volume 1932, pp. 119–129.
3. Wróbel, L.; Sikora, M.; Michalak, M. Rule Quality Measures Settings in Classification, Regression and Survival Rule Induction—An Empirical Approach. *Fundam. Inform.* **2016**, *149*, 419–449. [CrossRef]

4. Nguyen, H.S.; Ślęzak, D. Approximate Reducts and Association Rules-Correspondence and Complexity Results. In *RSFDGrC 1999*; Zhong, N., Skowron, A., Ohsuga, S., Eds.; Springer: Berlin/Heidelberg, Germany, 1999; Volume 1711, pp.137–145.

5. Moshkov, M.J.; Piliszczuk, M.; Zielosko, B. Greedy Algorithm for Construction of Partial Association Rules. *Fundam. Inform.* **2009**, *92*, 259–277. [CrossRef]

6. Żabiński, K.; Zielosko, B. Algorithm based on eav model. *Entropy* **2021**, *23*, 14. [CrossRef]

7. Zielosko, B.; Żabiński, K. Selected approaches for decision rules construction-comparative study. In *Knowledge-Based and Intelligent Information & Engineering Systems: Proceedings of the 25th International Conference KES-2021, Szczecin, Poland, 8–10 September 2021*; Watróbski, J., Salabun, W., Toro, C., Zanni-Merk, C., Howlett, R.J., Jain, L.C., Eds.; Elsevier: Szczecin, Poland, 2021; Volume 192, pp. 3667–3676.

8. Alsolami, F.; Amin, T.; Moshkov, M.; Zielosko, B.; Żabiński, K. Comparison of heuristics for optimization of association rules. *Fundam. Inform.* **2019**, *166*, 1–14. [CrossRef]

9. Guyon, I.; Gunn, S.; Nikravesh, M.; Zadeh, L. (Eds.) *Feature Extraction: Foundations and Applications*; Studies in Fuzziness and Soft Computing; Springer: Berlin/Heidelberg, Gernamy, 2006; Volume 207.

10. Stańczyk, U.; Zielosko, B.; Jain, L.C. Advances in Feature Selection for Data and Pattern Recognition: An Introduction. In *Advances in Feature Selection for Data and Pattern Recognition*; Stańczyk, U., Zielosko, B., Jain, L.C., Eds.; Springer International Publishing: Cham, Switzerland, 2018; pp. 1–9.

11. Reif, M.; Shafait, F. Efficient feature size reduction via predictive forward selection. *Pattern Recognit.* **2014**, *47*, 1664–1673. [CrossRef]

12. Pawlak, Z.; Skowron, A. Rough sets and Boolean reasoning. *Inf. Sci.* **2007**, *177*, 41–73. [CrossRef]

13. Ślęzak, D.; Wróblewski, J. Order based genetic algorithms for the search of approximate entropy reducts. In *RSFDGrC 2003*; Wang, G., Liu, Q., Yao, Y., Skowron, A., Eds.; Springer: Berlin/Heidelberg, Gernamy, 2003; Volume 2639, pp. 308–311.

14. Chen, Y.; Zhu, Q.; Xu, H. Finding rough set reducts with fish swarm algorithm. *Knowl.-Based Syst.* **2015**, *81*, 22–29. [CrossRef]

15. Pawlak, Z.; Skowron, A. Rudiments of rough sets. *Inf. Sci.* **2007**, *177*, 3–27. [CrossRef]

16. Zielosko, B.; Stańczyk, U. Reduct-based ranking of attributes. In *Knowledge-Based and Intelligent Information & Engineering Systems: Proceedings of the 24th International Conference KES-2020, Virtual Event, 16–18 September 2020*; Cristani, M., Toro, C., Zanni-Merk, C., Howlett, R.J., Jain, L.C., Eds.; Elsevier: Amsterdam, The Netherlands, 2020; Volume 176, pp. 2576–2585.

17. Zielosko, B.; Piliszczuk, M. Greedy Algorithm for Attribute Reduction. *Fundam. Inform.* **2008**, *85*, 549–561.

18. Yang, Y.; Chen, D.; Wang, H. Active Sample Selection Based Incremental Algorithm for Attribute Reduction With Rough Sets. *IEEE Trans. Fuzzy Syst.* **2017**, *25*, 825–838. [CrossRef]

19. Raza, M.S.; Qamar, U. Feature selection using rough set-based direct dependency calculation by avoiding the positive region. *Int. J. Approx. Reason.* **2018**, *92*, 175–197. [CrossRef]

20. Wang, C.; Shi, Y.; Fan, X.; Shao, M. Attribute reduction based on k-nearest neighborhood rough sets. *Int. J. Approx. Reason.* **2019**, *106*, 18–31. [CrossRef]

21. Ferone, A. Feature selection based on composition of rough sets induced by feature granulation. *Int. J. Approx. Reason.* **2018**, *101*, 276–292. [CrossRef]

22. Błaszczyński, J.; Słowiński, R.; Szeląg, M. Sequential covering rule induction algorithm for variable consistency rough set approaches. *Inf. Sci.* **2011**, *181*, 987–1002. [CrossRef]

23. Sikora, M.; Wróbel, L.; Gudyś, A. GuideR: A guided separate-and-conquer rule learning in classification, regression, and survival settings. *Knowl.-Based Syst.* **2019**, *173*, 1–14. [CrossRef]

24. Fürnkranz, J. Separate-and-Conquer Rule Learning. *Artif. Intell. Rev.* **1999**, *13*, 3–54. [CrossRef]

25. Valmarska, A.; Lavrač, N.; Fürnkranz, J.; Robnik-Šikonja, M. Refinement and selection heuristics in subgroup discovery and classification rule learning. *Expert Syst. Appl.* **2017**, *81*, 147–162. [CrossRef]

26. Kotsiantis, S.B. Decision Trees: A Recent Overview. *Artif. Intell. Rev.* **2013**, *13*, 261–283. [CrossRef]

27. Nguyen, H.S. Approximate Boolean reasoning: Foundations and applications in data mining. In *Transactions on Rough Sets V*; Peters, J.F., Skowron, A., Eds.; Springer: Berlin/Heidelberg, Gernamy, 2006; Volume 4100, pp. 334–506.

28. Stańczyk, U.; Zielosko, B.; Żabiński, K. Application of Greedy Heuristics for Feature Characterisation and Selection: A Case Study in Stylometric Domain. In Proceedings of the Rough Sets–International Joint Conference, IJCRS 2018, Quy Nhon, Vietnam, 20–24 August 2018; Nguyen, H.S., Ha, Q., Li, T., Przybyla-Kasperek, M., Eds.; Springer: Berlin/Heidelberg, Gernamy, 2018; Volume 11103, pp. 350–362.

29. Amin, T.; Chikalov, I.; Moshkov, M.; Zielosko, B. Relationships Between Length and Coverage of Decision Rules. *Fundam. Inform.* **2014**, *129*, 1–13. [CrossRef]

30. Stańczyk, U.; Zielosko, B. Heuristic-based feature selection for rough set approach. *Int. J. Approx. Reason.* **2020**, *125*, 187–202. [CrossRef]

31. Zielosko, B.; Żabiński, K. Optimization of Decision Rules Relative to Length Based on Modified Dynamic Programming Approach. In *Advances in Feature Selection for Data and Pattern Recognition*; Intelligent Systems Reference Library; Stańczyk, U., Zielosko, B., Jain, L.C., Eds.; Springer: Berlin/Heidelberg, Gernamy, 2018; Volume 138, pp. 73–93.

32. Shang, Y. A combinatorial necessary and sufficient condition for cluster consensus. *Neurocomputing* **2016**, *216*, 611–616. [CrossRef]

33. Tan, P.; Steinbach, M.; Karpatne, A.; Kumar, V. *Introduction to Data Mining*; Pearson: London, UK, 2019.

34. Świeboda, W.; Nguyen, H.S. Rough Set Methods for Large and Spare Data in EAV Format. In Proceedings of the 2012 IEEE RIVF International Conference on Computing Communication Technologies, Research, Innovation, and Vision for the Future, Ho Chi Minh City, Vietnam, 27 February–1 March 2012; pp. 1–6.

35. Breiman, L.; Friedman, J.H.; Olshen, R.A.; Stone, C.J. *Classification and Regression Trees*; Chapman and Hall/CRC: Boca Raton, FL, USA, 1984.

36. Agrawal, R.; Srikant, R. Fast algorithms for mining association rules in large databases. In *VLDB*; Bocca, J.B., Jarke, M., Zaniolo, C., Eds.; Morgan Kaufmann: San Francisco, CA, USA, 1994; pp. 487–499.

37. Kowalski, M.; Stawicki, S. SQL-Based Heuristics for Selected KDD Tasks over Large Data Sets. In Proceedings of the Federated Conference on Computer Science and Information Systems, Wrocław, Poland, 9–12 September 2012; pp. 303–310.

38. Sarawagi, S.; Thomas, S.; Agrawal, R. Integrating Association Rule Mining with Relational Database Systems: Alternatives and Implications. *Data Min. Knowl. Discov.* **2000**, *4*, 89–125. [CrossRef]

39. Ślęzak, D. Rough Sets and Bayes Factor. In *Transactions on Rough Sets III*; Peters, J.F., Skowron, A., Eds.; Springer: Berlin/Heidelberg, Germany, 2005; pp. 202–229.

40. Mitchell, T.M. *Machine Learning*; McGraw-Hill: New York, NY, USA, 1997.

41. Zielosko, B.; Stańczyk, U.; Żabiński, K. Ranking of attributes—Comparative study based on data from stylometric domain. In *Knowledge-Based and Intelligent Information & Engineering Systems: Proceedings of the 26th International Conference KES-2022, Verona, Italy, 7–9 September 2022*; Cristani, M., Toro, C., Zanni-Merk, C., Howlett, R.J., Jain, L.C., Eds.; Elsevier: Verona, Italy, 2022; Volume 207, pp. 2737–2746.

42. Dua, D.; Graff, C. UCI Machine Learning Repository, 2017. University of California, Irvine, School of Information and Computer Sciences. Available online: http://archive.ics.uci.edu/ml (accessed on 23 March 2022).

Disclaimer/Publisher's Note: The statements, opinions and data contained in all publications are solely those of the individual author(s) and contributor(s) and not of MDPI and/or the editor(s). MDPI and/or the editor(s) disclaim responsibility for any injury to people or property resulting from any ideas, methods, instructions or products referred to in the content.

Article

Processing Real-Life Recordings of Facial Expressions of Polish Sign Language Using Action Units

Anna Irasiak [1], Jan Kozak [2,*], Adam Piasecki [3] and Tomasz Stęclik [3]

[1] Deartament of Pedagogy, Jan Dlugosz University in Czestochowa, Al. Armii Krajowej 13/15, 42-200 Czestochowa, Poland
[2] Department of Machine Learning, University of Economics in Katowice, 1 Maja 50, 40-287 Katowice, Poland
[3] Łukasiewicz Research Network—Institute of Innovative Technologies EMAG, Leopolda 31, 40-189 Katowice, Poland
* Correspondence: jan.kozak@ue.katowice.pl

Abstract: Automatic translation between the national language and sign language is a complex process similar to translation between two different foreign languages. A very important aspect is the precision of not only manual gestures but also facial expressions, which are extremely important in the overall context of a sentence. In this article, we present the problem of including facial expressions in the automation of Polish-to-Polish Sign Language (PJM) translation—this is part of an ongoing project related to a comprehensive solution allowing for the animation of manual gestures, body movements and facial expressions. Our approach explores the possibility of using action unit (AU) recognition in the automatic annotation of recordings, which in the subsequent steps will be used to train machine learning models. This paper aims to evaluate entropy in real-life translation recordings and analyze the data associated with the detected action units. Our approach has been subjected to evaluation by experts related to Polish Sign Language, and the results obtained allow for the development of further work related to automatic translation into Polish Sign Language.

Keywords: action units; automatic translation; sign language; entropy of real data

Citation: Irasiak, A.; Kozak, J.; Piasecki, A.; Stęclik, T. Processing Real-Life Recordings of Facial Expressions of Polish Sign Language Using Action Units. *Entropy* **2023**, *25*, 120. https://doi.org/10.3390/e25010120

Academic Editors: Amelia Carolina Sparavigna and Gholamreza Anbarjafari

Received: 16 November 2022
Revised: 30 November 2022
Accepted: 30 December 2022
Published: 6 January 2023

Copyright: © 2023 by the authors. Licensee MDPI, Basel, Switzerland. This article is an open access article distributed under the terms and conditions of the Creative Commons Attribution (CC BY) license (https://creativecommons.org/licenses/by/4.0/).

1. Introduction

Legislation currently in force in the European Union and around the world requires that people with disabilities be treated equally and be provided with unrestricted communication and access to information [1]. In the Polish context, two documents have been enacted over the past few years that specifically regulate information and communication accessibility, and digital accessibility for people with disabilities, including Deaf people [2,3]. The documents mentioned above provided the impulse to search for technological solutions to remove or minimize existing barriers in this area.

It is worth mentioning that, for many Deaf people, a sign language is the one first and dominant in everyday life while the national language is the second one and the level of personal proficiency in it varies [4]. In addition, the lack of a universally-valid written form of a sign language promotes the use of digital technologies and, at the same time, overcomes the barrier of this lack. It allows Deaf people to communicate at a distance that just a few decades ago was impossible or greatly hindered. A number of studies about a sign language recognition, generation and translation are currently underway. Their purpose is to help break down barriers for a sign language users in everyday life. In this regard, the topics of an ongoing research generally concern the translation of a national language into a sign language as input using text, sound, or image [5,6]. There is also emerging research into reverse translation, an example of which is a solution described in [7]; if that is applied the system can recognise sign language poses and translate through avatars in the form of talking faces. A lot of work is also focused on developing bidirectional communication capabilities by creating solutions that translate spoken languages into sign languages and can also recognise sign languages as in the works of [8–10]. In addition, in the aspect of Polish Sign Language, research work has been conducted to facilitate the communication of a Deaf person who communicates using a sign language with a person who does not know such a way of conversation [11–15].

Many of the works indicate the great importance of non-manual components in sign communication. This language does not rely only on manual gestures but also on facial expressions and other non-manual markers. It poses a major challenge to researchers working on the topic of sign language analysis and synthesis. The examples of proposed solutions for dealing with existing difficulties in designing facial expression recognition systems are provided by research into Japanese and Brazilian Sign Language [16,17].

In addition, many research projects are being conducted in the area of sign language synthesis and developing solutions that practically use developed systems, especially for the development of signing avatars, which is also the subject of a project we are currently conducting. In [18], the author indicates three interrelated threads related to the best way to portray the linguistic and paralinguistic information expressed on a signer's face. First, it is a linguistic approach to facial expressions, and because of including that an avatar must be required to communicate intelligibly; second, computer graphics, which should provide the right tools and technologies. A third theme addresses the topic of sign language representation systems from the point of view of their ability to represent non-manual signs and facial expressions. Non-manual signs, facial expressions, and the generation of synthetic emotions have also been addressed in papers [19–23]. Those articles also describe efforts to improve the quality, realism, and facial expression in sign language animation.

In Poland, additional potential for research in the above-mentioned areas is provided by extensive corpus-based research into Polish Sign Language conducted at the University of Warsaw by the Section for Sign Linguistics [24]. In the project of corpus research, for a period of 10 years, approximately 565 h of frontal-view recordings of individual signers have been collected. What is important is that these elicited recordings were obtained from 150 Deaf PJM signers from all over the country, and the group of informants included people of various ages, places of origin, and gender in equal proportions. On the one hand, the corpus thus developed provides an invaluable source of foundational data for use in ongoing research into the recognition of facial expressions during the broadcast of a message in Polish Sign Language. On the other hand, the development of automatic systems for recognizing faces in footage will enable linguistic research in new areas and will greatly speed up the search for and selection of non-manual data.

All research efforts related to the development of tools for the recognition of sign language (both sign language gestures and non-manual components) and the proper reproduction of this language in the form of images in motion aim to develop fully-fledged automatic sign language technologies and to enable free communication between hearing and Deaf people.

Digital solutions for sign communication and translation are being developed for the fields of medicine [25], security and transportation [26,27] and public administration [28]. Many efforts have also been made in the field of education. Among others, a Turkish project was described in [29], in which the benefits of using a 3D Avatar in the educational process of Deaf children were presented. For the purpose of the experiment, an avatar was created and a test was performed using it to compare the educational effectiveness of the avatar with text-based educational tools. The results indicate that avatar-based tutoring was more effective in assessing the child's knowledge of certain words in a sign language. 3D avatars are also being used to teach specific electrical engineering concepts in Portuguese Sign Language (LIBRAS) [30], to present content from a Mexican history textbook for elementary 4th grade in Mexican Sign Language [31] or to create digital math educational materials in American Sign Language and Arabic Sign Language [32,33]. Thanks to automated tools, it is also possible to learn the basics of sign language (ASL) on one's own. A computer system has been developed in which, based on a neural network, it is possible to classify fingerspelling alphabet letters recorded with a webcam [34].

In the area of education, one can also consider the potential of using different types of dictionaries such as Arabic [35] or Indian [36] and automatic translators such as Paula, the Avatar of English Sign Language, being developed at DePaul University [37].

The technological solutions currently under development must address the problem comprehensively, considering the current state of knowledge about Deaf communities and culture, their needs and experiences, and actively including Deaf sign language users in the research. It is also necessary to be aware of the complexity of the construction and functioning of sign languages and to consider this fact in developing related technological tools. A review of the presented publications showed that researchers are aware of the existing problems and are making every effort to deal with them.

The presented solution can be applied, among other things, in linguistics (corpus research, annotation, construction of sign language dictionaries), didactics (teaching of mimic expression during sign language communication and verification of its correctness), comparative social research (search and comparison, mimic characteristics in given sign communities), or creation of technical solutions for sign language visualization.

Therefore, in this work, we propose the use of action units as support units for the automatic annotation of facial expressions in the sign language translation process. The main aim of this work is to evaluate entropy in real translation recordings and analyze the data associated with the detected action units. This is an important contribution because it is not possible to annotate every frame of the recording. Our work, among other things, is an analytical contribution, allowing us to check the loss of information in real recordings of sign language signers labelled with the use of action units. This paper is also concerned with the analysis of the relationship between the different action units in real data sets—recordings of a sign language signer. We also present the whole process carried out from recording to the annotation of action units. Our discussions are supported by the expertise of Polish Sign Language experts and their indications of the possible application of the detected action units in a larger automatic sign language translation project.

The article is organized as follows: In the next section, we present the theoretical background of the research. The literature review is discussed as well. The third section presents our research methodology. The fourth section includes the numerical experiments and the generated results. The discussion is included after the numerical experiments. The two last sections include a short summary and details of future research.

2. Background

The research described in the paper has been undertaken within the *Avatar2PJM* (Project: Framework of an automatic translator into the Polish Sign Language using the avatar mechanism, The National Centre for Research and Development, GOSPOSTRATEG-IV/0002/2020) project. The goal of the project is to develop a framework that would allow to translate of utterances in the Polish language to Polish Sign Language using an avatar and artificial intelligence methods. The innovative character of the solution lies in consideration of emotions and non-verbal elements of the utterance in the visualization of gestures. The basis for launching the research is the need to develop a solution that will increase the activity of the deaf and contribute to the liquidation of social barriers that these people face. These problems can be overcome by providing the deaf with the tool to support communication in their native language (Polish Sign Language, PSL). The project has been commissioned by the Chancellery of the Prime Minister and will be carried out by the Łukasiewicz Research Centre—Institute of Innovative Technologies EMAG and the Institute of the Polish Language of the Polish Academy of Sciences.

The project assumes the development of a method to translate the Polish language into Polish Sign Language along with a mechanism to control the avatar application. The sign language avatar is a computer representation (animation) of language phenomena. Thanks to good video-recorded reference material, it is possible to animate any described utterance. That is why one of the research stages dealt with acquiring the largest possible number of video recordings of a sign language translator within a Motion Capture (MoCap) session. MoCap is a technique of recording the three-dimensional movements of an actor. It is used in computer games and imitates the natural movements of objects or people in a very realistic way in order to achieve a natural effect. In the case of sign language avatars, MoCap allows copying the signs of the sign language and increasing the comprehension of the communicated content because, from the animator's perspective, the uttered signs of the sign language consist of geometrical poses and movements.

A sign language message consists of sign language signs and different additional information because what is expressed physically results from coexisting linguistic and non-linguistic processes. While producing computer-generated animations, the emotional context of the utterance is taken into account, along with such phenomena as proper lips movements or voiceless speech, which are performed during sign language messages. What is particularly important here is the sign language interpreter's facial expressions and the information they convey. Such elements are also significant in the context of data indispensable to developing the translation module. The material acquired during the MoCap session is used to feed the animation module and acquire a set of input data for the translation module based on machine learning methods. To make it possible, it is necessary to submit the set of recordings to the annotation process. Annotation describes particular elements of sing-language signs in particular time intervals of the sign duration. In addition, the process describes singular signs, dictionary-based interpretations (lemmas, lexemes), and information, e.g., about non-manual elements of a sign. Because one of the key annotated elements is the sign language interpreter's face, and the process is very time-consuming, the researchers attempted to examine the possibility of automatic recognition of the translator's mimic poses. Such automatic annotation would significantly improve and speed up the annotator's work. This paper describes the partial results of research undertaken in this domain.

One of the expected effects of the project is pilot testing in selected on-line information services run by the administration. A common use of automatic translation mechanisms in public-service internet systems will be a constructive step to improve the digital availability of public administration. In addition, the project team will examine the career potential of the deaf as well as their satisfaction with contacting public administration before and after the application of the virtual translator. This will allow determining social and economic barriers faced by the deaf while contacting the administration and moving on the job market. The career potential of the deaf will be analyzed and data acquisition methods will be determined to achieve the highest possible professional activation result. The results of the project will allow to permanently liquidate barriers encountered by Polish Sign Language users.

2.1. Polish Sign Language—The Role of a Facial Expression

Polish Sign Language (pl. polski język migowy, PJM), like other sign languages, is an autonomous, standard, and fully-fledged natural language, which means it constitutes a two-class system of conventional characters for universal communication. The physical nature of a sign language text, which makes it different from phonic languages, is not vocal-auditory but visual-spatial [38].

An utterance in PJM is composed of manual and non-manual signals. What is essential, the former one play both expressive and linguistic functions in natural sign languages. The expressive function consists of presenting, by articulators such as eyes, eyebrows, mouth, head, and shoulders, various emotional states while performing certain signs (e.g., expressing sadness while showing the SAD sign). However, more important is the linguistic function of non-manual signals, which proves the autonomy of a sign language and allows to distinguish PJM from the phonic Polish language [39]. The linguistic function of non-manual components in PJM consists of several aspects: (1) their phonological construction within signed words, (2) their lexical functioning as independent non-manual signs, (3) their grammatical functioning at the morphological level, and (4) their syntactic functioning in distinguishing signed sentences [40,41]. In the next sections of this chapter, these four aspects of the linguistic function of non-manual elements in PJM will be presented.

2.1.1. Non-Manual Component within Sign Words

At the phonological level, non-manual signals may be obligatorily embedded as an additional element in some signs. It is important to note that the sign of sign language consists of three sublexical parameters that are the sign equivalents of phonemes (cheremes). They are hand configuration, the location of sign articulation, and the movement performed during articulation [42]. Robert Battison [43] added two more properties: palm orientation and non-manual elements to this classification.

Considering the above-mentioned elements, three types of signs are distinguished: manual, multi-modal, and non-manual. Manual signs have only three basic parameters: hand configuration, location, and movement. Apart from these three parameters, multi-modal signs contain the fourth one—a non-manual signal. Non-manual signs are articulated using only non-manual signals without the use of hands and, therefore, can be self-realized at the lexical level and will be discussed in the next subsection [44].

Non-manual features in sign languages include facial actions and expressions, head movements and positions, shoulders, and the position of an upper body as a whole. The area of the body below the waist (hips, legs, and feet) very rarely serves as an active articulator. In our research, attention will be focused on the aspects of a facial expression which involve the eyebrows, eyelids, eye gaze, cheeks, nose, lips, and jaw. These parts of the face can assume the role of independent articulators or can be used simultaneously with the head, shoulders, and the whole body or with manual components [45,46]. Each of the indicated parts of the face can make appropriate movements, which, for Polish Sign Language, were included in the classification of Piotr Tomaszewski, presented below in the form of a diagram (Figure 1) (In our research and further in this paper, we concentrate only on facial expression, and we overlooked other "places" such as head, shoulders, and body—torso—movements.) In it, no other parts of the face, such as the forehead, are marked because they are redundant. For example, due to the anatomical structure, raising the eyebrow always produces a frown of the forehead effect so this latter part of the face is predictable. Each part of the face described as a category of "a place" feature has a set of ways of expressing opinions, feelings, and meanings, defined as "a setting" feature. The most numerous group of settings is boasted of by the mouths, which, both in PJM and in other sign languages, also play an articulatory function with the use of certain signs [41].

Therefore, various mouth actions and their combinations are classified into at least two clearly identifiable types of mouth patterns. These are "mouthings", which are said to be derived from the surrounding spoken language, and "mouth gestures", formed within a sign language and, thus, inherent to it [47,48]. Furthermore, different mouth configurations form the basis to create "minimal pairs", which means that these kinds of signals can distinguish different sign words, e.g., pairs of words such as (BIEGLE "fluent" i.e., in a sentence "Ja migam biegle") and (SZKODA "too bad" i.e., in a sentence "Nie udało mi się wygrać, szkoda") are distinguished by different mouth configurations: round vs. stretched [40].

2.1.2. Independent Non-Manual Sign Words

Non-manual signals, e.g., independent non-manual signs, can be used at the lexical level. They refer to signs which do not require the use of the hands at all but employ, in the articulation, only non-manual elements such as specific facial expressions or head movements. An example of a sign using only a facial expression is a sign (NMS: ZGADZA_SIE "That's right…"), which is articulated by wrinkling the nose, and the examples of signs using only head movements are signs (TAK "yes") and (NIE "no") [40].

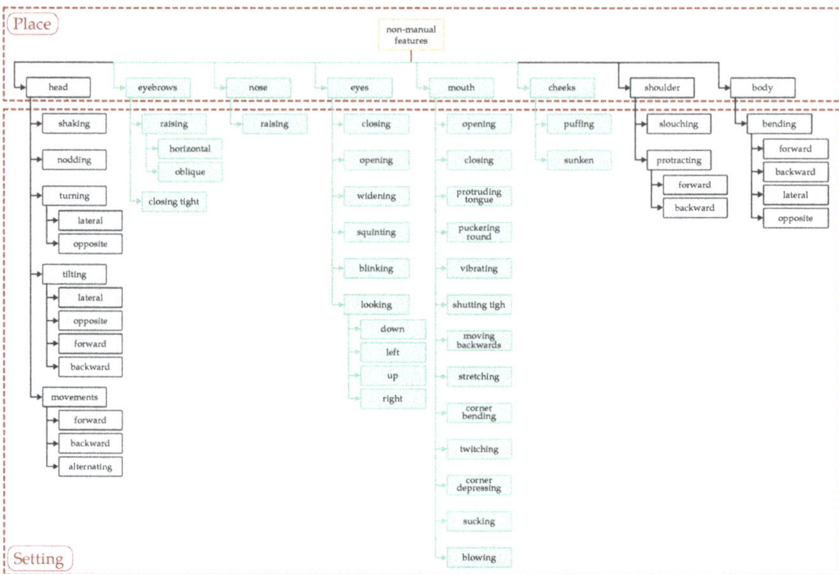

Figure 1. Place and settings features of non-manual components in PJM signs.

2.1.3. Role of Non-Manual Signals in Modifying Some Sign Words

Non-manual signals can also be superimposed on a single sign word or a sequence of words, fulfilling a grammatical function, e.g., to form comparative or superlative adjective forms. Non-manual factors may constitute a meaning-modifying (enriching) feature. In order to shorten the statement, the procedure of a simultaneous use of a feature is used, during which, instead of the sequence of characters ("nominal" + "attributive"), only the sign of the object with a non-manually assigned feature is used (e.g., sign DZIEWCZYNA "girl", NMS: a neutral facial expression, and sign ŁADNA_DZIEWCZYNA "pretty girl", NMS: smiling face, eyes squinting). A facial expression also allows to express the intensity of a feature. It is a way of intensifying or weakening the meaning of adjectives (e.g., distinguishes signs ŁADNY "pretty", NMS: calm, smiling face, and BARDZO_ŁADNY "very pretty", NMS: smiling face, squinting eyes, a slight head movement to the left), and it is also a way of modifying the meanings of verbs (e.g., CHCIEĆ "want", NMS: a slight nod, lips tightened, and BARDZO_CHCIEĆ "want very much", NMS: firmly nod, lips tighten, looking up). The handshape of these lexemes remains the same, the only determinant of intensity is the non-manual component [46,49].

2.1.4. Syntactic Functions of Non-Manual Signals in Distinguishing Signed Sentences

Non-manual signals in the syntactic function are essential, especially when distinguishing sentences. Due to the possibility of using non-manual grammatical components, different kinds of sentences with different clauses are formed. Below are a few kinds of sentences and examples of their non-manual components.

-   yes/no questions—PJM they can be answered simply by confirming or denying the entire sentence. They are marked in sign language by a slight forward tilt of the head and raising the eyebrows during the whole sentence. That is the only non-manual form that distinguishes the corresponding statement from the question.
-   "wh" questions—PJM the group of question words used for this purpose includes who, what, where, when, why, what kind of, how many and which. In this kind of utterances, the grammatical non-manual component consists of lowered eyebrows and squinted eyes that occur either over the entire wh-question or solely over a wh-phrase that has moved to a sentence-final position.

- negative sentences—a kind of sentence into which signs indicating negation are added. As a non-manual component, there is a relatively slow side-to-side head shake that co-occurs with a manual sign of negation, and the eyes may squint or close.
- conditional sentences—they contain subordinate sentences that express the conditions of implementing proposals included in superordinate clauses. In sign languages, subordinate sentences are formed by raised eyebrows, wide eyes, head forward (or back) and tilted to the side, followed by a pause after which the eyebrows and head return to neutral position [39,40,50].

A facial expression, which is a grammatical exponent, imitates the natural facial expressions accompanying the formation of the aforementioned types of sentences. It is also possible to create sign sentences that combine some of the above sentence types [46,49].

2.2. Action Units

Action units (AUs) define facial muscle activity so that it is possible to indicate activity that affects facial expressions (facial appearance). The origins of action units are related to the facial coding system proposed in 1978 in the work [51]. In this system, all visually perceivable facial expressions are described. The mentioned expressions are divided according to muscle movements in the following steps.

In the following years, attempts were made to detect units of action automatically, but mainly these were approaches related to a specific expression (happiness, sadness, etc.). However, at the beginning of the 19th century, with the increase in computational capabilities and thus the development of machine learning algorithms and computer vision, work was taken on more complex AU [52].

With the development of more machine learning methods, including deep learning, automatic AU detection became more and more precise [53–55]. This makes it possible to apply models learned to detect AU to real-world problems (including real-time detection). In general, however, many of the tools only detect a limited number of AUs. In this way, the responsible action units shown in Figure 2 are most often recognised. In this case, we have additionally subdivided the detected action units due to aspects related to sign language (this is a subdivision for the execution of a movement with the corresponding part of the face).

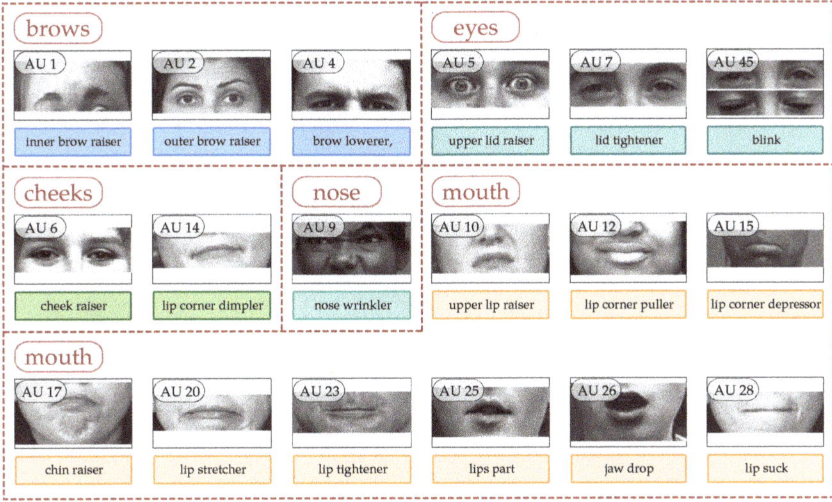

Figure 2. Action units with sign language-compatible facial parts—images from [56].

In this work, we aim to analyze the application of AU to the real-world problem of automatic facial expression annotation in the sign language translation process. Therefore, we focus on using existing algorithms for AU detection [57] and rely only on the AU detected by this tool, as described above.

3. Research Methodology

This article aims to analyze action units in the context of their use in automatic text-to-sign language translation (using a specially prepared avatar that signs appropriate sequences in sign language and performs body movements and facial expressions). Our approach is one of avatar control and concerns facial expressions, so we focus only on the face of the signer.

For the application of action units (AUs), analyses related to the entropy of specific AUs depending on the recording, as well as correlation, were applied. The aim is to create rules to correlate specific AU sequences with facial expression elements important for sign language annotation.

In this work, we analyze the real recordings related to the *Avatar2PJM* project described in Section 2. Consequently, our work is also related to image processing. Figure 3a shows one frame of the actual recording. In our case, we only consider facial expressions, so the identification of the face itself and the removal of the background must be made (Figure 3b). Only the image prepared in this way is used to find action units.

The next step of the analysis carried out is AU detection. At this stage of the work, we are using an off-the-shelf and tested tool for researchers working on computer vision and machine learning for AU analysis—OpenFace 2.0 [57]. The approach used gives better accuracy for detecting face landmarks and face action units than the OpenFace [58] tool. The method is based on linear support vector machine learning. However, it has been shown in the work [57] that the results obtained are similar to methods based on deep learning. It is worth noting that, in the presented solution, AU is detected in two ways. The first one (called presence) is about detecting whether AU is found in a given frame of the video (0—not found, 1—found), while the second one (called intensity) determines the intensity of occurrence of a given AU (from 0.0 to 1.0, where the closer to 0.0, the lower the intensity and vice versa). It should be noted that both models were trained on different learning datasets (described in more detail in [57]), so different results are possible.

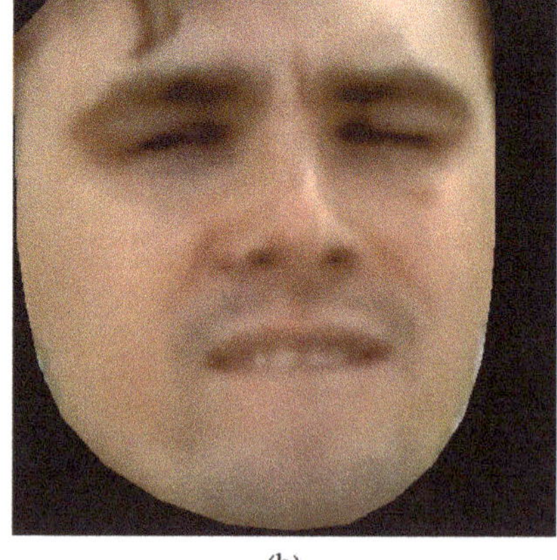

(a)   (b)

Figure 3. The real data used in this work: (**a**) Image before face extraction. (**b**) Image after face extraction and background removal.

The use of OpenFace 2.0 allows for real-time analysis – which is essential for our project. Additionally, it is based on approaches [59–61], but for our work, the most important issue is the quality of AU detection. In this case, the authors in the [57] paper showed that the solution we used shows better results (reported as Pearson correlation coefficient) than other popular solutions based on, among others, the convolutional neural network.

The result of this part of the work was to obtain a table consisting of columns describing each frame of the recording with 40 features related to, among others: frame number, time stamp, and AUs. Depending on the method, these are different AUs for presence: 1, 2, 4, 5, 6, 7, 9, 10, 12, 14, 15, 17, 20, 23, 25, 26, 28, and 45; and for intensity: 1, 2, 4, 5, 6, 7, 9, 10, 12, 14, 15, 17, 20, 23, 25, 26 and 45. As our work is related to the analysis of the applicability of AUs in automatic sign language translation, the AU results have been grouped, as suggested by Polish Sign Language practitioners, due to the part of the face. They are matched with the corresponding recording frame, as shown in Figure 4.

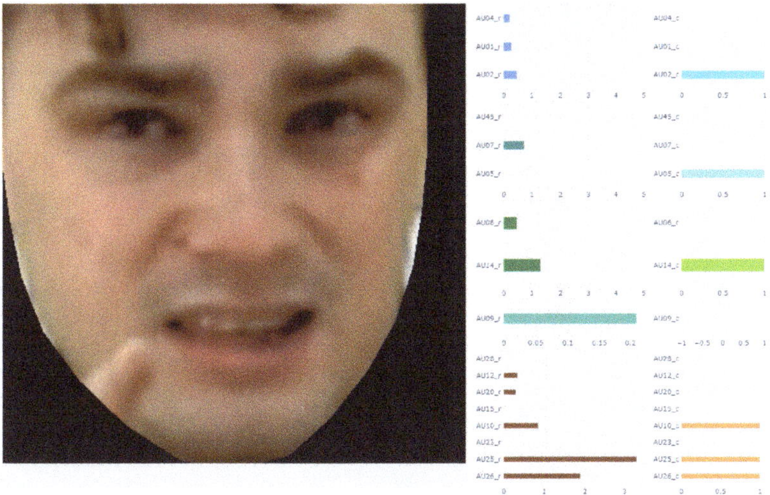

Figure 4. Matching AU with the corresponding recording frame.

It should be noted that the recordings used in our analyses consisted of 25 frames per 1 s. Labeling facial expressions consistent with Polish Sign Language proved to be impossible based on a single frame of the recording. Therefore, in the next step, it was proposed to prepare average AU values for one second of recording—this gave 25 frames in one comparison. The results of this approach are presented in Figure 5.

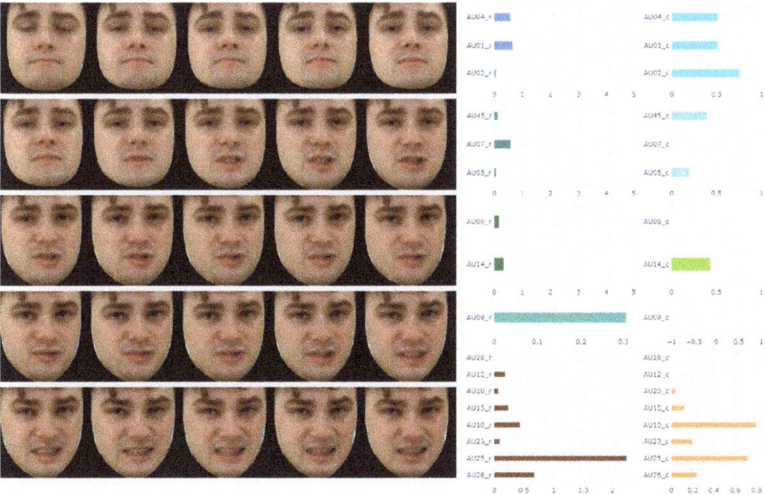

Figure 5. Matching AU with the corresponding one-second (5 frames) of the recording.

The Polish Sign Language experts showed that the facial expression changes too often for one second of recording. This is also evident in Figure 5. Therefore, analyses were also carried out for a smaller number of frame of the recordings. As a result, finally, the average AU values for 5 consecutive recording frames were determined. The final result of the data preparation is presented in Figure 6.

Figure 6. Matching AU with the corresponding 5 frames of the recording.

4. Results

The experiments aimed to test the relationship between AUs in real sign language recordings. In order to do this, recordings made for the *Avatar2PJM* project were analyzed—a total of several hundred megabytes of recordings (25 frames per second). During our analysis, a number of possibilities were tested—we also made comparisons between the recordings of individual utterances in relation to the material as a whole. This was all done to enable us to find correlations between AU and Polish Sign Language signs in the future—during annotations for automatic translation.

All recordings were processed according to the steps described in Section 3. Consequently, tables were created for each set of recordings describing:

- each frame of the recording;
- the average of 5 consecutive frames of the recording (with the frames averaged offset, so first frames [1, 2, 3, 4 and 5] then [2, 3, 4, 5 and 6] etc. up to $[n-4, n-3, n-2, n-1$ and $n]$, where n is the number of all frames in the recording;
- the average of 25 consecutive frames in the recording (analogous to the 5-frame approach).

For the results presented in this work, tables were created containing respectively: 20,211, 20,127 and 19,665 rows. This is the result of combining several different recordings, hence offsets reduce the number of frames (and thus indirectly the rows in the table). In each row, AU-related labels have been extracted—depending on the method chosen (presence or intensity).

4.1. Degree of Variability of the Data

As a result of our analyses, we used entropy (see Equation (1), where we are dealing only with binary AU values, where p is the probability of one of these values occurring—the probability that an AU occurs or does not occur in the recording) as a measure of variability in the data. To do this, we determined the entropy value for each of the designated AUs over all the recordings. In this way, we are able to decide on how much the value of each AU changes—an alternative to this is the histogram. Still, by using entropy, we can indicate a gain or loss in variability, depending on our approach (single frame, 5 frames and 25 frames analysis).

$$E(p) = -p \log_2 p - (1 - p) \log_2(1 - p). \tag{1}$$

It should be noted that for the present approach, determining the entropy value was not a problem—there are only two values of each feature (of each AU) in these data. In the case of the intensity approach, on the other hand, we proposed to normalize the data so that all values less than or equal to 0.5 were labeled, 0, and greater than 0.5 were labeled 1. Ultimately, in both cases, there are two possible values of the feature so that when the entropy is 0.0, the data are maximally ordered (there is only one AU value). When the entropy is 1.0, the data are maximally unordered (i.e., an AU is found as many times as it is not found).

Tables 1 and 2 record the entropy values for each AU according to analyzing each frame separately, averaging over 5 frames and averaging over 25 frames (analogous to the description in Section 3). As can be seen, for the present approach, there are essentially no differences in the entropy of each AU between the situation when each frame is analyzed separately and five frames in sequence. Thus, it can be said that averaging the AUs with an offset of five frames, i.e., $\frac{1}{5}$ s, allows for a better analysis of the prepared data (see Figure 6) by the expert while maintaining an identical distribution of information.

The situation is slightly different when averaging 25 frames, i.e., analyzing the entire second. In this case, for as many as 10 AUs out of 18, there is a change of more than 0.04, with a change of more than 0.1 once (in the case of AU_{12}, the lip corner puller). In only two cases is the change close to 0 (this is the case for AU_4 and AU_5, i.e., information related to the elevation or lowering of the eyebrows). This indicates a change in the information in the data for the vast majority of AUs.

Similar correlations are found for the intensity approach, although in this case differences also appear when changing from 1 frame to 5 frames. This has to do with normalization; even so, the difference does not exceed 0.025 (in the case of AU_5, so raising the eyelids is precisely 0.0241). In total, only in 5 cases does the difference exceed 0.01.

Similarly, when comparing 1 frame with 25 frames, the differences are higher than for the present approach. In this case, only twice are they smaller than 0.02 (not once are they smaller than 0.01). In contrast, in as many as 8 cases (out of 17) the difference exceeds 0.1, where for AU_5 it is more than 0.18, and for AU_{17} (chin raise) it exceeds 0.21.

Table 1. Entropy value for each of the action units for the present approach.

Action Unit	1 Frame	5 Frames	25 Frames
AU_1	0.7351	0.7354	0.6723
AU_2	0.9407	0.9398	0.9231
AU_4	0.9996	0.9997	0.9999
AU_5	0.9984	0.9985	0.9990
AU_6	0.2660	0.2659	0.1987
AU_7	0.3489	0.3469	0.2768
AU_9	0.4081	0.4089	0.3309
AU_{10}	0.9666	0.9670	0.9617
AU_{12}	0.5575	0.5561	0.4554
AU_{14}	0.8885	0.8876	0.8420
AU_{15}	0.9057	0.9063	0.8772
AU_{17}	0.9659	0.9641	0.9338
AU_{20}	0.8956	0.8932	0.8329
AU_{23}	0.8357	0.8328	0.7647
AU_{25}	0.9172	0.9183	0.8900
AU_{26}	0.8238	0.8252	0.7472
AU_{28}	0.0684	0.0669	0.0265
AU_{45}	0.7198	0.7206	0.6368

Our research indicates that it is possible to use the approach shown in Figure 6 to support the work of the expert when investigating AU mapping rules in facial expressions used in Polish Sign Language. Therefore, in the following steps, it is possible to focus on the approach related to the analysis of five frames of recordings. In this case (and the present approach), it can be observed that for some AUs the entropy value is very high (above 0.8 and often close to 1.0), but in some cases, the entropy is low. For example, for AU_{28} (lip suction), the entropy is only 0.0669, so the repeatability is very high. It is also possible to distinguish AU_6, AU_7 and AU_9, i.e., action units related to the cheeks and nose.

Interestingly, the intensity approach is characterized by slightly different values—this has to do with the normalization of the values but also with the different ways of training for each approach. Here, the lowest of the entropies relates to AU_5 and is 0.3 (in the present approach, it was close to 1.0). It can be seen that for the intensity approach, a certain repetition of AU is noticeable in more cases.

Table 2. Entropy value for each of the action units for the intensity approach.

Action Unit	1 Frame	5 Frames	25 Frames
AU_1	0.7721	0.7649	0.7506
AU_2	0.3691	0.3581	0.3152
AU_4	0.9768	0.9765	0.9873
AU_5	0.3236	0.2995	0.1399
AU_6	0.4370	0.4264	0.3061
AU_7	0.9050	0.9073	0.9274
AU_9	0.4542	0.4327	0.3109
AU_{10}	0.8671	0.8671	0.8927
AU_{12}	0.5216	0.5001	0.4184
AU_{14}	0.6244	0.6150	0.5585
AU_{15}	0.6726	0.6562	0.6267
AU_{17}	0.9199	0.9023	0.7031
AU_{20}	0.6407	0.6231	0.5620
AU_{23}	0.5439	0.5208	0.4196
AU_{25}	0.9904	0.9833	0.8872
AU_{26}	0.9957	0.9983	0.9777
AU_{45}	0.6477	0.6419	0.5226

4.2. Correlation of Action Units

Figures 7 and 8 present correlation heat maps for both approaches using five frames. Based on expert knowledge and experiments from Section 4.1, in this section, we present only the analyses for five frames.

As we can see, there are also differences between the two approaches used in the case of correlation. In the case of the present approach, the correlation between AUs is mainly noticeable between AU_{25} and AU_{26} (mouth opening and jaw lowering—this is a well-known relationship), but the situation is more interesting for the correlation between AU_4 (eyebrow lowering) and AU_{45} (blink) and AU_6 (cheek raiser) and AU_{12} (lip corner puller). This indicates that more correlations can be found beyond the classical correlations during sign language recordings, where facial expressions are very significant and often emphasized. It is also possible to notice a considerable lack of correlation between, for example, AU_4 and AU_5 or AU_{17} and AU_{25}, i.e., opposite facial expressions. This confirms that the use of AU in the sign language translation approach may be more relevant due to the high emphasis on facial expressions by sign language speakers.

In the intensity approach, on the other hand, there is a correlation between AU in a much higher number of cases. This is a signal to experts that when using the intensity approach, AUs are more likely to occur simultaneously. It should be noted that for these analyses, the AU values were not normalized in any way. The highest correlation concerning the other AUs is shown by AU_6. A significant correlation of AU_6 is seen with respect to AU_7, AU_9, AU_{10}, AU_{12} and AU_{14}. Similarly, as with the present approach, a significant lack of correlation occurs between AU_{17} and AU_{25}.

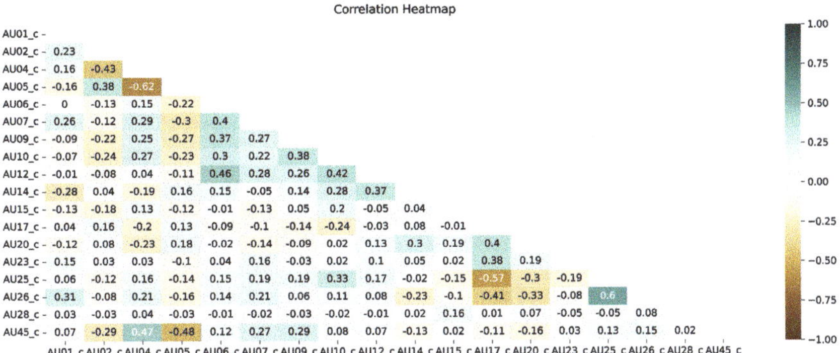

Figure 7. Correlation heat map for the presence approach and the use of an average of 5 frames.

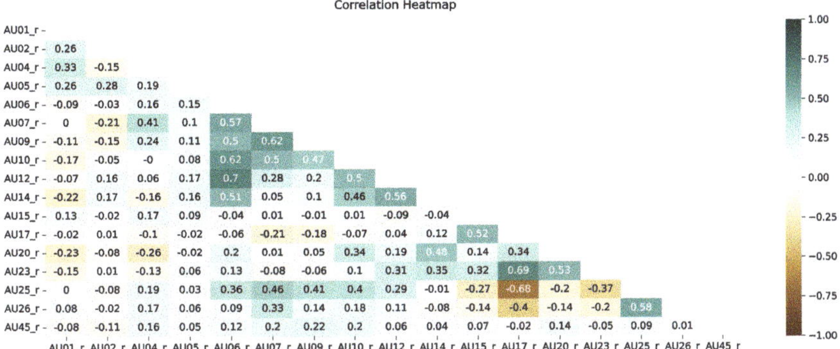

Figure 8. Correlation heat map for the intensity approach and the use of an average of 5 frames.

5. Discussion

In our work, we have seen an opportunity to use action units to mark facial expressions during sign language translation. However, to enable the work of experts, it is necessary to adapt the real data, in the form of recordings, for expert analysis. To this end, we proposed a comparison of AUs found from two different approaches with real video recordings. A comparison of the entropy of these data shows that it is possible to use five more frames of footage simultaneously (with averaged AU values). This makes the determination of facial expressions more precise and allows for a deeper analysis of the expert's knowledge of AU values.

It has also been shown that the correlation between the individual AUs in the case of recordings of sign language signers shows other relationships in addition to the classical ones. This is due to the high intensity of facial expressions and the particular emphasis of facial expressions on individual gestures. Accordingly, the use of AU labeling is justified in the context of future facial expression labeling during automatic sign language annotation.

An effective AU recognition program will be able to be successfully used in linguistics. In particular, in corpus studies of Polish Sign Language. It will make it possible, for example, to search corpus texts for the occurrence of a given facial expression or element of facial expression and to analyze this occurrence in connection with the sign signs at which the expression appears. This solution will also make it possible to build sign language dictionaries (in particular, Polish Sign Language, which is of interest to us), in which content searching can be carried out based on the indicated facial expression element or combinations of indicated features of individual facial parts forming the selected facial expression system (Search-by-video sign language dictionaries).

Using such a solution will benefit linguistic research but will also make it possible to develop and implement new methods to teach the correct reading of mimicry in PJM-signaled communication by deaf people, as well as to teach proper mimicry expression in conjunction with signaled manual communication.

Due to the implementation of the Avatar2PJM project, the team's further work on developing the tool for automatic translation into Polish Sign Language will focus on using it to automate the annotation of Polish Sign Language in ELAN. The inspiration for the work on automated annotation came from the very involved and time-consuming work of annotating a large corpus of sign language texts needed for the next stages of project work. Therefore, we began analyzing the possibility of automating this process. The planned end result of the work is the possibility of automatically generating records in ELAN with a precise indication of time intervals and annotating predefined non-manuals in them. The basis for showing non-manuals in a given time interval is the achievement of a sufficiently high AU detection rate in a given time interval of a recording with sign language content. Non-manuals annotated in ELAN would be identified by the researcher as a closed catalogue before the automatic annotation process begins as "controlled vocabulary" with respect to the individual assumptions of a given project. An example of a catalogue of non-manual names for the Polish Sign Language is Figure 1, where "place" denotes the annotated part of the signer's face (annotated in separate ELAN layers due to the potential for overlapping intervals) and "settings" are lists of possible annotated changes in the layout of that facial area.

The research carried out so far shows the potential and scope for using AU in automatic annotation, but the execution of such a tool requires further research and implementation work. The analyses performed so far have been conducted on real-life recordings, but our trials demonstrate that the tool under development will be able to successfully annotate recordings of a very different nature, such as video excerpts, found data from sign corpus, social media materials, etc. These recordings should be prepared in advance and meet the indicated criteria, e.g., regarding the quality/resolution of the recordings. Also important is its ability to be used to annotate different sign languages, as the method of annotation can be determined by the researcher himself and adapted to the non-manuals present in a given sign language. Once developed, the automatic annotation model will therefore be replicable for different sign languages.

In the model presented here, it is necessary to map AU detection in individual facial areas to the corresponding names of non-manuals specified for a given sign language. On the other hand, the tangible result of the implementation of the automatic annotation tool is the ability to quickly acquire a large number of annotated recordings in terms of the non-manuals present in them, which creates a large input database for the development of the automatic translation tool. The final result of the automatic annotation tool of sign language non-manuals is ready-made data files that are input for the automatic translation tool.

We are also aware of some limitations in the use of the described method and the need for further analysis of the possibility of using AU in recognizing facial expressions in Polish Sign Language.

In our study, we used 18 AUs responsible for recognizing muscle activity in different parts of the face. After expert analysis, it can be concluded that this number of AUs is insufficient to mark all facial expressions that are used during communication in Polish Sign Language. Especially complex seems to be the aspect of mouth actions. Hanke [62] points to an extensive list of 59 "mouth gestures" recognized in research on British, Dutch, and German Sign Language. This should be included the large variety of "mouthings" as non-manuals of sign language related to articulation in spoken language. Therefore, it will be necessary to find solutions dedicated specifically to this area of faces. On the other hand, we note that the developed method allows for the marking of crucial facial expressions, thus, providing a chance for at least partial automation of work during the translation or annotation of sign language.

6. Conclusions

This work aimed to analyze actual sign language recordings in terms of the use of action units in the automatic translation of the text into Polish Sign Language. This is part of the $Avatar2PJM$ project, in which correct translation into sign language with appropriate facial expressions is an important aspect.

In this work, we analyzed the entropy of action units in real recordings and its change when averaged—which has to do with the labeling of recordings by experts. Marking at $\frac{1}{25}$ s is impossible, but $\frac{1}{5}$ s is already a sufficient time range for experts. It was shown that the entropy does not change significantly when considering five frames of recording, allowing further work on the design.

The correlation between each action unit and the frequency of occurrence of each was also determined. This is valuable information for experts who are working on finding rules to map known action units in facial expressions required for translation into Polish Sign Language.

Future work will develop rules to label facial expressions based on detected action units. In the case of the intensive approach, we will also examine the use of certain margins when determining intensity—known from the three-way decision theory. In addition, it is worth considering the approach of using machine learning directly to find the repetition of the labeling (done by the experts) without using action units. For this purpose, however, a sufficiently large dataset should be prepared.

Author Contributions: Conceptualization, J.K. and A.I.; methodology, J.K. and A.I.; software, T.S.; validation, T.S. and J.K.; formal analysis, J.K.; investigation, A.I., J.K., A.P. and T.S.; resources, A.I., J.K., A.P. and T.S.; data curation, T.S. and J.K.; writing—original draft preparation, A.I., J.K., A.P. and T.S.; writing—review and editing, A.I., J.K., A.P. and T.S.; visualization, J.K.; supervision, J.K.; project administration, J.K. and A.P.; funding acquisition, A.P. All authors have read and agreed to the published version of the manuscript.

Funding: This research was funded by The framework of an automatic translator into the Polish Sign Language using the avatar mechanism, The National Centre for Research and Development, grant number GOSPOSTRATEG-IV/0002/2020.

Conflicts of Interest: The authors declare no conflict of interest.

References

1. Convention on the Rights of Persons with Disabilities: Resolution. UN General Assembly, A/RES/61/106. 2007. Available online: https://www.un.org/en/development/desa/population/migration/generalassembly/docs/globalcompact/A_RES_61_106.pdf (accessed on 15 November 2022).
2. Sejm of the Republic of Poland. Act on Ensuring Access for Persons with Special Needs of 19 July 2019. *J. Laws* **2019**, *2019*, 1696.
3. Sejm of the Republic of Poland. Act on Digital Accessibility of Websites and Mobile Applications of Public Entities of 4 April 2019. *J. Laws* **2019**, *2019*, 848.
4. Marschark, M.; Tang, G.; Knoors, H. Bilingualism and bilingual deaf education. In *Perspectives on Deafness*; Oxford University Press: New York, NY, USA, 2014.
5. Roelofsen, F.; Esselink, L.; Mende-Gillings, S.; De Meulder, M.; Sijm, N.; Smeijers, A. Online Evaluation of Text-to-sign Translation by Deaf End Users: Some Methodological Recommendations (short paper). In Proceedings of the 1st International Workshop on Automatic Translation for Signed and Spoken Languages (AT4SSL), Virtual, 20 August 2021; pp. 82–87.
6. San-Segundo, R.; Barra, R.; D'Haro, L.; Montero, J.M.; Córdoba, R.; Ferreiros, J. A spanish speech to sign language translation system for assisting deaf-mute people. In Proceedings of the Ninth International Conference on Spoken Language Processing, Pittsburgh, PA, USA, 17–21 September 2006.
7. Mazumder, S.; Mukhopadhyay, R.; Namboodiri, V.P.; Jawahar, C. Translating sign language videos to talking faces. In Proceedings of the Twelfth Indian Conference on Computer Vision, Graphics and Image Processing, Jodhpur, India, 19–22 December 2021; pp. 1–10.
8. Cormier, K.; Fox, N.; Woll, B.; Zisserman, A.; Camgöz, N.C.; Bowden, R. Extol: Automatic recognition of british sign language using the bsl corpus. In Proceedings of the 6th Workshop on Sign Language Translation and Avatar Technology (SLTAT) 2019, Hamburg, Germany, 29 September 2019.

9. Saggion, H.; Shterionov, D.; Labaka, G.; Van de Cruys, T.; Vandeghinste, V.; Blat, J. SignON: Bridging the gap between sign and spoken languages. In Proceedings of the Annual Conference of the Spanish Association for Natural Language Processing: Projects and Demonstrations (SEPLN-PD 2021) Co-Located with the Conference of the Spanish Society for Natural Language Processing (SEPLN 2021), Málaga, Spain, 21–24 September 2021; pp. 21–25.
10. Xiao, Q.; Qin, M.; Yin, Y. Skeleton-based Chinese sign language recognition and generation for bidirectional communication between deaf and hearing people. *Neural Netw.* **2020**, *125*, 41–55. [CrossRef] [PubMed]
11. Oszust, M.; Wysocki, M. Polish sign language words recognition with Kinect. In Proceedings of the 2013 6th International Conference on Human System Interactions (HSI), Sopot, Poland, 6–8 June 2013; pp. 219–226.
12. Romaniuk, J.; Suszczańska, N.; Szmal, P. Thel, a language for utterance generation in the thetos system. In *Proceedings of the Language and Technology Conference*; Springer: Berlin/Heidelberg, Germany, 2011; pp. 129–140.
13. Warchoł, D.; Kapuściński, T.; Wysocki, M. Recognition of fingerspelling sequences in polish sign language using point clouds obtained from depth images. *Sensors* **2019**, *19*, 1078. [CrossRef] [PubMed]
14. Kapuscinski, T.; Wysocki, M. Recognition of signed expressions in an experimental system supporting deaf clients in the city office. *Sensors* **2020**, *20*, 2190. [CrossRef]
15. Kowalewska, N.; Łagodziński, P.; Grzegorzek, M. Electromyography Based Translator of the Polish Sign Language. In *Proceedings of the International Conference on Information Technologies in Biomedicine*; Springer: Berlin/Heidelberg, Germany, 2019; pp. 93–102.
16. da Silva, E.P.; Costa, P.D.P.; Kumada, K.M.O.; De Martino, J.M. Facial action unit detection methodology with application in Brazilian sign language recognition. *Pattern Anal. Appl.* **2022**, *25*, 549–565. [CrossRef]
17. Yabunaka, K.; Mori, Y.; Toyonaga, M. Facial expression sequence recognition for a japanese sign language training system. In Proceedings of the 2018 Joint 10th International Conference on Soft Computing and Intelligent Systems (SCIS) and 19th International Symposium on Advanced Intelligent Systems (ISIS), Toyama, Japan, 5–8 December 2018; pp. 1348–1353.
18. Wolfe, R.; McDonald, J.; Johnson, R.; Moncrief, R.; Alexander, A.; Sturr, B.; Klinghoffer, S.; Conneely, F.; Saenz, M.; Choudhry, S. State of the Art and Future Challenges of the Portrayal of Facial Nonmanual Signals by Signing Avatar. In *Proceedings of the International Conference on Human-Computer Interaction*; Springer: Berlin/Heidelberg, Germany, 2021; pp. 639–655.
19. Gonçalves, D.A.; Baranauskas, M.C.C.; dos Reis, J.C.; Todt, E. Facial Expressions Animation in Sign Language based on Spatio-temporal Centroid. *Proc. ICEIS* **2020**, *2*, 463–475.
20. Huenerfauth, M. Learning to generate understandable animations of American sign language. In Proceedings of the 2nd Annual Effective Access Technologies Conference, Rochester Institute of Technology, Rochester, NY, USA, 15 June 2014.
21. Johnson, R. Towards enhanced visual clarity of sign language avatars through recreation of fine facial detail. *Mach. Transl.* **2021**, *35*, 431–445. [CrossRef]
22. Kacorri, H. TR-2015001: A Survey and Critique of Facial Expression Synthesis in Sign Language Animation. *CUNY Academic Works*. 2015. Available online: https://academicworks.cuny.edu/gc_cs_tr/403 (accessed on 15 November 2022)
23. Smith, R.G.; Nolan, B. Emotional facial expressions in synthesised sign language avatars: A manual evaluation. *Univers. Access Inf. Soc.* **2016**, *15*, 567–576. [CrossRef]
24. Kuder, A.; Wójcicka, J.; Mostowski, P.; Rutkowski, P. Open Repository of the Polish Sign Language Corpus: Publication Project of the Polish Sign Language Corpus. In Proceedings of the LREC2022 10th Workshop on the Representation and Processing of Sign Languages: Multilingual Sign Language Resources, Palais du Pharo, France, 20–25 June 2022; pp. 118–123.
25. De Maria Marchiano, R.; Di Sante, G.; Piro, G.; Carbone, C.; Tortora, G.; Boldrini, L.; Pietragalla, A.; Daniele, G.; Tredicine, M.; Cesario, A.; et al. Translational research in the era of precision medicine: Where we are and where we will go. *J. Pers. Med.* **2021**, *11*, 216. [CrossRef]
26. Ebling, S.; Glauert, J. Building a Swiss German Sign Language avatar with JASigning and evaluating it among the Deaf community. *Univers. Access Inf. Soc.* **2016**, *15*, 577–587. [CrossRef]
27. Martin, P.M.; Belhe, S.; Mudliar, S.; Kulkarni, M.; Sahasrabudhe, S. An Indian Sign Language (ISL) corpus of the domain disaster message using Avatar. In Proceedings of the Third International Symposium in Sign Language Translations and Technology (SLTAT-2013), Chicago, IL, USA, 18–19 October 2013; pp. 1–4.
28. Zwitserlood, I.; Verlinden, M.; Ros, J.; Van Der Schoot, S.; Netherlands, T. Synthetic signing for the deaf: Esign. In Proceedings of the Conference and Workshop on Assistive Technologies for Vision and Hearing Impairment (CVHI), Granada, Spain, 29 June–2 July 2004.
29. Yorganci, R.; Kindiroglu, A.A.; Kose, H. Avatar-based sign language training interface for primary school education. In Proceedings of the Workshop: Graphical and Robotic Embodied Agents for Therapeutic Systems, Los Angeles, CA, USA, 20 September 2016.
30. Lima, T.; Rocha, M.S.; Santos, T.A.; Benetti, A.; Soares, E.; Oliveira, H.S.D. Innovation in learning—The use of avatar for sign language. In *Proceedings of the International Conference on Human-Computer Interaction*; Springer: Berlin/Heidelberg, Germany, 2013; pp. 428–433.
31. Barrera Melchor, F.; Alcibar Palacios, J.C.; Pichardo-Lagunas, O.; Martinez-Seis, B. Speech to Mexican Sign Language for Learning with an Avatar. In *Proceedings of the Mexican International Conference on Artificial Intelligence*; Springer: Berlin/Heidelberg, Germany, 2020; pp. 179–192.

32. Hayward, K.; Adamo-Villani, N.; Lestina, J. A Computer Animation System for Creating Deaf-Accessible Math and Science Curriculum Materials. In *Proceedings of the Eurographics*; Education Papers; The Eurographics Association: Norrköping, Sweden, 2010; pp. 1–8.
33. Shohieb, S.M. A gamified e-learning framework for teaching mathematics to arab deaf students: Supporting an acting Arabic sign language avatar. *Ubiquitous Learn. Int. J.* **2019**, *12*, 55–70. [CrossRef]
34. Rajendran, R.; Ramachandran, S.T. Finger Spelled Signs in Sign Language Recognition Using Deep Convolutional Neural Network. *Int. J. Res. Eng. Sci. Manag.* **2021**, *4*, 249–253.
35. Aliwy, A.H.; Ahmed, A.A. Development of arabic sign language dictionary using 3D avatar technologies. *Indones. J. Electr. Eng. Comput. Sci.* **2021**, *21*, 609–616.
36. Sugandhi, P.K.; Kaur, S. Online multilingual dictionary using Hamburg notation for avatar-based Indian sign language generation system. *Int. J. Cogn. Lang. Sci.* **2018**, *12*, 120–127.
37. Filhol, M.; Mcdonald, J. Extending the AZee-Paula shortcuts to enable natural proform synthesis. In Proceedings of the Workshop on the Representation and Processing of Sign Languages, Miyazaki, Japan, 12 May 2018.
38. Świdziński, M. Jak Głusi przyswajają język: O językach migowych i miganych. In *W Język Migowy We Współczesnym Szkolnictwie na świecie i w Polsce*; Grzesiak, I., Ed.; Wydawnictwo—Stanisław Sumowski: Malbork, Poland, 2007; pp. 16–24.
39. Tomaszewski, P.; Rosik, P. Sygnały niemanualne a zdania pojedyncze w Polskim Języku Migowym: Gramatyka twarzy. *Porad. Językowy* **2007**, *1*, 33–49.
40. Tomaszewski, P.; Farris, M. Not by the hands alone: Functions of non-manual features in Polish Sign Language. In *Studies in the Psychology of Language and Communication*; Matrix: Warsaw, Poland, 2010; pp. 289–320.
41. Tomaszewski, P. *Fonologia Wizualna Polskiego języka Migowego*; Matrix: Warsaw, Poland, 2010.
42. Stokoe Jr, W.C. Sign language structure: An outline of the visual communication systems of the American deaf. *J. Deaf Stud. Deaf Educ.* **2005**, *10*, 3–37. [CrossRef]
43. Battison, R. Phonological deletion in american sign language. *Sign Lang. Stud.* **1974**, *5*, 1–19. [CrossRef]
44. Tomaszewski, P. Lingwistyczny opis struktury polskiego języka migowego. In *Język Jako Przedmiot badań Psychologicznych. Psycholingwistyka ogólna i Neurolingwistyka*; Okuniewska, K.H., Ed.; Wydawnictwo Szkoły Wyższej Psychologii Społecznej: Warsaw, Poland, 2011.
45. Mikulska, D. Elementy niemanualne w słowniku i tekście Polskiego Języka Migowego. In *Studia nad Kompetencją Językową i Komunikacją Niesłyszących*; Polski Komitet Audiofonologii: Warszawa, Poland, 2003; pp. 79–98.
46. Crasborn, O.A. *Nonmanual Structures in Sign Language*; Elsevier: Oxford, UK, 2006; pp. 668–672.
47. Braem, P.B.; Sutton-Spence, R. *The Hands Are The Head of The Mouth. The Mouth as Articulator in Sign Languages*; Signum Press: Hamburg, Germany, 2001.
48. Mohr, S. *Mouth Actions in Sign Languages. An Empirical Study of Irish Sign Language*; De Gruyter Mouton: Berlin, Germany, 2014.
49. Fabisiak, S. Przejawy imitacyjności w systemie gramatycznym polskiego języka migowego. *LingVaria* **2010**, *1*, 183.
50. Michael, N.; Yang, P.; Liu, Q.; Metaxas, D.N.; Neidle, C. A Framework for the Recognition of Nonmanual Markers in Segmented Sequences of American Sign Language. In Proceedings of the British Machine Vision Conference, Dundee, UK, 29 August–2 September 2011; pp. 1–12.
51. Ekman, P.; Friesen, W.V. Facial action coding system. *Environ. Psychol. Nonverbal Behav.* **1978**. [CrossRef]
52. Lien, J.J.J.; Kanade, T.; Cohn, J.F.; Li, C.C. Detection, tracking, and classification of action units in facial expression. *Robot. Auton. Syst.* **2000**, *31*, 131–146. [CrossRef]
53. Jaiswal, S.; Valstar, M. Deep learning the dynamic appearance and shape of facial action units. In Proceedings of the 2016 IEEE Winter Conference on Applications of Computer Vision (WACV), Lake Placid, NY, USA, 7–10 March 2016; pp. 1–8.
54. Breuer, R.; Kimmel, R. A deep learning perspective on the origin of facial expressions. *arXiv* **2017**, arXiv:1705.01842.
55. Nadeeshani, M.; Jayaweera, A.; Samarasinghe, P. Facial emotion prediction through action units and deep learning. In Proceedings of the 2020 2nd International Conference on Advancements in Computing (ICAC), Malabe, Sri Lanka, 10–11 December 2020; Volume 1, pp. 293–298.
56. Cohn, J.F.; Ambadar, Z.; Ekman, P. Observer-based measurement of facial expression with the Facial Action Coding System. *Handb. Emot. Elicitation Assess.* **2007**, *1*, 203–221.
57. Baltrusaitis, T.; Zadeh, A.; Lim, Y.C.; Morency, L.P. Openface 2.0: Facial behavior analysis toolkit. In Proceedings of the 2018 13th IEEE International Conference on Automatic Face & Gesture Recognition (FG 2018), Xi'an, China, 15–19 May 2018; pp. 59–66.
58. Baltrušaitis, T.; Robinson, P.; Morency, L.P. Openface: An open source facial behavior analysis toolkit. In Proceedings of the 2016 IEEE Winter Conference on Applications of Computer Vision (WACV), Lake Placid, NY, USA, 7–10 March 2016; pp. 1–10.
59. Wood, E.; Baltrusaitis, T.; Zhang, X.; Sugano, Y.; Robinson, P.; Bulling, A. Rendering of eyes for eye-shape registration and gaze estimation. In Proceedings of the IEEE international conference on computer vision, Washington, DC, USA, 7–13 December 2015; pp. 3756–3764.
60. Zadeh, A.; Chong Lim, Y.; Baltrusaitis, T.; Morency, L.P. Convolutional experts constrained local model for 3d facial landmark detection. In Proceedings of the IEEE International Conference on Computer Vision Workshops, Venice, Italy, 22–29 October 2017; pp. 2519–2528.

61. Zhang, K.; Zhang, Z.; Li, Z.; Qiao, Y. Joint face detection and alignment using multitask cascaded convolutional networks. *IEEE Signal Process. Lett.* **2016**, *23*, 1499–1503. [CrossRef]
62. Hanke, T.; Marshall, I.; Safar, E.; Schmaling, C.; Bentele, S.; Blanck, D.; Dorn, R.; Langer, G.; von Meyenn, A.; Popescu, H.; et al. Visicast deliverable d5-1: Interface definitions, 2002. *ViSiCAST Project Report.* 2002. Available online: https://www.visicast.cmp.uea.ac.uk/Papers/ViSiCAST_D5-1v017rev2.pdf (accessed on 15 November 2022).

Disclaimer/Publisher's Note: The statements, opinions and data contained in all publications are solely those of the individual author(s) and contributor(s) and not of MDPI and/or the editor(s). MDPI and/or the editor(s) disclaim responsibility for any injury to people or property resulting from any ideas, methods, instructions or products referred to in the content.

MDPI
St. Alban-Anlage 66
4052 Basel
Switzerland
Tel. +41 61 683 77 34
Fax +41 61 302 89 18
www.mdpi.com

Entropy Editorial Office
E-mail: entropy@mdpi.com
www.mdpi.com/journal/entropy

www.ingramcontent.com/pod-product-compliance
Lightning Source LLC
LaVergne TN
LVHW070152170726
843515LV00007B/1905